Mathematics
for
Management

Mathematics for Management

RICHARD C. LUCKING
The Management Centre
University of Bradford

A Wiley–Interscience Publication

JOHN WILEY & SONS
Chichester · New York · Brisbane · Toronto

British Library Cataloguing in Publication Data:

Lucking, Richard C.
Mathematics for management.
1. Management—Mathematics
I. Title
510′.2′4658 HD30.25 80-40127

ISBN 0 471 27779 7 (cloth)
ISBN 0 471 27781 9 (paper)

Typeset by Preface Ltd, Salisbury, Wilts.
and printed by The Pitman Press, Bath, Avon.

To
Charles, Pat, and Nicola

Acknowledgements

The text has been usefully improved by the comments and criticisms of several of my colleagues at Bradford, and I would particularly like to acknowledge the assistance of Richard Finn, Ali Mosleshirazi, and Chris Stewart. Thanks must also go to the several hundred students whose uninhibited inventiveness in misinterpretation have helped me to improve my own skills in setting the odd unambiguous question. Last, but emphatically not least, my sincere thanks to Jo Baxter for her unstinted enthusiasm and patience in the typing of the manuscript.

Contents

PART TWO

PART THREE

Preface

The last few years have witnessed both a rapid growth in the number of courses offered by universities and polytechnics in the fields of management, economics, and business studies, and an increasing emphasis within these areas on quantitative analysis and the development of numerate skills. Even so, most such degree courses accept students with very diverse backgrounds in mathematics, and first-year classes may well include several students with a recent good 'A' level pass, and others with an average 'O' level taken several years previously. Introductory courses in mathematics thus need to bring the less well qualified student up to the standard of the more advanced students in selected topics, as well as develop the latter's capabilities more in the direction of mathematics applied to subjects less technical than the physical sciences. This text is based on the first-year course in mathematics offered to undergraduate students at the University of Bradford Management Centre, and was developed in the first instance specifically for this course. Despite this particular origin, however, the level of the text is appropriate for recommendation to students with at least mathematics at 'O' level standard, following courses in Business Studies, Economics, Finance, and related social sciences with some quantitative input. Material is also included which should prove useful for understanding the mathematics involved in introductory courses in statistics in the same subject areas (although the book does not attempt to provide a full course in basic statistics, which is better left to a separate specialist volume). Likewise at the postgraduate level, the text is appropriate for those students following MBA, MSc, or research programmes in management whose academic backgrounds have been in non-quantitative subjects.

In line with the usual structure of many degree courses, this text aims to help the student acquire and develop one of the basic analytical skills (mathematics) early in the course, which would then be exercised in relation to appropriate classes of problems (such as operational research or economics) in a subsequent stage of the degree programme. The emphasis is therefore on helping the social science type student become a better mathematician, the techniques covered being selected with a view to those types of application likely to be faced subsequently. This emphasis on developing mathematical skills is in contrast to many texts where the mathematical ability of the reader is assumed, and is exercised to greater or lesser degree on classes of problems, rather than being explicitly developed as skills.

The applied nature of the subject is emphasized in this text through the choice of worked and set examples which run in parallel with the main mathematical techniques discussed. After the introductory chapter, which reviews basic mathematical notation, the text proceeds through five main classes of techniques:

1. Basic set theory, leading into relations, functions, and graphs.
2. Linear relationships.
3. Curvilinear relationships, leading to the development of the differential and integral calculus.
4. Matrix algebra, which is developed from a second look at simultaneous linear equations.
5. Mathematical series.

This material is developed through examples of some applications, so for example:

1. An introduction to linear programming, using graphical representation, is presented after the section on linear relationships.
2. Several aspects of mathematical models are discussed in conjunction with the section on nonlinear relationships.
3. Various discounting techniques in finance are explained and developed from first principles in the section on mathematical series.
4. The theme of optimization under constraints, common to several applications, is emphasized in order to demonstrate their element of similarity.

The chart illustrates the many connections within the material covered. The various 'pure' mathematical inputs toward the left-hand side of the diagram are employed in the applications listed towards the right-hand side.

Besides worked examples within the text, three further sets of examples are included. Simpler examples for the student are included at the end of each chapter (with answers provided at the end of the book); intermediate problems, suitable for tutorial work, also follow each chapter (without answers); and more advanced problems are included at the end of each of the six main sections of the book, the solutions to this latter group being explained more fully, and providing in effect an extension to the material of the main text.

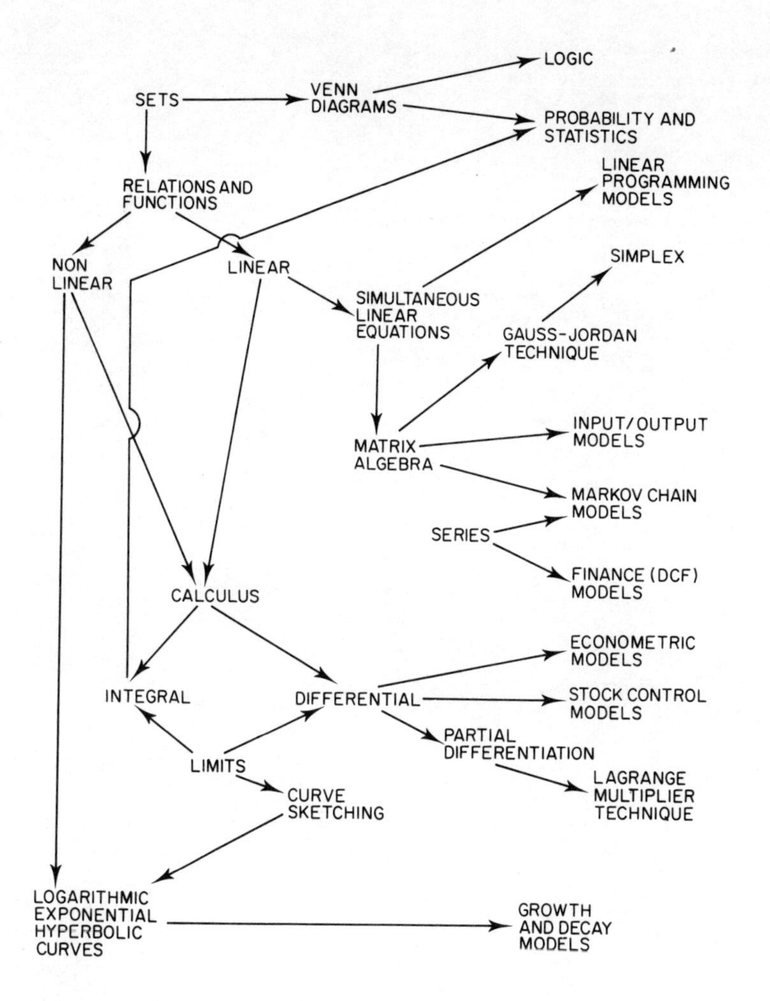

LOGIC
SETS
VENN DIAGRAMS
PROBABILITY AND STATISTICS
LINEAR PROGRAMMING MODELS
RELATIONS AND FUNCTIONS
SIMPLEX
NON LINEAR
LINEAR
SIMULTANEOUS LINEAR EQUATIONS
GAUSS-JORDAN TECHNIQUE
INPUT/OUTPUT MODELS
MATRIX ALGEBRA
MARKOV CHAIN MODELS
SERIES
FINANCE (DCF) MODELS
CALCULUS
ECONOMETRIC MODELS
INTEGRAL
DIFFERENTIAL
STOCK CONTROL MODELS
PARTIAL DIFFERENTIATION
LIMITS
LAGRANGE MULTIPLIER TECHNIQUE
CURVE SKETCHING
LOGARITHMIC EXPONENTIAL HYPERBOLIC CURVES
GROWTH AND DECAY MODELS

PART ONE

Introduction

This introductory chapter begins by summarizing the basic algebraic rules and notation which will be used and assumed throughout the remainder of the text. Although it should be adequate as a basis for revision, it is deliberately very concise, and any reader who finds himself having great difficulties with this section is advised to refer in the first instance to an introductory text on algebra and arithmetic.

Subsequent sections address themselves to the task of developing the readers' commonsense in spotting errors and avoiding illegal mathematical operations. Aspects of approximation are covered, as well as intuitive or geometric reasoning to assist in the manipulation of algebraic formulae.

CHAPTER 1

Basic Notation

The phrase 'the language of mathematics' suggests that a mathematical expression can, in principle, be translated into an English language form. Or, putting it another way, the underlying logic of a mathematical operation can be expressed either by means of symbols (i.e. usual mathematical notation) or by a statement in English (or Chinese, or French, etc.). However, mathematical notation is

(a) Brief. Complicated operations in terms of logic can be expressed very succinctly.
(b) Universal. A unique language, with one set of rules or grammar which is universally accepted.
(c) Rigorous. English can sometimes be vague, or ambiguous. A mathematical expression should only have one possible interpretation.

For example, the following would be extremely difficult to write in English, and would certainly not be brief

$$3ab\,\sqrt{xy}/[(\log x + 6 + \mathrm{e}^2)^{-2}]^{1/3} + \sum_{i=4}^{7} i^2 \geq 6 \times 10^{-8} \sum_{i=2}^{5} x_i.$$

Firstly, therefore, we must understand the conventions of the notation, and how an expression like that above should be interpreted.

1.1 ARITHMETIC OPERATIONS

The four basic operations in arithmetic are addition, subtraction, multiplication, and division, which are written in symbol form as +, −, ×, and /. When combining several operations, the multiplication and division should always be carried out before addition and subtraction, i.e. × and / before + and −. Otherwise, operations are performed in sequence from left to right; for example,

$$5 \times 12 + 9 - 3 \times 2/6 + 7 = 60 + 9 - 6/6 + 7 = 60 + 9 - 1 + 7 = 75.$$

1.2 BRACKETS

When brackets, () or [] etc., appear in a statement, all operations inside the bracket are performed *before* those outside the bracket; for example,

$$5 \times (6 - 8) + 4/2 = 5 \times (-2) + 4/2 = -5 \times 2 + 4/2 = -10 + 2 = -8.$$

When brackets are *nested*, the innermost set of operations is carried out first, for example,

$$\begin{aligned}(3 \times (5 - 2) + 7 \times (8 - 4))/2 &= (3 \times 3 + 7 \times 4)/2 = (9 + 28)/2 \\ &= 37/2 = 18{\cdot}5.\end{aligned}$$

1.3 MODULUS SIGN

The modulus sign is used to mean *the positive value of*, and is used in the following way (note the similar precedence to brackets):

$$|-6| = 6$$
$$|-3(4) + 1| = |-12 + 1| = |-11| = 11.$$

1.4 ALTERNATIVE NOTATIONS

The multiplications sign (×) is not always used. It can be replaced by a full stop type dot (.), thus:

$$3 \times 4 = 3.4 = 12.$$

When brackets are employed, or alphabetic letters are used, then it may be omitted altogether.

$$3 \times 4 = 3.4 = 3(4) = 12 \quad \text{(but } \textit{not} \text{ 34)}$$

However,

$$\mathrm{y} \times \mathrm{z} = \mathrm{y.z} = \mathrm{yz} = \mathrm{y(z)}$$

or

$$2 \times \mathrm{y} = 2.\mathrm{y} = 2\mathrm{y} = 2(\mathrm{y}).$$

The division sign is often written as ÷, rather than /.

1.5 EQUALITIES AND INEQUALITIES

If we want to put a statement such as *the weight of an apple is 110 g* into mathematical notation then we might first define a quantity x in the following way:

Let x = weight of an apple in grams, then $x = 110$.

This is an example of an *equality*, and if we operate on both sides of the equation in the same way, the equality is not affected; for example,

Multiply both sides by 2:	$2x = 220$
Add any number to each side:	$2x + 91 = 220 + 91 = 311$
Take the *reciprocal* of both sides:	$\frac{1}{x} = \frac{1}{110}$.

An *inequality* is a statement such as *the weight of an apple is less than 110 g*. Here we have to introduce some more symbols

$>$ means *is greater than*
$<$ means *is less than*
$\geqslant$ means *is greater than or equal to*
$\leqslant$ means *is less than or equal to*

so, if x is defined to be the weight of an apple in grams, as before,

$x \leqslant 110$ means the weight is less than or equal to 110 g.

We can also compare numbers, so that $50 > 13$ should be obvious, for example. However, for negative numbers we must be more careful and the easiest way to see this is to imagine all numbers (positive and negative) to be drawn on a *number line* across the page as below

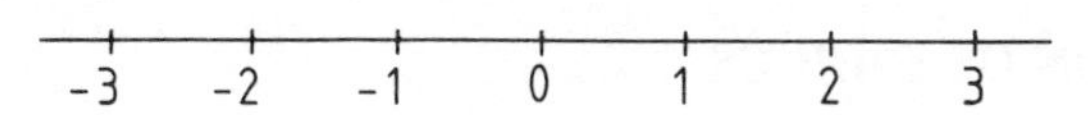

When two numbers are compared, the one on the right is the greater and the one on the left is the lesser; for example,

$50 > 13$	$0{\cdot}1 < 0{\cdot}6$
$-50 < -13$	$-0{\cdot}6 < -0{\cdot}1$
$-0{\cdot}1 < -0{\cdot}001$	$-3 < 0$.

We cannot alter numbers on both sides of an inequality as freely as for an equality; for example,

	$3 < 8$	
Adding to both sides	$3 + 5 < 8 + 5$	$8 < 13$
Multiplying by a positive number	$4 \times 3 < 4 \times 8$	$12 < 32$

but,

Taking reciprocals	$\frac{1}{3} > \frac{1}{8}$	$0{\cdot}33 > 0{\cdot}125$
Multiplying by a negative number	$-4 \times 3 > -4 \times 8$	$-12 > -32$.

Finally, we can also place an upper and a lower limit upon the range of values for some quantity; for example,

$100 < x \leqslant 120$ (i.e. more than 100, but less than or equal to 120).

1.6 ADDITIONAL NOTATION

Rarer, but useful, symbols include

$\simeq$ or $\cong$ or $\doteqdot$ mean *approximately equal to*

$\neq$ means *not equal to*

$\not>$ means *not greater than*

$\propto$ means *proportional to*

1.7 POWERS AND INDICES

y^2 is read *y squared* and means $(y \times y)$
y^3 is read *y cubed* and means $(y \times y \times y)$
y^4 is read *y to the fourth* and means $(y \times y \times y \times y)$
y^5 is read *y to the fifth* and means $(y \times y \times y \times y \times y)$
y^n is read *y to the* n^{th} and means n y's multiplied together.

These are all examples of the *index* notation.

For example y^4, read *y to the fourth* can be expressed as *y raised to the fourth power*, and the number 4 is the *index* or the *exponent*.

1.8 ADDITION OF INDICES

$$\begin{aligned} y^3 \times y^2 &= (y \times y \times y) \times (y \times y) \\ &= (y \times y \times y \times y \times y) \\ &= y^5. \end{aligned}$$

So, when multiplying powers of the same number (y), *add* the indices; for example,

$$y^2 \times y^4 \times y^5 = y^{11}$$
$$y^3 \times y^7 = y^{10}$$
$$y \times y^2 = y^3 \quad \text{(N.B. } y \text{ can always be written as } y^1\text{)}.$$

These rules can be extended to division, when we can introduce the convention of a *negative* power; for example,

$$\frac{y^6}{y^2} = \frac{y \times y \times y \times y \times y \times y}{y \times y} = y^4 = y^{(6-2)}.$$

So, for division, we should *subtract* indices. Therefore

$$\frac{y^2}{y^6} = \frac{y \times y}{y \times y \times y \times y \times y \times y} = \frac{1}{y^4} = y^{(2-6)} = y^{-4}$$

so that

$$y^{-4} = \frac{1}{y^4}.$$

Similarly,

$$y^{-7} = \frac{1}{y^7} \quad y^{-1} = \frac{1}{y}.$$

Note also that

$$y^5 \times y^{-5} = y^{5-5} = y^0$$

but

$$y^5 \times y^{-5} = y^5 \times \frac{1}{y^5} = 1$$

hence

$$y^0 = 1 \quad \text{for } any\ y.$$

1.9 MULTIPLICATION OF INDICES

We can also raise powers to a power; for example;

$$\begin{aligned}(y^2)^3 &= (y \times y) \times (y \times y) \times (y \times y)\\ &= y \times y \times y \times y \times y \times y\\ &= y^6\end{aligned}$$

and

$$\begin{aligned}(y^{-2})^3 &= y^{-2} \times y^{-2} \times y^{-2}\\ &= \frac{1}{y \times y} \times \frac{1}{y \times y} \times \frac{1}{y \times y}\\ &= 1/(y \times y \times y \times y \times y \times y \times y)\\ &= 1/y^6 = y^{-6}.\end{aligned}$$

So, when raising a power to a power, *multiply* the indices.

1.10 ROOTS

We know that $5 \times 5 = 25$ or $5^2 = 25$. That is, *5 squared equals 25*. We could say instead that *the square root of 25 is 5*.

$$5 = \sqrt{25} \text{ (square root)}, \ 5^2 = 25$$

Similarly,

$$5 = {}^3\sqrt{125} \text{ (cube root)}, \ 5^3 = 125$$
$$5 = {}^4\sqrt{625} \text{ (fourth root)}, \ 5^4 = 625.$$

We can now show that the operation of taking a root can be expressed instead as a power. Suppose we take $5^4 = 625$. Let us raise each side to a power $\frac{1}{4}$, without as yet knowing what a power of $\frac{1}{4}$ means, then

$$(5^4)^{1/4} = (625)^{1/4}$$

or

$$5^1 = (625)^{1/4} \quad \text{since we must multiply powers together (see Section 1.9)}$$

but

$$5 = \sqrt[4]{625} \quad \text{from above.}$$

So, if we are to be consistent with the laws of powers already proved, we must interpret a power of $\frac{1}{4}$ as equivalent to a fourth root. So,

$$y^{1/2} = \sqrt{y}$$
$$y^{1/3} = \sqrt[3]{y}$$
$$y^{1/4} = \sqrt[4]{y} \quad \text{and so on.}$$

One further point should be made about the root sign $\sqrt{\ }$. Strictly, the square root of 25 can take either of two values: $+5$ or -5 since

$$(+5)^2 = 25$$

and

$$(-5)^2 = 25.$$

To avoid confusion, the usual convention is to take the symbol $\sqrt{\ }$ to mean the *positive* root, and then the alternative signs are introduced explicitly; for example,

$$\pm\sqrt{25} = \pm 5.$$

1.11 POWERS OF TEN

Since

$$\begin{aligned}(ab)^3 = (a \times b)^3 &= (a \times b) \times (a \times b) \times (a \times b) \\ &= (a \times a \times a) \times (b \times b \times b) \\ &= a^3 b^3.\end{aligned}$$

In general, $(ab)^n = a^n b^n$ and so similarly $(ab)^{1/2} = a^{1/2} b^{1/2}$, so that $\sqrt{ab} = \sqrt{a}\sqrt{b}$.

This result is very useful, as it enables us to put awkward numbers into *exponential form* before taking roots. Now,

$$10^1 = 10 \qquad 10^{-1} = 0{\cdot}1$$
$$10^2 = 100 \qquad 10^{-2} = 0{\cdot}01$$
$$10^3 = 1000 \qquad 10^{-3} = 0{\cdot}001$$

$10^4 = 10000$ $\quad\quad$ $10^{-4} = 0{\cdot}0001$

etc. $\quad\quad$ etc.

So we can express very large, or very small, numbers in the following ways:

$$35921 = 3{\cdot}5921 \times 10^4 = 3{\cdot}5921 \times 10^4$$

$$67124396 = 6{\cdot}7124396 \times 10^7 = 67{\cdot}124396 \times 10^6$$

$$0{\cdot}000031 = 3{\cdot}1 \times 10^{-5} = 31 \times 10^{-6}$$

$$0{\cdot}000296 = 2{\cdot}96 \times 10^{-4} = 2{\cdot}96 \times 10^{-4}.$$

The form in the middle column is the exponential form. This has been adjusted in the right-hand column to give an even power of 10, multiplied by a number between 1 and 99. It is this form which is most useful for taking the square root; for example,

$$\sqrt{67124396} = \sqrt{67{\cdot}124396 \times 10^6} = \sqrt{67{\cdot}124396} \times \sqrt{10^6}$$
$$= 8{\cdot}193 \times 10^3 = 8193$$

$$\sqrt{0{\cdot}000296} = \sqrt{2{\cdot}96 \times 10^{-4}} = \sqrt{2{\cdot}96} \times \sqrt{10^{-4}}$$
$$= 1{\cdot}72 \times 10^{-2} = 0{\cdot}0172.$$

Note that

$$\sqrt{10^6} = (10^6)^{1/2} = 10^3 \quad \text{and} \quad \sqrt{10^{-4}} = (10^{-4})^{1/2} = 10^{-2}.$$

So for cube roots we would want the power of 10 to be divisible by 3; for example,

$$\sqrt[3]{10^{-9}} = (10^{-9})^{1/3} = 10^{-3}.$$

1.12 LOGARITHMS

$$8 = 2^3 \text{ and } 16 = 2^4.$$

So, if we call 2 the *base*, then we must raise the base to power 3 to achieve 8, and to power 4 to achieve 16.

We can define a logarithm in the following way:

$$3 = \log_2 8 \qquad 4 = \log_2 16$$

(note the suffix 2 indicating the base that we are using). Now,

$$8 \times 16 = 128 \quad \text{but also} \quad 2^3 \times 2^4 = 2^7.$$

That is,

$$3 + 4 = 7 \quad \text{or} \quad \log_2 8 + \log_2 16 = \log_2 128.$$

So, instead of multiplying two numbers (8 and 16) we can instead *add* the logarithms of the two numbers, and then *antilog* to get back to the product. This is often preferable, since addition is an easier operation than multiplication.

The whole process becomes:

$8 \times 16 = ?$	... the original problem
But $\log_2 8 = 3$, $\log_2 16 = 4$	... from log tables
$3 + 4 = 7$	... Addition operation
Antilog(7) = 128	... from antilog tables, base 2
So that $8 \times 16 = 128$	... problem solved.

Two common bases are used (as 2 is *not* very usual)

(i) $e \cong 2{\cdot}718$	natural or Napierian logs
(ii) 10	logs to base 10.

Base 10 is the usual choice for calculations, as the following example shows:

$\log_{10} 2 = 0{\cdot}3010$	$\log_{10} 0{\cdot}2 = \bar{1}{\cdot}3010$
$\log_{10} 20 = 1{\cdot}3010$	$\log_{10} 0{\cdot}02 = \bar{2}{\cdot}3010$
$\log_{10} 200 = 2{\cdot}3010$	$\log_{10} 0{\cdot}002 = \bar{3}{\cdot}3010$
etc.	etc.

The $\bar{2}$, or *bar 2* implies that the *decimal* part is positive and the *integer* part is negative. So that strictly

$$\bar{2}{\cdot}3010 = -1{\cdot}6090$$

but it is convenient in logarithm calculations to retain the form $\bar{2}{\cdot}3010$.

When base e is used (natural logarithms) then an alternative notation is often used:

$$\log_e x \text{ or } \ln x.$$

For base 10, the base is often left out explicitly, so that:

$$\log x \text{ means } \log_{10} x.$$

1.13 LOGARITHMS AND POWERS

Logarithms are also useful for powers, and hence roots; for example,

$$2^3 = \text{Antilog}(3 \times 3{\cdot}010) = \text{Antilog}(0{\cdot}9030) = 8{\cdot}0$$

since $\log_{10} 2 = 0{\cdot}3010$. Or

$$\sqrt[6]{0{\cdot}008303} = ? \quad \text{Now } \log_{10}(0{\cdot}008303) = \bar{3}{\cdot}9192$$

$$\begin{aligned} \therefore \sqrt[6]{0{\cdot}008303} &= (0{\cdot}008303)^{1/6} = \text{Antilog}(\bar{3}{\cdot}9192 \times 1/6) \\ &= \text{Antilog}((\bar{6} + 3{\cdot}9192) \times 1/6) \\ &= \text{Antilog}(\bar{1}{\cdot}6532) \\ &= 0{\cdot}4500. \end{aligned}$$

1.14 MANIPULATION OF LOGARITHMS

If $A = n^a$ and $B = n^b$ then, by definition of a logarithm,

$$a = \log_n A \quad \text{and} \quad b = \log_n B \quad \text{(any base, } n\text{)}$$

This is usually the basic step underlying most of the rules of logarithms.

(i) $\log_n AB = \log_n A + \log_n B$

Proof

$$AB = n^a n^b = n^{a+b}$$
$$\therefore\ a + b = \log_n AB \text{ by definition of a logarithm.}$$

But

$$a + b = \log_n A + \log_n B$$
$$\therefore\ a + b = \log_n AB = \log_n A + \log_n B.$$

(ii) $\log(A/B) = \log A - \log B$ (proof similar to (i))

(iii) $\log A^m = m \log A$

Proof

$$A = n^a, \text{ so } A^m = (n^a)^m = n^{am}$$

$$\therefore\ am = \log_n(A^m) \quad \text{by definition of a logarithm.}$$

But

$$a = \log_n A$$
$$\therefore\ am = m \log A = \log(A^m).$$

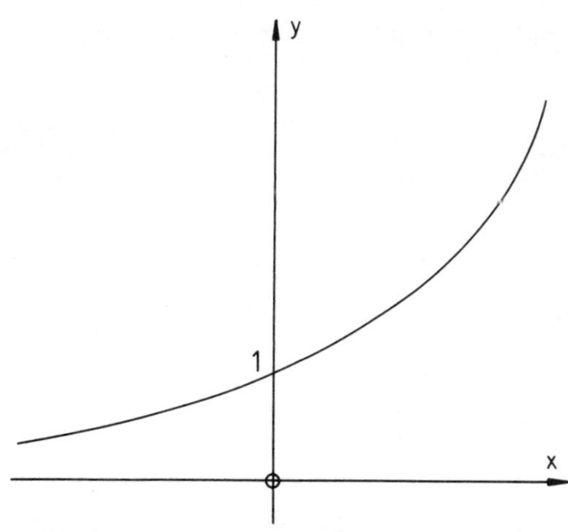

Figure 1.1(a) The exponential form $y = n^x$

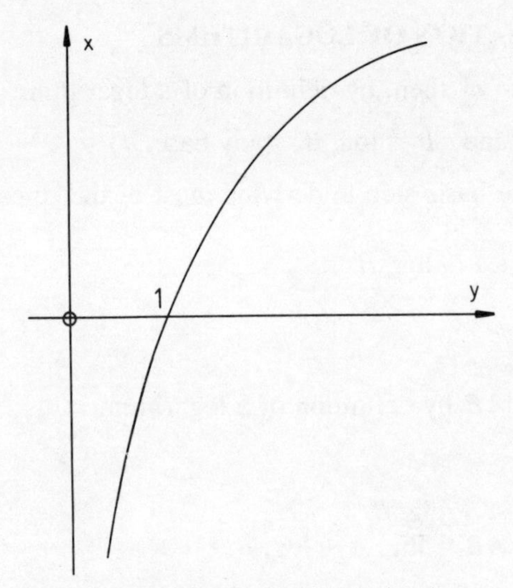

Figure 1.1(b) The logarithmic form $x = \log_n y$

Graph of the logarithmic function

$y = n^x$ and $x = \log_n y$ are equivalent statements and so they have the same shape on a graph.
Either
turn through 90° anti-clockwise, and reflect in x-axis
or
reflect in the line $x = y$ to achieve, by either move, $x = \log_n y$.

1.15 THE SUMMATION SIGN Σ (sigma)

Sigma is the Greek letter s, and the capital form denotes the operation *the sum of* in mathematics. Consider the list of numbers:

$$x_1, x_2, x_3, x_4, x_5$$

which have suffices 1, 2, 3, 4, and 5 and are read as *x-one, x-two, x-three, x-four, x-five*. Do not confuse the three similar expressions:

$2x$, read *two-x*, which means $2 . x$
x^2, read *x-squared*, which means $x . x$
x_2, read *x-two*, which means the second number in some list such as that above.

We can very neatly abbreviate the sum of a list such as that above using the capital sigma

$$x_1 + x_2 + x_3 + x_4 + x_5 = \sum_{i=1}^{5} x_i.$$

In words, *take the sum of all the subscripted x's, from subscript (or suffix) 1 up to and including subscript 5*. Similarly,

$$\sum_{i=2}^{4} x_i = x_2 + x_3 + x_4$$

$$\sum_{i=2}^{4} x^i = x^2 + x^3 + x^4$$

$$\sum_{i=1}^{3} (x_i^2 + 1) = (x_1^2 + 1) + (x_2^2 + 1) + (x_3^2 + 1)$$

$$\sum_{i=2}^{5} 4x_i^3 = (4x_2^3) + (4x_3^3) + (4x_4^3) + (4x_5^3)$$

$$= 4(x_2^3 + x_3^3 + x_4^3 + x_5^3) = 4 \sum_{i=2}^{5} x_i^3.$$

1.16 SENSE AND NONSENSE

In the sections which follow, we attempt to give the reader some insight into operations and calculations which are mathematically unwise or illegitimate. There is always a danger that putting anything incorrect into print will be more readily fixed in one's memory than all the correct statements put together. However, there is some value in doing this, since restricting discussion to what is allowed can fail to demonstrate explicitly those operations which, although illegal, are tempting to the less experienced mathematician. Woven into the text of what follows, therefore, are several statements on what you *should not do*, mathematically. In order to minimize the risk of these being absorbed uncritically by a casual glance, all incorrect mathematical statements are clearly labelled as such.

1.17 APPROXIMATIONS

Particularly in the age of electronic calculators, it is tempting to deliver an answer to a question which is as precise as possible. So firstly we make a distinction between *precision* and *accuracy*.

Precision is concerned with the degree of exactness which is claimed by virtue of, for example, the number of decimal places which are given. The following three heights of the same person are increasingly precise in their description.

1·8 m   1·832 m   1·8324578 m

Accuracy is concerned with the question as to whether these measurements are correct, within the degree of precision claimed. The true height of the same person might well be 1·784 m (to 3 decimal places), and the increasing precision of the first set of figures is then of no value.

Accuracy is therefore often concerned with the question as to whether or not our instrument of measurement is operating in the way we think it is (such as a clock running at the right pace, a metre rule which is a metre in length, etc.). Precision is concerned with the degree of detail which we can consistently read from the instrument (such as a clock to within one second, or a rule to within one millimetre).

The result of an arithmetic calculation, therefore, should have a precision which reflects that of the original data used. We must take account, also, of the way in which errors (or implicit errors, by virtue of using imprecise figures) combine. Suppose A, B, C, etc. are a set of figures we wish to combine in calculation. Suppose, likewise, that a, b, c are the corresponding errors (or maximum likely errors) which we would associate with these numbers. If

$$A \pm B = C$$

then

$$a^2 + b^2 = c^2$$

gives us some idea of the size of the error in C, given that in A and B; for example,

$$A = 1{\cdot}42 \qquad (a \doteqdot 0{\cdot}01)$$
$$B = 0{\cdot}091 \qquad (b \doteqdot 0{\cdot}001)$$
$$C = A + B = 1{\cdot}42 + 0{\cdot}091$$
$$\therefore C = 1{\cdot}511.$$

But

$$c^2 = (0{\cdot}01)^2 + (0{\cdot}001)^2 = 0{\cdot}000101$$
$$c \doteqdot 0{\cdot}01.$$

Notice that for a summation (and similarly for subtraction), the error with the largest *absolute value* is the most important. The result implies that it would be over-optimistic to give C in any form more precise than

$$C = 1{\cdot}51.$$

For products, suppose

$$D \times E = F$$

then

$$\left(\frac{d}{D}\right)^2 + \left(\frac{e}{E}\right)^2 = \left(\frac{f}{F}\right)^2$$

and in this case the squares of the *relative* errors add; for example,

$$D = 1{\cdot}42 \qquad (d \doteqdot 0{\cdot}01)$$
$$E = 0{\cdot}09 \qquad (e \doteqdot 0{\cdot}01)$$
$$F = D \times E = 1{\cdot}42 \times 0{\cdot}09$$
$$\therefore F = 0{\cdot}1278$$
$$\left(\frac{f}{F}\right)^2 = \left(\frac{0{\cdot}01}{1{\cdot}42}\right)^2 + \left(\frac{0{\cdot}01}{0{\cdot}09}\right)^2 = 0{\cdot}0123$$
$$\therefore \frac{f}{F} = 0{\cdot}111 \doteqdot \frac{1}{9}.$$

Notice that for a multiplication (and similarly for division), the error with the largest *relative value* is the most important. Even though the absolute values of the errors d and e were deemed the same, the error in E was far more significant and accounts for most of the error in F.

In fact

$$\frac{f}{F} \doteqdot \frac{e}{E} \doteqdot \frac{1}{9} \quad \text{(about 11\%)}.$$

Hence, F should be quoted in a form no more precise than

$$F = 0{\cdot}13.$$

The reader may wonder where we obtained the values for a, b, d, and e. These were taken from the given precisions for the four numbers A, B, D, and E. Since $A = 1{\cdot}42$, we have assumed that this quantity is known to this precision and no better. The greatest doubt we could have about this number (in other words, the maximum probable error) is 0·01, or 1 in the second decimal place. Likewise for the other numbers. If this assumption is incorrect, and we actually know A more precisely, then our calculation for c would need revision.

For example, if in fact

$$A = 1{\cdot}420000$$

then $a = 0{\cdot}000001$ at most, and thus c will be affected.

A brief summary

Our explanation has been fairly detailed, but in practice, estimation of precision can be a fairly quick process. To illustrate this, consider the following calculation.

$$X = \frac{312 \times (0{\cdot}12 + 4{\cdot}1)}{(161 + 94{\cdot}5) \times 0{\cdot}0012} = 4294{\cdot}325.$$

If the figures are given only as precisely as they are known then we can estimate the precision of X.

Error in (0·12 + 4·1)	is approximately 0·1, or 2%
Error in 312	is approximately 1, or 0·3%
Error in (161 + 94·5)	is approximately 1, or 0·4%
Error in 0·0012	is approximately 0·0001, or 8%.

For the final multiplication and division operations, *relative* errors combine, and the largest of these is clearly 8%. So, if we only know the original figures to the precision stated, we should not quote X more precisely than

$$X = 4300.$$

(Even the figure 3 is a little generous, since we have really indicated a precision of a little over 2%.)

Rounding-off decimals

When rounding-off numbers or decimals, account is taken of the first significant figure to be removed. If it lies between 0 and 4 (inclusive), then the final remaining figure is unaltered. If the first significant figure to be removed is between 5 and 9 (inclusive), then the final remaining figure is increased by 1. For example,

1·8324578 becomes, on successive rounding:
1·832458
1·83246
1·8325
1·833
1·83
1·8
2.

1.18 ELECTRONIC CALCULATORS

When a calculation of some complexity is undertaken on an electronic calculator, it is useful if one can ensure that:

(a) the calculation has been undertaken successfully and the correct answer has been obtained;
(b) the calculation has been undertaken as efficiently as possible, with as great a precision and speed as practicable.

The first condition can be satisfied to some extent if one develops a facility for quick approximations, to check that the answer is of the correct order of magnitude. The second condition can be achieved if the operations are executed in an intelligent sequence. We discuss both these aspects in this section, and assume that the calculator available is of a fairly basic type, includes the operations $+$, $-$, $\times$, $\div$, X^2, $1/X$, $\sqrt{X}$, and that there is a single memory facility.

Quick approximations

The art of quick approximation is often to drop the precision, to simplify the calculation. For example, to take our earlier example

$$\frac{312 \times (0{\cdot}12 + 4{\cdot}1)}{(161 + 94{\cdot}5) \times 0{\cdot}0012}$$

this is approximately

$$\frac{300 \times 4}{250 \times 0{\cdot}001} = \frac{1200 \times 1000}{250} = 4800$$

which agrees with our more precise calculation (4294) in terms of order of magnitude.

Taking another calculation, familiar in statistics:

$$\frac{0{\cdot}12 - 0{\cdot}10}{\sqrt{\dfrac{(0{\cdot}011)^2}{20} + \dfrac{(0{\cdot}012)^2}{50}}} \doteqdot \frac{0{\cdot}02}{\sqrt{\dfrac{(0{\cdot}01)^2}{20} + \dfrac{(0{\cdot}01)^2}{50}}}$$

$$= \frac{0{\cdot}02}{(0{\cdot}01)\sqrt{\dfrac{1}{20} + \dfrac{1}{50}}} = \frac{2}{\sqrt{\dfrac{70}{1000}}} = 2 \times \sqrt{\frac{10000}{700}}$$

$$\doteqdot 2 \times \frac{100}{25} \quad (\text{since } 25^2 = 625 \doteqdot 700)$$

$$= 8.$$

Exact calculation gives 6·69, which is in good enough agreement with our rough estimate.

Sequence of calculation

With only a single memory, it is more important to carry out any calculation in such a way that intermediate values need not be written down, if at all possible. We shall consider the two calculations above, again, to demonstrate this and also to show how to avoid loss of precision.

To calculate

$$\frac{0{\cdot}12 - 0{\cdot}10}{\sqrt{\dfrac{(0{\cdot}011)^2}{20} + \dfrac{(0{\cdot}012)^2}{50}}}$$

It is generally wiser to commence with the denominator, the result of which may be entered into the memory, ready to divide the numerator.

The sequence of operations (with brief explanation) might then be

ENTRY or KEY	CONSEQUENCE
0·011	entered and . . .
X^2	squared
÷ 20 =	to give $(0{\cdot}011)^2/20$
M+	entered into memory.
0·012	entered and . . .
X^2	squared
÷ 50 =	to give $(0{\cdot}012)^2/50$
M+	entered into memory also
MR	total memory recalled and . . .
$\sqrt{X}$	square-rooted to give denominator
MC	clear memory of residual content
M+	enter denominator into memory.
0·12	entered
− 0·10 =	to give numerator
÷ MR =	to divide numerator by denominator and obtain final result

However, as this sequence stands, precision has been reduced very considerably. This has occurred because, during the calculation of $(0{\cdot}011)^2/20$ (and its counterpart), we reached a very small number (0·00000605) so that much of the storage contained (non-significant) zeros. It is quite likely, in fact, to be rounded-off by the calculator, a point which may not be noticed by the operator. To avoid this, it is wiser in this example to calculate instead

$$\frac{0{\cdot}12 - 0{\cdot}10}{\sqrt{\dfrac{11^2}{20} + \dfrac{12^2}{50}}}.$$

We have thus increased the denominator by a factor of

$$\sqrt{(10^3)^2} \quad \text{or} \quad 10^3.$$

The final calculated result is then a factor of 10^3 too small, so we should scale this upwards by a factor of 1000 to compensate. With this procedure, the calculation is at all times undertaken within the limits of precision of the calculator.

Similar problems can arise with extended calculations of this type:

$$\frac{312 \times 4{\cdot}22}{255{\cdot}5 \times 0{\cdot}0012}.$$

Wherever possible, the order of calculation should be determined by the need to avoid excessively large, or excessively small, numbers. This often implies that multiplication and division operations should be alternated, to keep the intermediate figures within limits. For this example, a suggested order of calculation would be:

312	entered
÷ 255·5 =	to give 312/255·5
× 4·22 =	to give 312/255·5 × 4·22
÷ 0·0012 =	to give (312/255·5) × (4·22/0·0012).

This is preferable (generally) to a sequence where all the multiplications are undertaken prior to all the divisions, particularly in cases such as the following

$$\frac{0{\cdot}00123 \times 0{\cdot}00456 \times 0{\cdot}00789}{0{\cdot}00321 \times 0{\cdot}00654 \times 0{\cdot}00987}.$$

If all the multiplications are undertaken before any divisions for this example, using a typical eight-digit calculator, the result zero is achieved (rather than 0·2136) due solely to problems of rounding-off.

1.19 POPULAR MISCONCEPTIONS

In this section we attempt to arm the reader against some of the more popular errors which creep in to the work of the unwary.

Functions of the type $(a + b)^n$

An important (or at least, frequent) class of errors would fall into the general category of putting

$(a + b)^n = a^n + b^n$. INCORRECT

For example

$(a + b)^2 = a^2 + 2ab + b^2$ CORRECT

not

$(a + b)^2 = a^2 + b^2$ INCORRECT

as can be seen from Figure 1.2. The area of the total square must be $(a + b) \times (a + b) = (a + b)^2$, whereas $a^2 + b^2$ only gives the shaded region. So, all cross-terms must be included thus:

Likewise

$(a + b + c)^2 = a^2 + b^2 + c^2 + 2ab + 2ac + 2bc$ CORRECT

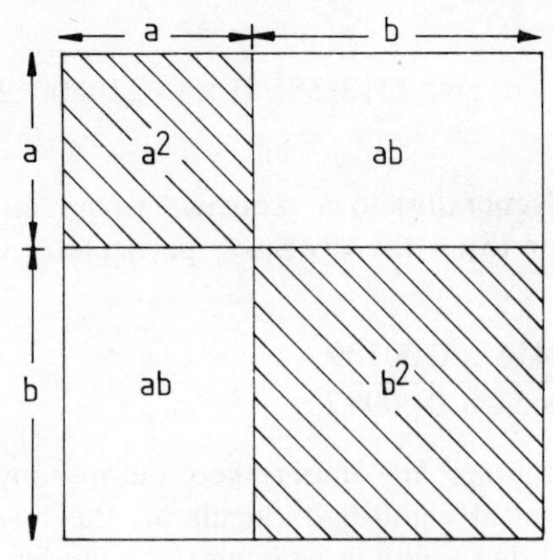

Figure 1.2 Geometric interpretation of $(a + b)^2$

not

$$(a + b + c)^2 = a^2 + b^2 + c^2. \qquad \text{INCORRECT}$$

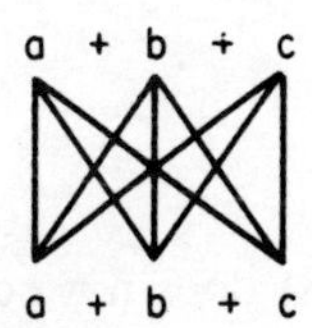

gives 9 terms altogether, whereas $a^2 + b^2 + c^2$ gives only the shaded region in Figure 1.3.

Another variation on this error might be higher powers:

$$(a + b)^3 = a^3 + 3a^2b + 3ab^2 + b^3 \qquad \text{CORRECT}$$

not

$$(a + b)^3 = a^3 + b^3 \qquad \text{INCORRECT}$$

(a three-dimensional figure of a cube could demonstrate this in similar manner to the above cases).

Or fractional powers:

$$\sqrt{a + b} = (a + b)^{1/2} \qquad \text{CORRECT}$$

not

$$\sqrt{a + b} = \sqrt{a} + \sqrt{b} \qquad \text{INCORRECT}$$

not

$$(a + b)^{1/2} = a^{1/2} + b^{1/2}. \qquad \text{INCORRECT}$$

If in doubt about the validity of such formulae, a useful trick is often to

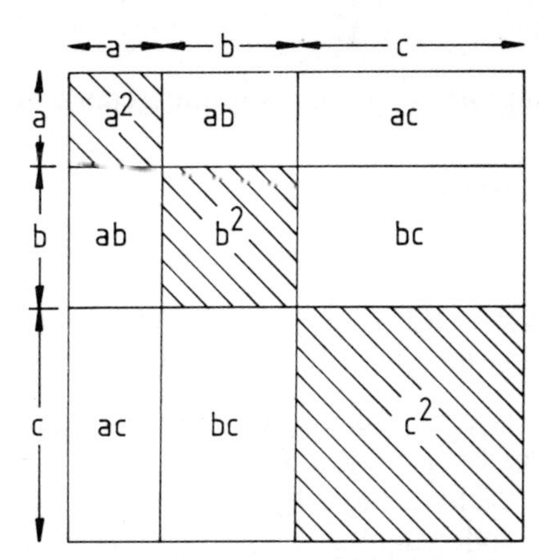

Figure 1.3 Geometric interpretation of $(a + b + c)^2$

refer to real numbers (rather than the algebraic form) and, quite simply, see if it works. For example,

$$\sqrt{9+16} = \sqrt{25} = 5$$

but

$$\sqrt{9} + \sqrt{16} = 3 + 4 = 7$$

and it is immediately apparent that the root of the sum (5) does not equal the sum of the roots (7).

Much of Section 1.7, on power notation, was developed by using specific powers and trying to find the general rules which should apply. But, if in doubt, a real example can often help immediately. It is perhaps easier to see that

$$(a^3)^2 = a^3 \times a^3 = a \times a \times a \times a \times a \times a = a^6 = a^{3 \times 2}$$

rather than the general form

$$(a^m)^n = a^{m \times n}.$$

So, at least try and see if a doubtful algebraic formula makes arithmetic sense with actual numbers, instead of alphabetic symbols.

Faulty interpretations

A common error is to introduce or, more likely, to ignore brackets and alter the meaning of an expression. For example,

$2x^2$ and $(2x)^2$ are *not* the same

since

$$2x^2 = 2.x.x$$
$$(2x)^2 = 2x.2x = 2.x.2.x = 4x^2.$$

This is particularly important when using the general form of an expanded power, such as

$$(X + Y)^3 = X^3 + 3X^2Y + 3XY^2 + Y^3$$

whereas

$$\begin{aligned}(2x + 5y)^3 &= (2x)^3 + 3(2x)^2\,(5y) + 3(2x).(5y)^2 + (5y)^3 \\ &= 8x^3 + 60x^2y + 150xy^2 + 125y^3.\end{aligned}$$

Notice that when X is replaced by $2x$, as in the example above, then X^3 must be interpreted as $(2x)^3$, not as $2x^3$.

Two other errors worthy of note by virtue of their frequency rather than their subtlety are the following:

$\log x^m$ is not the same as $(\log x)^m$.

The distinction can probably be made clearer if we introduce brackets into the first function, thus;

$$\log(x)^m \neq (\log x)^m.$$

From Section 1.14 we have

$$\log(x)^m \equiv m \log x$$

so that putting

$$\log(x)^m \equiv (\log x)^m \qquad \text{INCORRECT}$$

would be tantamount to putting

$$m \log x \equiv (\log x)^m \qquad \text{INCORRECT}$$

or

$$mA \equiv A^m \qquad \text{INCORRECT}$$

(since we could always find a single number A which equalled $\log x$). This last form is more clearly nonsense, since it suggests that multiplication by m is identical to raising to the power m: that is,

$$Am = A^m \qquad \text{INCORRECT}$$

(in real numbers, that

2.3 is identical to 2^3

or

6 is identical to 8

which is perhaps the clearest example of an incorrect identity).

We have used the identity symbol, $\equiv$, above which means *is identically equal to*. It implies that any such equality always holds, regardless of the value of any variable involved, rather than indicating equality for a particular value of the variable. For example,

$$\log x^m \equiv m \log x$$

is true for any x, whereas

$$x + 2 = 5$$

is only true if x takes the value $+3$.

Finally in this section, we can perhaps point out the obvious just once more, and mention that

e^x and x^e

are not the same functions. This is a popular but invalid transposition when the subject of differentiation arises (Chapters 7 and 8). The distinction may be a little clearer if we choose any number rather than e (a little

mysterious since it disguises its very specific arithmetic value of 2·718 . . .).
So, for example

$$\underbrace{\underset{(3.3.3.\ldots)}{3^x}}_{x \text{ terms}} \quad \text{is not the same as} \quad \underbrace{\underset{(x.x.x)}{x^3}}_{3 \text{ terms}} .$$

Solution to a quadratic equation

We finally draw the reader's attention to the useful formula for the solution of a quadratic equation. If

$$ax^2 + bx + c = 0$$

then

$$x = \frac{-b \pm \sqrt{b^2 - 4ac}}{2a} . \tag{i}$$

This formula is introduced here because it is sometimes used incorrectly by the unwary for any similar polynominal type equation where there are only 3 coefficients. It *cannot* be used, for example, to solve the cubic:

$$ax^3 + bx^2 + c = 0.$$

It might be mentioned, however, that the formula can be useful for higher order equations if these can be reduced to the quadratic form by suitable substitution.

For example, if

$$ax^4 + bx^2 + c = 0$$

then if we put $X = x^2$, we have

$$aX^2 + bX + c = 0$$

and this can be solved for X (and hence x) using the formula (i).

Or, if we have

$$ax^3 + bx^2 + cx = 0$$

then x is a common factor throughout (so that $x = 0$ is a root) and we have

$$x(ax^2 + bx + c) = 0$$
$$\therefore\ x = 0 \text{ or } ax^2 + bx + c = 0$$

and we proceed using (i) above to find the other two solutions for x.

1.20 BASIC EXAMPLES FOR THE STUDENT

B1.1 Rewrite using the symbols $<, >, \leqslant, \geqslant$.

(a) x is not as great as 300
(b) x is at most 10
(c) x is more than 15 but not as high as 28
(d) x is not more than 150 but is at least 96
(e) x is not less than 50 but not more than 150.

B1.2 Evaluate the following

(a) $3\,|4-5|+3\,|4|+3(|4|-|-5|)$
(b) $3\times 2+6/2\times 7+4$
(c) $\dfrac{3-(-3-(4-5)-2)-7}{-(-(-1))-1}$.

B1.3 Reduce the following inequalities to their simplest form

(a) $x-1<1-x$
(b) $x+z\leqslant 2+(y-2+x)$
(c) $3(x-2)+4(1-x)\leqslant -3$.

B1.4 Find the square roots of the following numbers:
0·04, 0·4, 0·00004, 4000, 40000, 40.

B1.5 Find the cube roots of the following numbers:
0·027, 27, 27000, 0·008, 8000.

B1.6 Evaluate or simplify the following

(a) $(27^{1/3}\times 3^2\times 81^{1/2})^{1/5}$
(b) $(4^{3/2}\times 2^3\times 64^{1/2})^{1/9}$
(c) $(x^2y^4)^{3/2}$
(d) $(x+y)^0$
(e) $\sqrt{xy}\,(x^2y^3)^{1/2}y^{1/4}$
(f) $\sqrt{xy}/(x^2y^3)$.

B1.7 Use logarithms to evaluate

(a) 0·37/0·002
(b) $\sqrt[3]{4000}$
(c) $(0{\cdot}99)^{19}$
(d) $(0{\cdot}78/0{\cdot}0084)^{1/2}$
(e) $\sqrt[6]{0{\cdot}08361}$
(f) $(3{\cdot}0149\times 4{\cdot}213)^0$.

B1.8 If $x_1=-3,\ x_2=0,\ x_3=4,\ x_4=2$, evaluate the following:

(a) $\left(\sum_{i=1}^{3} x_i\right)^2$
(b) $\sum_{i=1}^{4}(ix_i+1)$
(c) $\sum_{i=3}^{4}(x_i^2-i^2)$
(d) $\sum_{i=1}^{3}(4x_i^2+3x_i+1)$.

1.21 INTERMEDIATE EXAMPLES FOR THE STUDENT

I1.1 Rewrite, using mathematical notation,

(a) x is not greater than 5 and not less than 2

(b) y is more than z, which is greater than 6, but both y and z are less than 12

(c) x is at least 20 and at most 80.

I1.2 Evaluate the following:

(a) $\dfrac{|6-(-4)|}{|(-3)-(-1)|}$ (b) $\dfrac{|-6(2+3)+5|}{|4-8|}$ (c) $\dfrac{(6-3)(4+1)}{6-(3+4)+1}$

(d) $(2 \times (2-6) + 9 \times 4(4-2))/10$ (e) $4 - 3/6 \times 2 + 5$.

I1.3 Solve the following equations and inequalities

(a) $2x - 3(x-1) = 4(2x+1) - 19$

(b) $2(x+2) - 3(2x-1) < 3(2-x)$

(c) $4(3x+7) \geqslant (x+12) - 4(x-4)$.

I1.4 Evaluate

(a) 4^3 (d) $2^{5/2} \times 2^{1/2}$

(b) $4^{1/2}$ (e) $10^{1/3} \times 10^2 \times 10^{2/3}$

(c) 4^0 (f) $16/16^{1/2}$.

I1.5 Find the square roots of the following numbers

(a) 0·81 (d) 20000

(b) 0·000081 (e) 0·000000002.

(c) 8·1

I1.6 Use logarithms to evaluate the following

(a) the cube root of 24·8

(b) The fourth root of 0·0028

(c) 0·95 raised to the sixth power.

I1.7 If $x_1 = 2, x_2 = 0, x_3 = -1, x_4 = 7, x_5 = -1$, evaluate the following

(a) $\sum_{i=2}^{5} x_i^2$ (c) $\sum_{i=1}^{4} x_i(x_{i+1})$

(b) $\sum_{i=1}^{5} ix_i$ (d) $\sum_{i=1}^{4} x_i(x_i + 1)$.

More Advanced Examples for Part One

A1.1 Express the following in its simplest form

$$\frac{(2x^2 + xy - 3y^2)\sqrt{x + y}}{\sqrt{x^2 - y^2}\sqrt{x - y}}.$$

A1.2 Express the following in its simplest form

$$\frac{(\sqrt[4]{xy})^{14}}{\sqrt{x^5}\sqrt{xy}\,(x^{1/4}y^{1/2})^6}.$$

A1.3 The mean value, $\bar{x}$, of a set of N observations is given by

$$\bar{x} = \frac{x_1 + x_2 + x_3 + \ldots + x_N}{N} = \frac{1}{N}\sum_{i=1}^{N} x_i.$$

The standard deviation of the observations, s, is given by

$$s^2 = \overline{(x_i - \bar{x})^2} = \frac{1}{N}\sum_{i=1}^{N}(x_i - \bar{x})^2.$$

Show that this formula for s can be simplified to the form often used for calculations:

$$s^2 = \overline{x^2} - (\bar{x})^2.$$

A1.4 (a) Show that

$$\log(x^m) = m \log x$$

for logarithms to any base.

(b) Sketch on the same graph the three curves

(i) $y = \log x$

(ii) $y = \log\left(\dfrac{1}{x}\right)$

(iii) $y - \log(x^2)$.

PART TWO

Introduction

Set theory, though a fairly recent development in mathematics, has proved to be a powerful one, particularly in terms of its ability to demonstrate greater unity among the various branches of mathematics. The equivalence of algebraic and geometric representations are emphasized in Chapter 3, for example, where functions and graphs are both developed from the basic notation of set theory. The value of algebraic and geometric representations as alternative expressions of the same problem runs as a minor theme throughout the book, in fact, since it is felt that an ability to define or interpret problems in alternative ways can assist our understanding and capacity to solve them, as well as strengthen our familiarity with the variety of techniques available.

Set theory is introduced from first principles in Chapter 2, and we aim here to assist the reader in formulating a rigorous mathematical statement from one expressed in everyday language. Set theory provides a particularly useful topic for starting with, since at the level with which we shall be concerned, most of the ideas are intuitively obvious. The importance of rigour in interpreting a language statement is emphasized in the section on basic logic, and the value of equivalent geometric figures is illustrated by the use of Venn diagrams, where the value of sets in logical classification becomes apparent. Venn diagrams are also useful in developing some of the basic ideas of probability, and this is demonstrated through examples.

CHAPTER 2

Sets and Logic

2.1 INTRODUCTION

In this chapter we are concerned with categories, or *sets* of objects. We begin with a brief consideration of *logic*, and the way it is used in drawing valid conclusions from a few simple statements of fact.

For example

I All students read books. Tony reads books. What can you conclude from these two statements?
 A. Some students are called Tony
 B. Tony is a student
 C. Tony could be Dutch
 D. Tony could be a student.

II If one eats too much, one becomes ill. This man is ill. What can you conclude from these two statements?
 A. This man has eaten too much
 B. This man might have some disease
 C. It can't be said that this man hasn't eaten too much
 D. You can't conclude anything.

III From a sample of 50 students, one half are English, and 21 have black hair. 15 are neither English nor have black hair. How many black-haired English are there in the group?

Notice that all the statements, and conclusions, make use of categories which are apparently well-defined: students, the ill, black-haired, and so on. By *well-defined* we mean here that a person

is either a student	or a non-student
and is either ill	or not-ill (well)
and is either black-haired	or non-black-haired

with no room for doubt within each pair of classes—we cannot allow one person to be both *ill* and *not-ill* simultaneously.

Examples I and II above will be examined more fully towards the end of this chapter (Section 2.8), but we solve example III in the following section to demonstrate some basic ideas of sets.

2.2 A NUMERICAL EXAMPLE

In example III above, we can imagine the entire population to be situated within a box, within which one group labelled *English* falls inside a circle. Non-English fall outside this circle but within the box (if they are part of the *population*) (Figure 2.1).

Similarly we could consider another group taken from the whole population—*black-haired* say—and include these within another circle (Figure 2.2). We now have a diagram which defines four groups of people:

A: Persons who are *neither* English *nor* black-haired
B: Persons who *are* English but *are not* black-haired
C: Persons who *are both* English and *are* black-haired
D: Persons who *are* black-haired but *are not* English.

We can solve the third problem III from this diagram, and to do so we shall introduce some simple notation using the modulus sign

$|\mathbf{A}|$ means the number of persons in group **A**
$|\mathbf{B}|$ means the number of persons in group **B**, and so on.

The original information in the problem can now be stated in the following way:

(1) $|\mathbf{A}| + |\mathbf{B}| + |\mathbf{C}| + |\mathbf{D}| = 50$ ... total sample (population)
(2) $|\mathbf{B}| + |\mathbf{C}| = 25$ ... two groups (**B** and **C**) include the English
(3) $|\mathbf{C}| + |\mathbf{D}| = 21$ ... two groups (**C** and **D**) have black hair
(4) $|\mathbf{A}| = 15$ ... one group is neither English nor black-haired
$|\mathbf{C}| = ?$ ... the problem, the number of black-haired English.

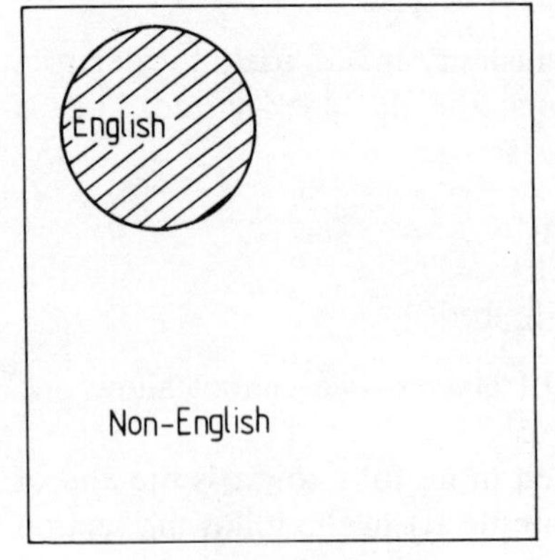

Figure 2.1

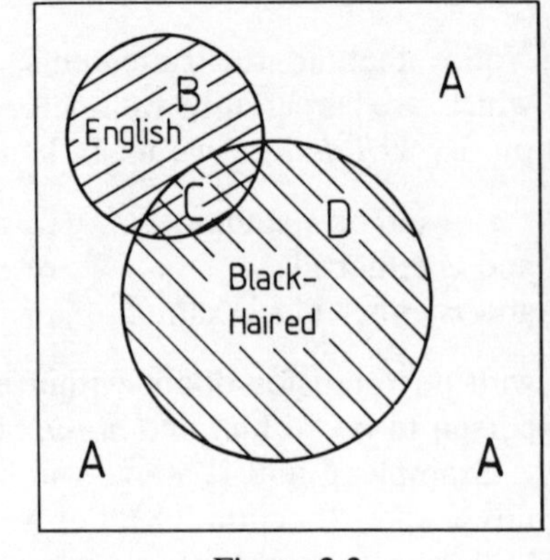

Figure 2.2

(1)–(2) gives $|\mathbf{A}| + |\mathbf{D}| = 25$, and (4) gives $|\mathbf{A}| = 15$, so that $|\mathbf{D}| = 10$. But also $|\mathbf{C}| + |\mathbf{D}| = 21$ so that $|\mathbf{C}| = 11$ ... the number of black-haired English.

Therefore there are 11 black-haired English in the group of 50.

2.3 FORMAL SET NOTATION AND THEORY

The example introduced some of the basic ideas behind set theory, but in order to proceed further we shall introduce some formal definitions and notation.

A **set** is any well-defined collection of objects (e.g. *students*) and is usually denoted by capital letters **A**, **B**, **C**, **X**, **Y**, **Z**, etc. Sets must be described in one of two ways (*tabulation* or by *defining property*) in such a way that no ambiguity exists. *All students* forms a well-defined set at any time, while *the fifty best students* leaves room for some doubt—what do we mean by *best* students?

An **element** is one object from a set (e.g. a particular student) and is usually denoted by lower-case letters a, b, c, x, y, z, etc. To show that a is an element of a set **A** we would write

$a \in \mathbf{A}$ $\quad$ a belongs to **A** $\quad$ or $\quad$ a is an element of **A**.

Also,

$a \notin \mathbf{A}$ $\quad$ a is not an element of **A**.

Tabulation can be used when it is possible and convenient to name and list every element of the set; for example,

$\mathbf{A} = \{4, 6, 8, 10, 12\} \qquad \mathbf{B} = \{\text{March, May}\}.$

Defining property can be used when all the elements of a set have some property in common, and this property can be stated as a defining rule; for example,

$$\mathbf{A} = \{x \mid x \text{ is an even number between 3 and 13}\}$$
$$\mathbf{B} = \{x \mid x \text{ is any month beginning with the letter M}\}.$$

The vertical line means *such that* so that the first statement above is read: **A** *is the set of all* x *such that* x *is an even number between 3 and 13*. In some cases we can only use the method of a defining property, such as when there is an infinite number of elements in the set. Where the elements have no common property, then tabulation must be used.

The **power** of a set **A** is written as $|\mathbf{A}|$, and is defined to be the number of elements in **A**. (We have used this definition already in solving problem III.)

The **null set** or **empty set** has no elements and is written as $\emptyset$. By definition $|\emptyset| = 0$. The null set should not be confused with the set {0} which contains one element (zero) and so has a power of one.

Subsets. A set **A** is said to be a subset of a set **B** if all the elements of **A** are also elements of **B**. We write

$\mathbf{A} \subset \mathbf{B}$ **A** is a subset of **B**

and

$\mathbf{A} \not\subset \mathbf{B}$ **A** is not a subset of **B**.

For example,

$\mathbf{A} = \{0, 1, 2, 3, 5\} \quad \mathbf{B} = \{0, 1\} \quad \mathbf{C} = \{2, 3, 4\}$

then

$\mathbf{B} \subset \mathbf{A}, \quad \mathbf{C} \not\subset \mathbf{A}.$

N.B. The null set is considered to be a subset of every set.

Proper subsets. According to our definition of subsets above, any set **A** is also a subset of itself. However, if set **A** is a subset of set **B**, and at least one element of **B** is not included within **A**, then **A** can be termed a *proper subset* of **B**. More formally, set **A** is a proper subset of set **B** if

$\mathbf{A} \subset \mathbf{B}$ and $\mathbf{A} \neq \mathbf{B}$.

Universe. The universe is the largest set or the complete set in a problem (e.g. the *whole population* in the earlier examples) and is denoted by the capital letter **U**. For example if we were considering days of the week as the elements in a set problem, then the *universe* would be all seven days

U = {Mon., Tues., Wed., Thurs., Fri., Sat., Sun.}.

Equality of sets implies all of the following (for two sets, **A** and **B**)

$\mathbf{A} = \mathbf{B} \qquad \mathbf{A} \subset \mathbf{B} \qquad \mathbf{B} \subset \mathbf{A} \qquad |\mathbf{A}| = |\mathbf{B}|$

i.e. **A** and **B** have identical elements, and are of the same power.

2.4 PICTORIAL REPRESENTATION: VENN DIAGRAMS

Sets can be represented pictorially as areas, usually circular (for convenience only) within which all the elements of the set are included. There is no relationship between the size of the circle and the power of the set (number of elements). The universe is usually represented by a rectangle enclosing all the relevant sets.

Examples

Two sets **A**, **B**.

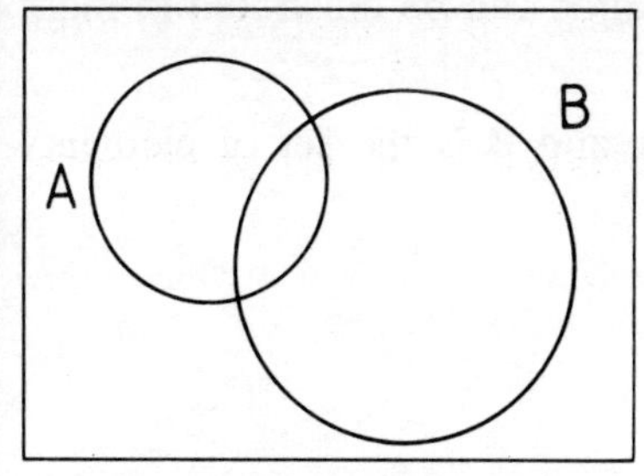

Figure 2.3(a) General case of two sets, with some elements common to both **A** and **B**; some elements exclusive to only either **A** or **B**; and some elements of the universe included in neither **A** nor **B**

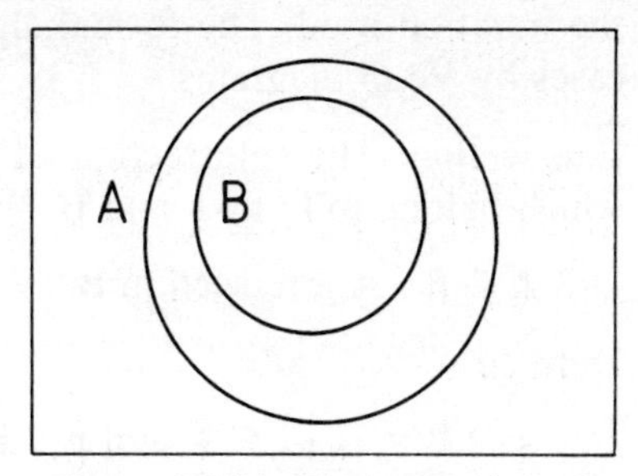

Figure 2.3(b) The case where **B** is a subset of **A**. In this case, or in the similar case where **A** is a subset of **B**, the two sets **A** and **B** are said to be *comparable*

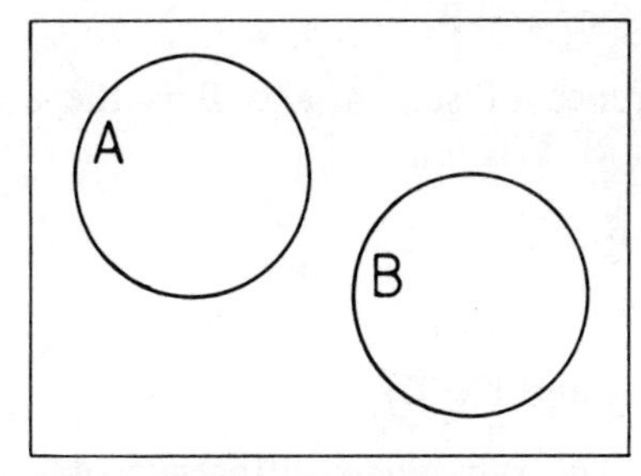

Figure 2.3(c) The case where **A** and **B** have no elements in common, and are said to be *disjoint*

Three sets **A**, **B**, and **C**.

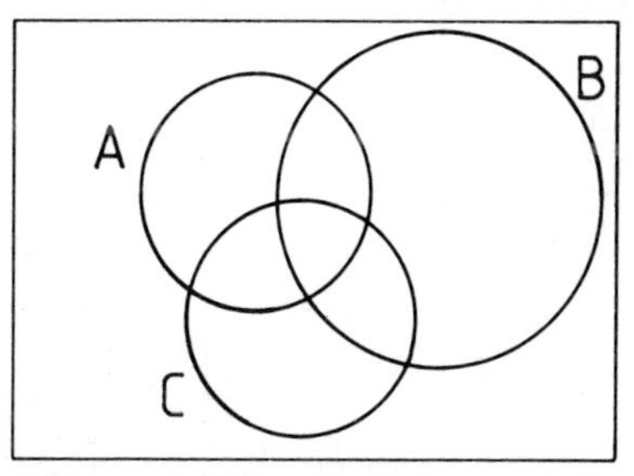

Figure 2.3(d) General case of three sets where elements belonging to all three, any two, only one, or none of the three sets are permitted

2.5 SET OPERATIONS: TWO SETS

We now consider the types of operations which can be applied to sets, and the notation used. The formal algebraic notation can be illustrated in most cases by Venn diagrams.

Intersection. The intersection of two sets **A** and **B** is the set of elements which belong to both **A** and **B**. Written:

$\mathbf{A} \cap \mathbf{B}$ **A intersection B.**

Defined

$$\mathbf{A} \cap \mathbf{B} = \{x \mid x \in \mathbf{A} \text{ and } x \in \mathbf{B}\}.$$

Union. The union of two sets **A** and **B** is the set of elements which belong to **A** or to **B** or to both **A** and **B**. Written:

$\mathbf{A} \cup \mathbf{B}$ **A union B.**

Defined:

$$\mathbf{A} \cup \mathbf{B} = \{x \mid x \in \mathbf{A} \text{ or } x \in \mathbf{B}\}.$$

Difference. The difference of sets **A** and **B** is the set of elements which belong to **A** but not to **B**. Written:

$\mathbf{A} - \mathbf{B}$ **A minus B.**

Defined:

$$\mathbf{A} - \mathbf{B} = \{x \mid x \in \mathbf{A} \text{ and } x \notin \mathbf{B}\}.$$

Symmetric difference. The symmetric difference of sets **A** and **B** is the set of elements which belong to **A** or to **B** but not to both. Written:

$\mathbf{A} \Delta \mathbf{B}$ symmetric difference of **A** and **B**.

Defined:

$$\mathbf{A} \Delta \mathbf{B} = (\mathbf{A} \cup \mathbf{B}) - (\mathbf{A} \cap \mathbf{B}).$$

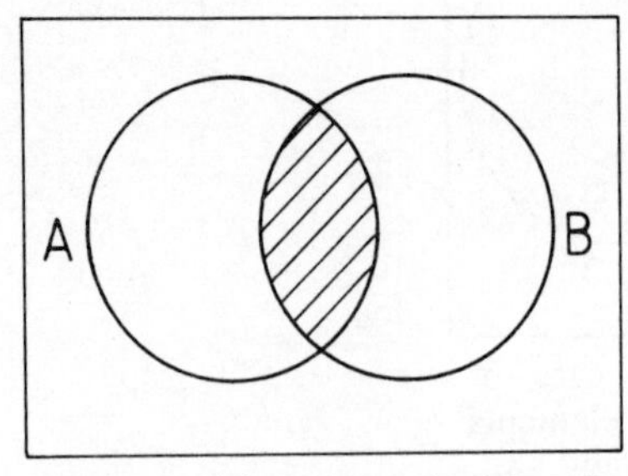

Figure 2.4(a) Intersection: $\mathbf{A} \cap \mathbf{B}$ is shaded

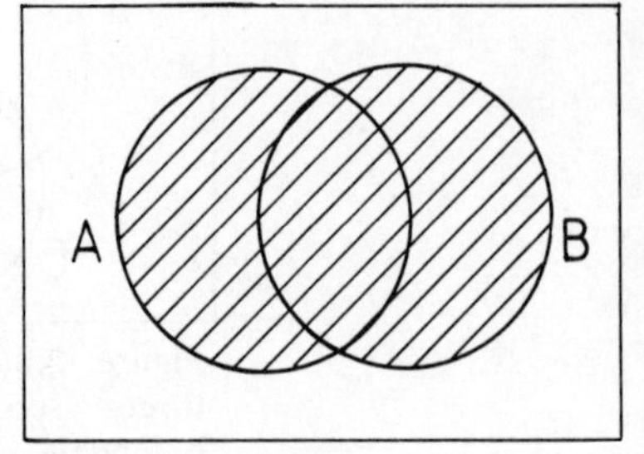

Figure 2.4(b) Union: $\mathbf{A} \cup \mathbf{B}$ is shaded

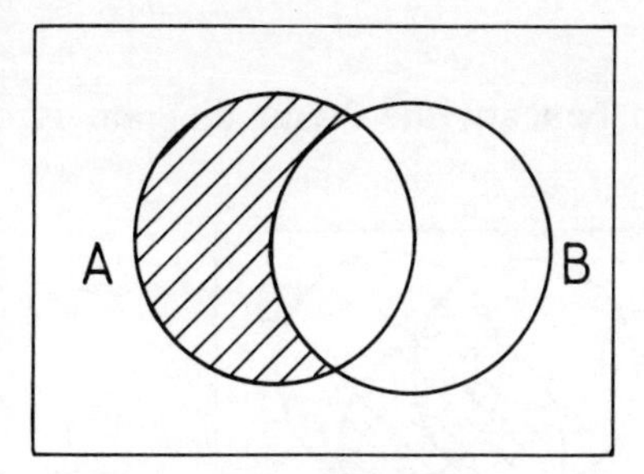

Figure 2.5(a) Difference: **A** − **B** is shaded

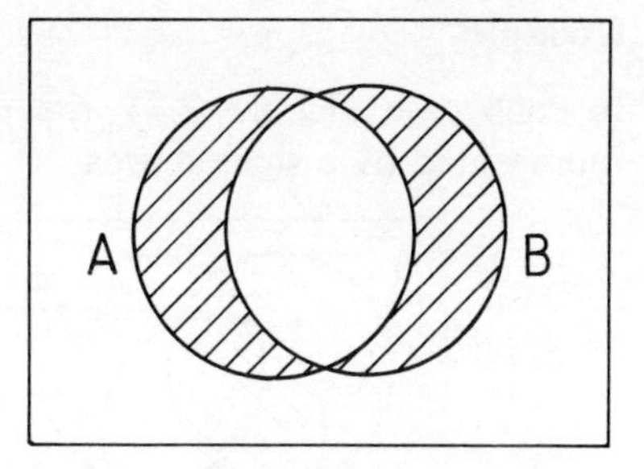

Figure 2.5(b) Symmetric difference: **A** Δ **B** is shaded

Complement. The complement of a set **A** is the set of all elements in the universe which do not belong to **A**. Written:

$\mathbf{A}'$ **A** dash or **A** complement.

Defined:

$$\mathbf{A}' = \{x \mid x \notin \mathbf{A}\}.$$

Note that $\mathbf{A}'$ could be defined by:

$$\mathbf{A}' = \mathbf{U} - \mathbf{A}.$$

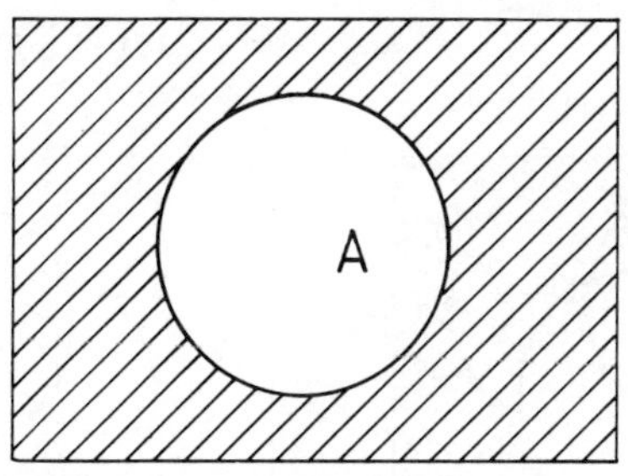

Figure 2.6 Complement: $\mathbf{A}'$ is shaded

2.6 SET OPERATIONS: THREE SETS OR MORE

The same set of operations (union, intersection, etc.) may be applied to groups of three or more sets, but brackets may be required to indicate which operation should be carried out first. The rules governing the use of brackets are the same as those governing the algebra of numbers. For example,

$$\mathbf{A} \cap (\mathbf{B} \cup \mathbf{C})$$

means the intersection of **A** with the union of **B** and **C**. The union operation is carried out before the intersection operation in this example.

Examples

In each case (Figure 2.7), the expression beneath the Venn diagram is represented by a shaded area.

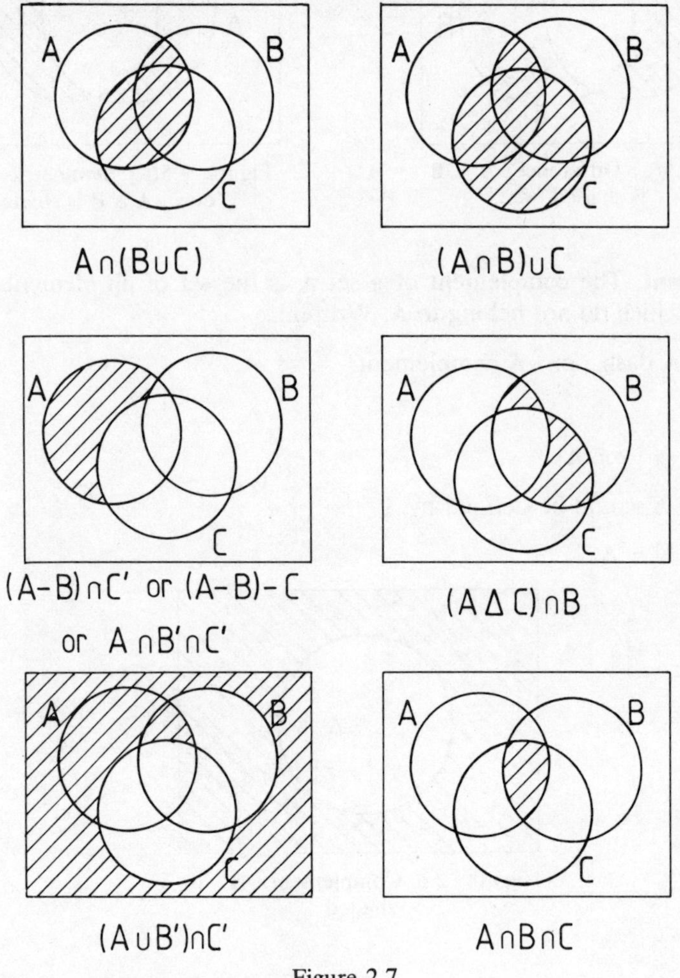

Figure 2.7

2.7 SETS IN COUNTING PROBLEMS

Example

A survey on the breakfast habits of a group of 50 people gives the following results:

29 people eat at least kippers
30 people eat at least kornflakes
17 people eat at least krispies

13 people eat at least kippers and krispies
11 people eat at least kornflakes and krispies
17 people eat at least kippers and kornflakes
8 people eat kippers, kornflakes and krispies.

How many of the people in the sample eat

(i) only one type of food?
(ii) only two of the three types of food?
(iii) none of the three types of food?

(Ignore breakfast foods other than kippers, kornflakes, or krispies.)

The problem is most easily solved using a Venn diagram (Figure 2.8). The solution to such problems always begins at the centre of the diagram, where all three sets intersect. The figure 8 can then be inserted directly, as the number of people eating all three types of food. We then proceed outwards from the centre. Thus, the statement *13 people eat at least kippers and krispies* gives the *total* figure for the intersection of the kippers and krispies sets. Since 8 have already been accounted for, this leaves 5 (i.e. 13 − 8) as the figure appropriate to the subset: (kippers and krispies but not kornflakes). In other words, the statement *13 people eat (at least) kippers and krispies* includes those people who also eat kornflakes, and those who do not. Often the words *at least* are omitted in the original statement of the problem, which makes this point less obvious.

So, we find that

(i) (7 + 10 + 1) people eat only 1 food

$\therefore$ 18 people eat only 1 food.

(ii) (9 + 3 + 5) people eat only 2 foods

$\therefore$ 17 people eat only 2 foods.

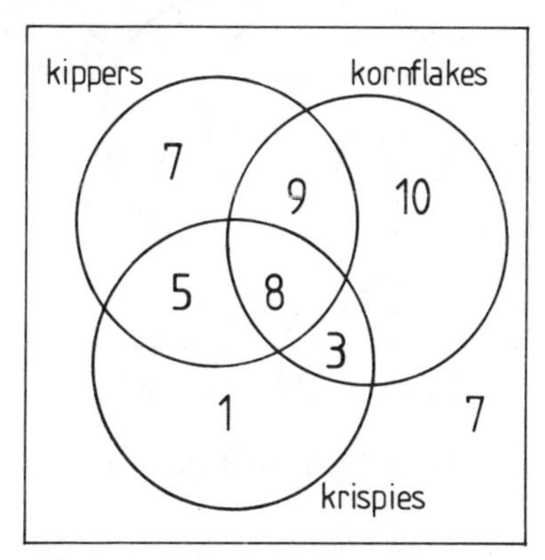

Figure 2.8 Solution to the breakfast counting problem

(iii) $|50 - (7 + 9 + 10 + 5 + 8 + 3 + 1)|$ people eat none of the 3 foods

$\therefore$ 7 people eat none of the 3 foods.

Derivation of general formulae for counting problems

The number of elements in $\mathbf{A}$ is $|\mathbf{A}|$; the number of elements in $\mathbf{B}$ is $|\mathbf{B}|$, etc. Therefore $|\mathbf{A} \cup \mathbf{B}| = |\mathbf{A}| + |\mathbf{B}| - |\mathbf{A} \cap \mathbf{B}|$, which can be seen from a Venn diagram as follows. The shaded region in Figure 2.9(a) represents

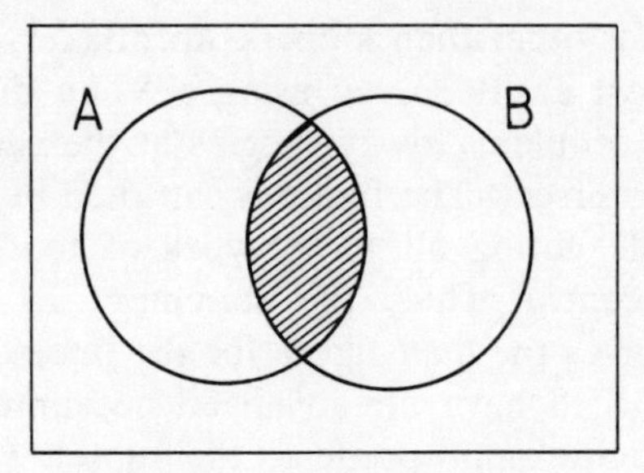

Figure 2.9(a) Two sets

the intersection of $\mathbf{A}$ and $\mathbf{B}$, i.e. $\mathbf{A} \cap \mathbf{B}$. Now if we add the number of elements in $\mathbf{A}$ (i.e. $|\mathbf{A}|$) to the number of elements in $\mathbf{B}$ (i.e. $|\mathbf{B}|$) we will have counted those elements in $(\mathbf{A} \cap \mathbf{B})$ twice. So, to count the total number of elements in $\mathbf{A}$, or in $\mathbf{B}$, or in both, we must add the number of elements in $\mathbf{A}$ to the number in $\mathbf{B}$, and subtract any that we have double counted. We then find that

$$|\mathbf{A} \cup \mathbf{B}| = |\mathbf{A}| + |\mathbf{B}| - |\mathbf{A} \cap \mathbf{B}|.$$

For three sets $\mathbf{A}$, $\mathbf{B}$, and $\mathbf{C}$, Figure 2.9(b), we find by a similar argument

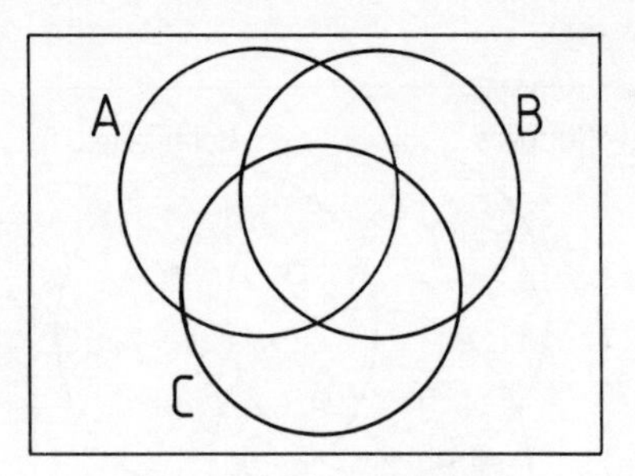

Figure 2.9(b) Three sets

that the number of elements in $\mathbf{A}$ or in $\mathbf{B}$ or in $\mathbf{C}$ or in any combination of $\mathbf{A}$, $\mathbf{B}$, and $\mathbf{C}$ is

$$|\mathbf{A} \cup \mathbf{B} \cup \mathbf{C}| = |\mathbf{A}| + |\mathbf{B}| + |\mathbf{C}| - |\mathbf{A} \cap \mathbf{B}| \\ - |\mathbf{A} \cap \mathbf{C}| - |\mathbf{B} \cap \mathbf{C}| + |\mathbf{A} \cap \mathbf{B} \cap \mathbf{C}|.$$

The same principles can be employed for generating formulae for more complex configurations of sets.

Example

A survey among a class of students shows that everyone

(i) is either a scientist or a poet, and
(ii) has either brown eyes or blue eyes.

30% have brown eyes and 45% are scientists. What can you say about the proportion of blue-eyed poets?

Referring to Figure 2.9(c), let **B** be the set of blue-eyed students and let **P** be the set of poets. **U** is the total class. The question is seeking information on $|\mathbf{B} \cap \mathbf{P}|$.

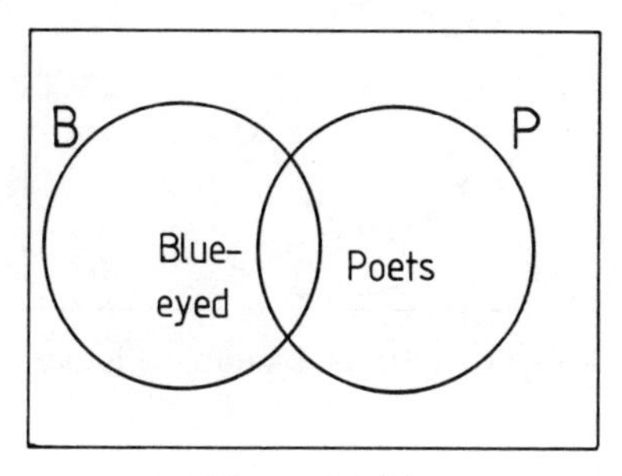

Figure 2.9(c)

Using percentages, $|\mathbf{B}| = 70$, $|\mathbf{P}| = 55$, $|\mathbf{U}| = 100$. Now

$|\mathbf{B} \cup \mathbf{P}| \leqslant 100$ blue eyed and/or poetic cannot exceed 100%

But

$$|\mathbf{B} \cup \mathbf{P}| = |\mathbf{B}| + |\mathbf{P}| - |\mathbf{B} \cap \mathbf{P}| = 70 + 55 - |\mathbf{B} \cap \mathbf{P}|$$
$$\therefore |\mathbf{B} \cap \mathbf{P}| = 125 - |\mathbf{B} \cup \mathbf{P}| \geqslant 125 - 100 = 25$$
$$\therefore |\mathbf{B} \cap \mathbf{P}| \geqslant 25$$

Also, since $|\mathbf{P}| = 55$, $|\mathbf{B}| = 70$ then $|\mathbf{B} \cap \mathbf{P}| \leqslant 55$

$$\therefore 25 \leqslant |\mathbf{B} \cap \mathbf{P}| \leqslant 55.$$

Therefore the proportion of blue-eyed poets lies between 25% and 55% inclusive.

Problem A2.4 in the more advanced examples for Part Two of this book uses the ideas of counting and sets to develop the basic laws of probability very simply. (The solution to this problem is provided at the end of the book.)

2.8 SETS IN LOGIC

We now return to the basic problems in logic posed in Section 2.1.

Example I

Let **B** be the set of all book-readers, let **S** be the set of all students, and let t be the *element* (or person) Tony. *All students read books* implies $\mathbf{S} \subset \mathbf{B}$. *Tony reads books* implies $t \in \mathbf{B}$.

From Venn diagram in Figure 2.10(a) it can be easily seen that *Tony* must fall within the circle labelled **B**, but may or may not fall within the circle **S**.

Conclusion: *Tony could be a student*.

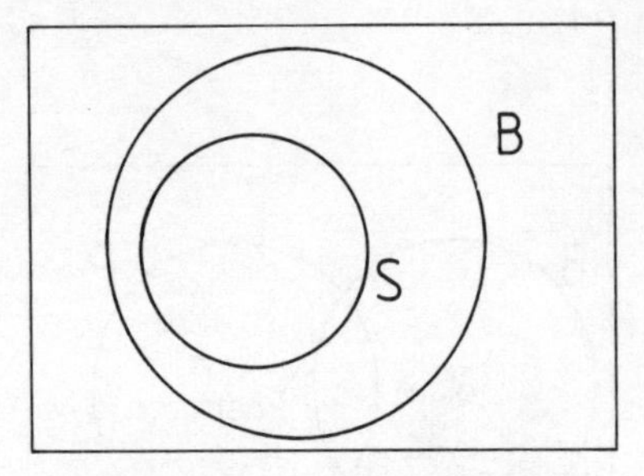

Figure 2.10(a) Example I logic

Example II

Let **I** be the set of ill people, let **E** be the set of people who have eaten too much, and let m be the element (or man) specified. *If one eats too much one becomes ill* implies $\mathbf{E} \subset \mathbf{I}$. *This man is ill* implies $m \in \mathbf{I}$.

From the Venn diagram in Figure 2.10(b) it can be seen that the man must fall within the circle labelled **I**, but *may* or *may not* fall within the circle **E**.

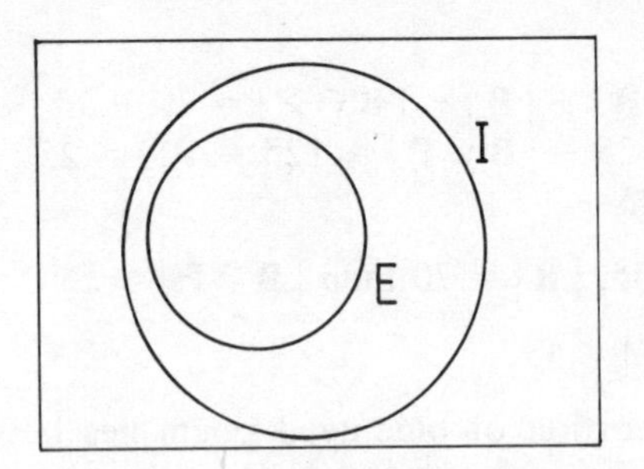

Figure 2.10(b) Example II logic

Now,

if he should fall *outside* circle **E**, he has *not* eaten too much
if he should fall *within* circle **E**, he has eaten too much.

We cannot distinguish between the two cases, through lack of evidence, and so all we can conclude is:

It can't be said that this man hasn't eaten too much.

2.9 LOGICAL CHAINS

Consider the statements:

Irrational men are never sane
Rational people never contradict themselves
Great men are always idealists
The philosophy and the actions of idealists always contradict.

What conclusions can we draw from these statements?

Let **R**, **S**, **C**, **G**, and **I** be the sets of Rational, Sane, Contradicting, Great, and Idealistic men respectively. (Note that for set theory to be valid we must be able to define each of these groups, allowing no room for ambiguity. This example is illustrative and we have assumed for the purposes of illustration that a suitable definition of *Rational*, *Sane*, etc. exists.)

The four statements therefore give us the four relations

(i) $\mathbf{R}' \subset \mathbf{S}'$
(ii) $\mathbf{R} \subset \mathbf{C}'$ or $\mathbf{C} \subset \mathbf{R}'$
(iii) $\mathbf{G} \subset \mathbf{I}$
(iv) $\mathbf{I} \subset \mathbf{C}$

and these can be formed into a *logical chain*

$$\mathbf{G} \subset \mathbf{I} \subset \mathbf{C} \subset \mathbf{R}' \subset \mathbf{S}'.$$

Major conclusion $\mathbf{G} \subset \mathbf{S}'$ Great men are always insane
Minor conclusions: $\mathbf{G} \subset \mathbf{R}'$ Great men are irrational
$\mathbf{G} \subset \mathbf{C}$ Great men contradict themselves
$\mathbf{I} \subset \mathbf{S}'$ Idealists are insane
$\mathbf{I} \subset \mathbf{R}'$ Idealists are irrational
$\mathbf{C} \subset \mathbf{S}'$ Those who contradict themselves are insane.

Finally, we could also write the logical chain as

$$\mathbf{S} \subset \mathbf{R} \subset \mathbf{C}' \subset \mathbf{I}' \subset \mathbf{G}'$$

an equivalent logical statement.

This would then give us the major conclusion

$\mathbf{S} \subset \mathbf{G}'$ Sane men are never great

which is equivalent to

$\mathbf{G} \subset \mathbf{S}'$ Great men are always insane.

Alternative minor conclusions follow similarly.

2.10 BASIC EXAMPLES FOR THE STUDENT

B2.1 Tabulate the following sets
- (a) $\{x \mid x \text{ is a letter in the word } \textit{management}\}$
- (b) $\{x \mid x^2 = 4\}$
- (c) $\{x \mid x \text{ is an integer and } 0 < x < 5\}$
- (d) $\{x \mid x(x + 1) = 0\}$.

B2.2 If $\mathbf{A} = \{-1, 0, 1\}$, are the following correct or incorrect?
- (a) $0 \in \mathbf{A}$
- (b) $\{0\} \in \mathbf{A}$
- (c) $\emptyset \subset \mathbf{A}$
- (d) $\emptyset \in \mathbf{A}$
- (e) $\{x \mid x^2 = 1\} \subset \mathbf{A}$
- (f) $\{x \mid x^2 = -1\} \subset \mathbf{A}$.

B2.3 In a survey of 100 families, the number that read the previous day's issues of various newspapers were found to be:

The Times	28
The Guardian	30
The Morning Star	42
Times and Guardian	8
Times and M. Star	10
Guardian and M. Star	5
All three	3

- (a) How many read none of the three papers?
- (b) How many read the Morning Star as their only newspaper?
- (c) How many read The Guardian and The Morning Star but not The Times?

B2.4 In Venn diagrams showing three overlapping sets **A**, **B**, and **C**, shade the areas
- (a) $\mathbf{A} \cup \mathbf{B} \cup \mathbf{C}$
- (b) $\mathbf{A} \cup (\mathbf{B} \cup \mathbf{C})$
- (c) $(\mathbf{A} \cap \mathbf{B}) \cup (\mathbf{A} \cap \mathbf{C})$
- (d) $(\mathbf{A} \cup \mathbf{B})' \cap \mathbf{C}$
- (e) $\mathbf{A}' \cap \mathbf{B}' \cap \mathbf{C}'$.

B2.5 $\mathbf{A} = \{x \mid x \text{ is an odd integer}\}$

$\mathbf{B} = \{x \mid x^3 = 8\}$

$\mathbf{C} = \{x \mid (x^2 - 9)(x^2 - 1) = 0\}$

$\mathbf{D} = \{2, 3, 4\}$

Illustrate the comparability of **A**, **B**, **C**, and **D** in a Venn diagram.

2.11 INTERMEDIATE EXAMPLES FOR THE STUDENT

I2.1 Describe the following sets in set notation and state the number of elements in each.

(a) The countries in the E.E.C.
(b) The outcomes when a coin is tossed.
(c) The positive numbers.
(d) The children of Queen Victoria who are still alive.

I2.2 A company has £1 000 000 available for investment. Four possible projects are to be considered, requiring the following amounts of money:

Project a: £200 000 Project b: £200 000
Project c: £800 000 Project d: £600 000

What subsets of $\{a, b, c, d\}$ could the company choose without exceeding the expenditure limit? What subsets require exactly £1 000 000?

I2.3 The report of an inspector who inspected a batch of 100 items gave the following number of defective items due to hardness, finish, and dimensional faults. Would you believe the inspector?

All three defects	5
Hardness and finish	10
Dimension and finish	8
Dimension and hardness	20
Finish	30
Hardness	23
Dimension	50

I2.4 **A**, **B**, and **C** are subsets of a universe **U**

$\mathbf{U} = \{1, 2, 3, 4, 5, 6, 7, 8\}$ $\mathbf{A} = \{1, 2, 3, 4\}$ $\mathbf{B} = \{3, 4, 5\}$
$\mathbf{C} = \{6, 7, 8\}$.

Are the following statements correct or incorrect?

(a) $\mathbf{B} \subset \mathbf{A}$ (c) $\mathbf{B} \subset \mathbf{A} \cup \mathbf{C}$
(b) $(\mathbf{B} \cap \mathbf{A}) \subset \mathbf{C}'$ (d) $\mathbf{C} \subset (\mathbf{A} \cap \mathbf{B})'$.

I2.5 Let **S** be a set of 52 playing cards. Within the pack let **D** be the set of diamonds, **C** the clubs, **Q** the queens, and **A** the set of aces. One card is drawn from the pack. Express the following events in terms of **D**, **C**, **Q**, and **A** and their unions, intersections, and complements.

(a) the card is a diamond or a club
(b) the card is an ace but not a club
(c) the card is the queen of diamonds
(d) the card is a queen or an ace but not a diamond
(e) the card is a club but not a queen or an ace.

How many elements are there in each of these events? Hence what are the probabilities of the events, assuming the card is drawn at random?

CHAPTER 3

Sets, Relations, Functions, and Graphs

3.1 NUMBERS

In this chapter we aim to illustrate the basic ideas of graphs and graphical representation. In particular the equivalence of algebraic notation and graphical (geometric) notation will be emphasized, and the latter will be introduced whenever appropriate to help familiarization with this form of representation. We will at first be using the terminology of set theory, introduced in the previous chapter, and shall be largely concerned with the set of all real numbers, **N**, and its subsets.

Firstly the significance of the description *real* numbers should be explained.

Real numbers The set of all real numbers, **N**, includes all negative and positive numbers, both rational and irrational (see below), from minus infinity through zero to plus infinity, though excluding infinity (+ or −) itself. They are distinguished from . . .

Imaginary numbers which can be most simply expressed as ib where b is real, and $i = \sqrt{-1}$, and . . .

Complex numbers which can be most simply expressed as $a + ib$ where a and b are real, and $i = \sqrt{-1}$.

We shall be concerned throughout this book with real numbers. Imaginary and complex numbers are very useful in dealing with phenomena where dynamic (particularly cyclic) considerations arise. They are thus used in many branches of the physical sciences, and in the analysis of economic and business cycles, for example, but this goes beyond the scope of this book.

The set of real numbers, **N**, has some important subsets:

Integers . . . . −3, −2, −1, 0, 1, 2, 3, ...

Natural numbers 1, 2, 3, 4, . . . (positive integers, excluding 0).

Rational numbers (from the word ratio, i.e. ratio-nal). These are all numbers which can be expressed as a ratio of two integers. We cannot necessarily express them as exact decimals, however, as *recurrence* may

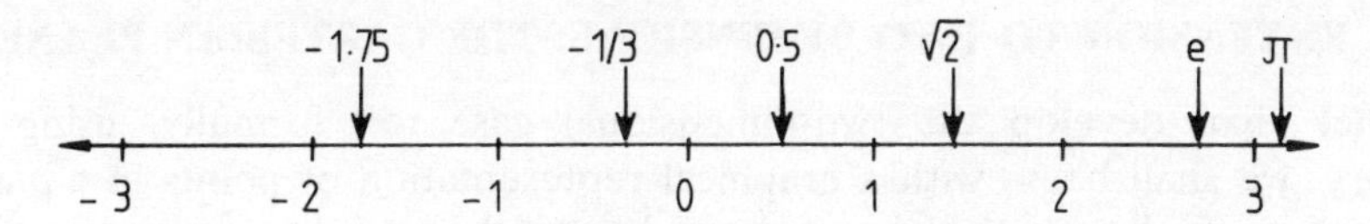

Figure 3.1 The number line showing some rational, irrational, and integral numbers and their corresponding points on the line

occur, leading to an infinite number of decimal places; for example,

17/4, 6/1, 1345/1000 or 4·25, 6·0, 1·345
1/3, 1/7 or $0{\cdot}\dot{3}$, $0{\cdot}\dot{1}4285\dot{7}$

(meaning 0·3333 . . . and 0·142857142857142 . . .).

Irrational numbers cannot be expressed exactly, either in decimal or in vulgar fraction form; for example,

$\pi \cong 3{\cdot}14159\ldots$
$e \cong 2{\cdot}71828\ldots$

π and e can both be defined in terms of converging infinite series.

3.2 PICTORIAL REPRESENTATION OF REAL NUMBERS AND INTERVALS

We can represent all real numbers on a line, the *number line*, as in Figure 3.1. Every real number (or element of the set **N**) can be represented by a single point on the line, and for every point on the line there is a corresponding real number. In other words there is a *one-to-one correspondence* between real numbers and points on the line.

Likewise, subsets of **N** which form *intervals*, or a continuous range, can be illustrated on the number line using the following conventions:

(i) All points *included* in the subset (or interval) are drawn as a continuous heavy line.
(ii) End points *included* in the subset are drawn as filled circles.
(iii) End points *excluded* from the subset are drawn as open, or empty, circles.

Examples

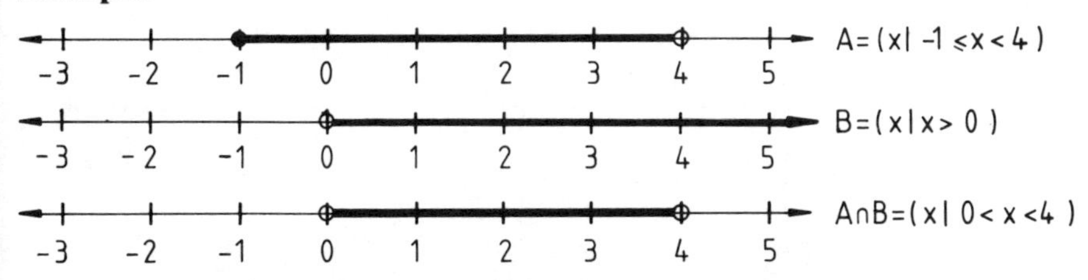

Figure 3.2 Subsets of **N** drawn as intervals on the number line

3.3 EXTENSION TO TWO DIMENSIONS: THE CARTESIAN PLANE

Rather than develop the two-dimensional case too formally, using set theory, we shall begin with a graphical representation of points in a *plane*, and refer only briefly to the accompanying set theory.

Firstly we draw two number lines at right angles to one another (when they are said to be *orthogonal*), and denote the distance along one of them in general as x, and along the other as y. These two lines or *axes* define a *plane* (or *cartesian plane*), in the plane of the paper, and intersect at the *origin* where we usually make $x = y = 0$.

Referring to Figure 3.3, we can now define the position of points such as p, q, r by reference to the two axes, and in this case find readily that

p is the point where $x = 1, y = 2$
q is the point where $x = 2, y = 1$
r is the point where $x = -2, y = -1$

It should now be obvious that we can denote the points p, q, and r by the abbreviated pairs of numbers thus:

p is the point $(1, 2)$
q is the point $(2, 1)$
r is the point $(-2, -1)$

where the *order* of the two numbers is important and the *signs* of the two numbers are important.

The usual conventions are that

(i) The x-coordinate precedes the y-coordinate
(ii) The x-axis is drawn horizontally, with positive values to the right
(iii) The y-axis is drawn vertically, with positive values in the upwards direction.

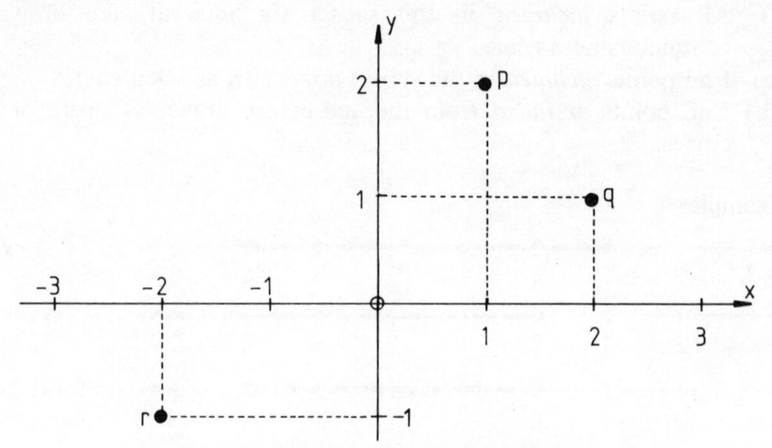

Figure 3.3 Points on the cartesian plane

In fact $(1, 2)$ is an example of what is sometimes called an *ordered pair*, and may often be treated as one element of a set of ordered pairs, as we shall show later. Also, it can be readily seen that there is a one-to-one correspondence between any point on the cartesian plane, and one ordered pair (a, b) where a and b are real numbers. a is the x-coordinate of the point, and b is the y-coordinate).

From the example of the points p, q, and r it should also be clear that

$$(a, b) = (c, d) \text{ only if } a = c \text{ and } b = d$$

and

$$(a, b) = (b, a) \text{ only if } a = b.$$

3.4 ORDERED PAIRS

Just as in the one-dimensional case, where individual points and intervals could be indicated on the number line, ordered pairs, either individually or grouped into intervals, may be drawn on the cartesian plane.

Notice the conventions used in Figures 3.4 to 3.6:

To denote *inclusion* in a set
- (i) Individual points are drawn as filled circles
- (ii) Lines (or boundaries of areas) are continuous
- (iii) Areas are shaded.

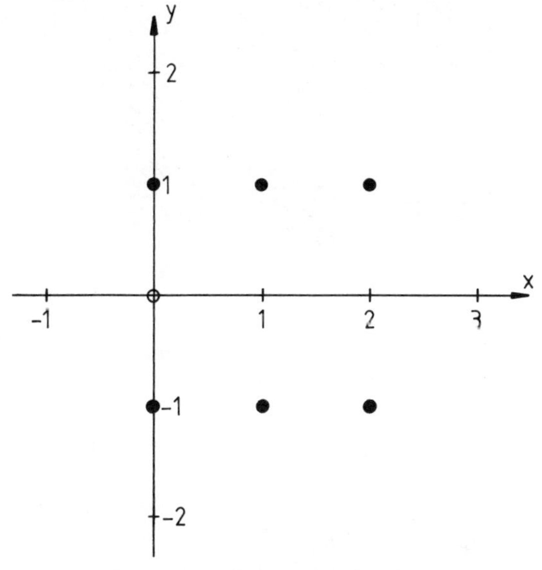

Figure 3.4 A set of six ordered pairs

$\mathbf{A} = \{(0, -1), \ (0, 1), \ (1, -1), \ (1, 1), \ (2, -1), (2, 1)\}$

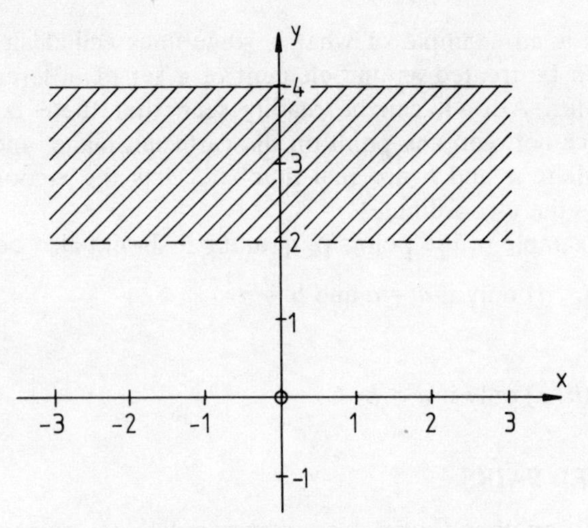

Figure 3.5 An infinite set of ordered pairs bounded only in the $\pm y$ directions

$$\mathbf{B} = \{(x, y) \mid x \in \mathbf{N},\ 2 < y \leqslant 4\}$$

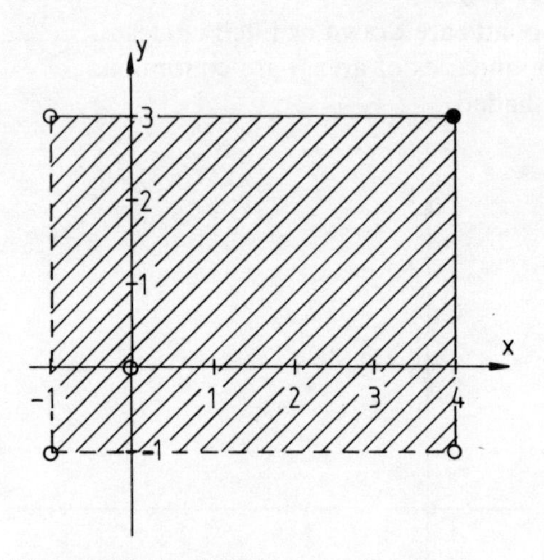

Figure 3.6 An infinite set of ordered pairs bounded in both x and y directions

$$\mathbf{C} = \{(x, y) \mid -1 < x \leqslant 4,\ -1 < y \leqslant 3\}$$

To denote *exclusion* in a set

(i) Individual points are drawn as empty circles
(ii) Lines (or boundaries) are dotted
(iii) Areas are left blank.

3.5 RELATIONS

A **Relation** is a set of ordered pairs, but by itself such a definition lacks much interest since *any* collection of ordered pairs on a graph (i.e. points and areas) can be included. We are usually more interested in those cases where some sort of well-defined relationship exists between the two parts (x and y) of the ordered pair.

Since a relation is a set we denote it by a capital letter, usually **R**, thus

$$\mathbf{R} = \{(x, y) \mid x \text{ is related to } y\}.$$

The set can then be tabulated, or described by a defining property.

Example

$$\mathbf{R} = \{(1, 2), (2, 4), (3, 6), (4, 8), (5, 10)\}$$

and

$$\mathbf{R} = \{(x, y) \mid y = 2x, x \text{ is a natural number less than } 6\}$$

are equivalent, and could both be represented by a series of five points on a graph (Figure 3.7).

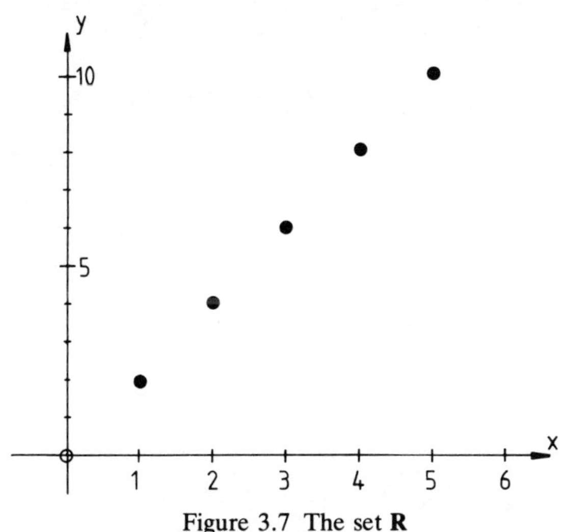

Figure 3.7 The set **R**

It should also be remembered that relations can be generated from the unions and/or intersections of simpler relations, since *relations are sets*.

Examples

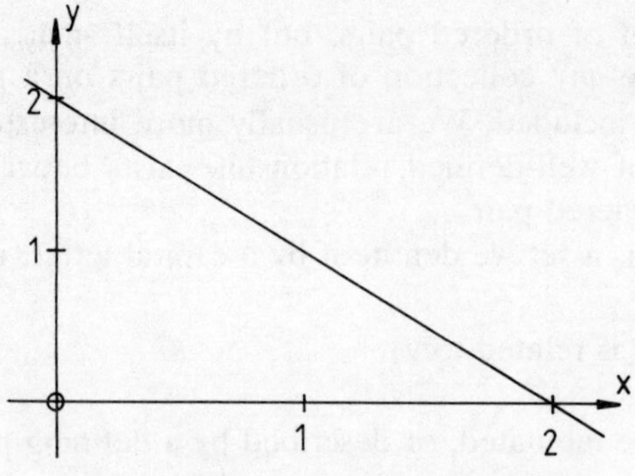

Figure 3.8 $\mathbf{R} = \{(x, y) \mid y + x = 2\}$

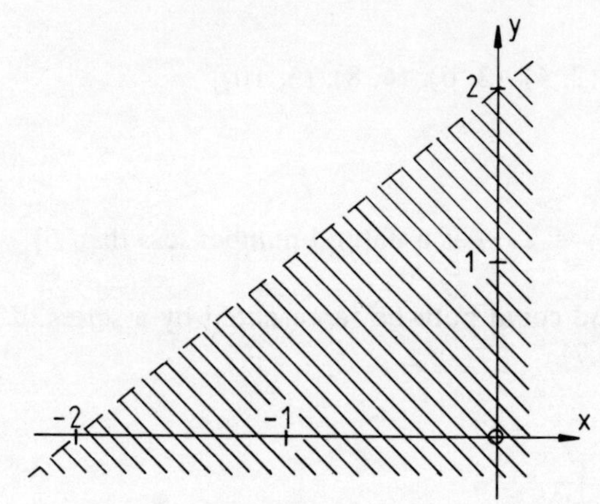

Figure 3.9 $\mathbf{R} = \{(x, y) \mid y < x + 2\}$

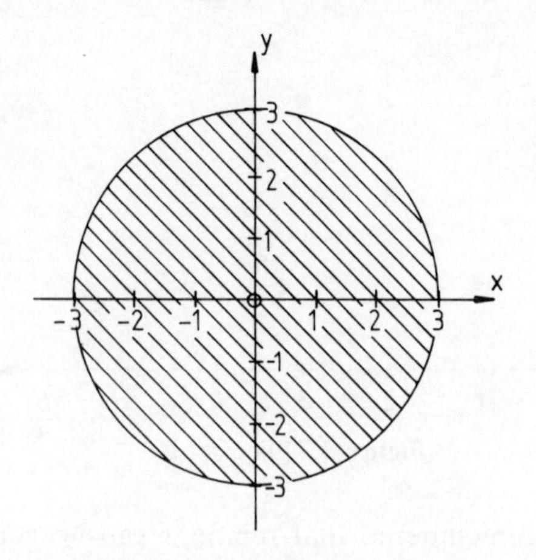

Figure 3.10 $\mathbf{R} = \{(x, y) \mid x^2 + y^2 \leqslant 9\}$

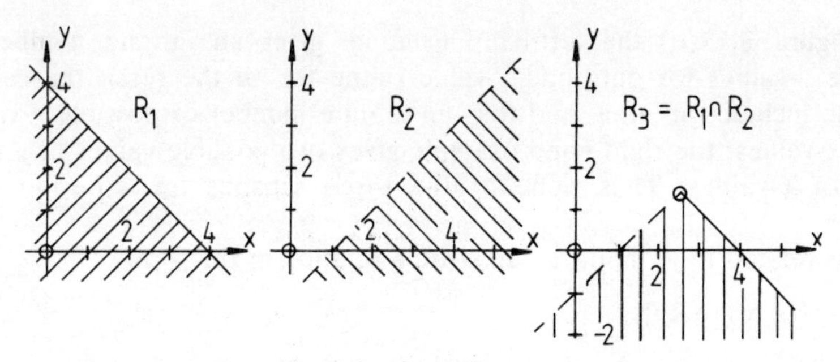

Figure 3.11 $\mathbf{R}_1 = \{(x, y) \mid x + y \leqslant 4\}$
$\mathbf{R}_2 = \{(x, y) \mid x - y > 1\}$
$\mathbf{R}_3 = \mathbf{R}_1 \cap \mathbf{R}_2 = \{(x, y) \mid x + y \leqslant 4 \text{ and } x - y > 1\}$

3.6 FUNCTIONS

An important class of relations are called *functions*, and can be simply defined as including all those relations where, for each value of x, there is at most one corresponding value of y.

In Figure 3.12(a), for each value of x there is at most one corresponding value of y, and hence they are all functions. Notice the careful way in which the closed and open circles have been included in the central example of Figure 3.12(a). If all circles had been closed, there would have been one value of x where the two linear segments would have overlapped, and thus have given two possible values of y. It would not then be a function.

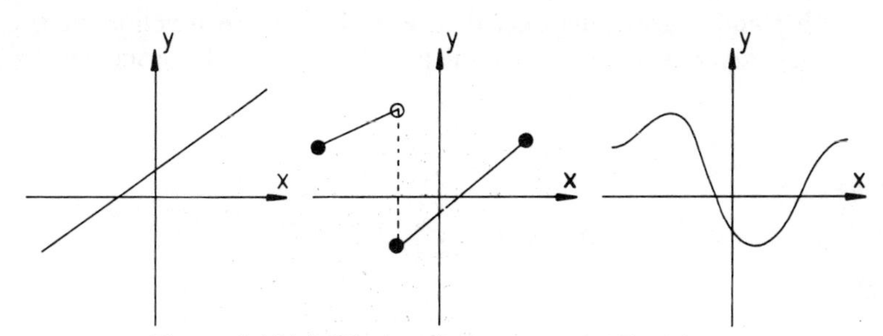

Figure 3.12(a) These relations are also functions

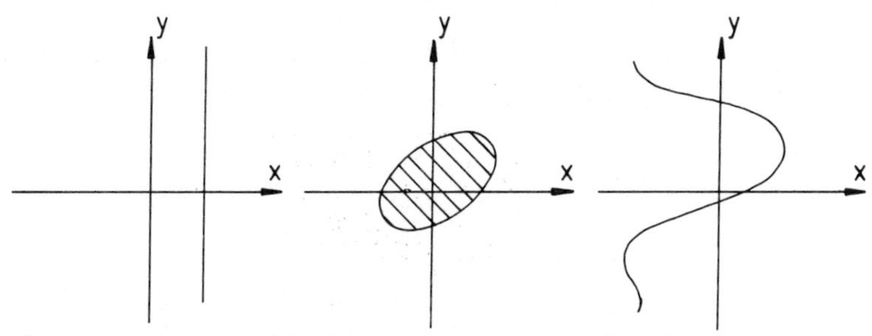

Figure 3.12(b) These relations are not functions

In Figure 3.12(b) the left-hand example gives an infinite number of possible y-values for only one x-value (none for all the rest); the central example includes an area, and thus an infinite number of possible y-values for all x-values; the right hand example gives two possible values of y for a range of x-values. Thus none of the three satisfies the criterion of a function.

If x is related to y, then we can write a relation in the form

$$F = \{(x, y) \mid x\mathbf{R}y\}$$

for example. This may be abbreviated to

$$x\mathbf{R}y.$$

For example,

$$x = y \qquad x > y.$$

If the relation is also a function then we can write

$$xfy \quad \text{or} \quad y = f(x)$$

The second form is very common, and reads *y is a function of x*.

3.7 INVERSE FUNCTIONS

If

$$y = f(x) \quad \text{and also} \quad x = g(y)$$

where both f and g are functions, then g is the *inverse function* of f, and f is the *inverse function* of g. It is then possible to use the notation $f = g^{-1}$, or $g = f^{-1}$.

Example

$$y = \mathrm{e}^x \qquad \text{or} \quad f = \{(x, y) \mid y = \mathrm{e}^x\}$$

so that

$$x = \log_\mathrm{e} y \quad \text{or} \quad g = \{(y, x) \mid x = \log_\mathrm{e} y\}$$
$$\text{or} \quad g = \{(x, y) \mid y = \log_\mathrm{e} x\}.$$

Notice the way in which the ordered pair (x, y) has been brought back into the conventional order in the expression for g. So, we have

$$f = \{(x, y) \mid y = \mathrm{e}^x\}$$

and

$$f^{-1} = \{(x, y) \mid y = \log_\mathrm{e} x\}.$$

3.8 GRAPHS

A pictorial representation on the cartesian plane of any relation is usually called a *graph*. A graph is therefore a pattern or distribution of points on the plane, each of which represents one ordered pair of the type (x, y). All the figures, from Figure 3.3 onwards, are therefore graphs of one form or another. The actual pattern made by the points can include random arrays or continuous areas, lines, and curves, depending on how the relationship between x and y is defined.

The full notation for a relation is rather tedious to use all the time, and so it is usually abbreviated in the following way

$$\mathbf{R} = \{(x, y) \mid y = 2x^2 + 7x + 3\} \quad \text{becomes} \quad y = 2x^2 + 7x + 3.$$

It should therefore be remembered that an equation such as $y = 2x^2 + 7x + 3$ can be considered as generating (and defining) a particular set of ordered pairs: for each value of x there is a corresponding value of y, thus giving (x, y).

In order to be able to draw a graph of a given relation or function we need to generate a table of ordered pairs and plot these on the cartesian plane. However in most cases the set will, strictly speaking, include an infinite number of ordered pairs. The usual practice is therefore to draw in a suitable number of points and join these by a smooth curve where appropriate. There are some pitfalls in this technique because, if we are careless, the smooth curve chosen may not be appropriate where the curve is peaked, or discontinuous or shows other signs of *bad behaviour*.

Examples

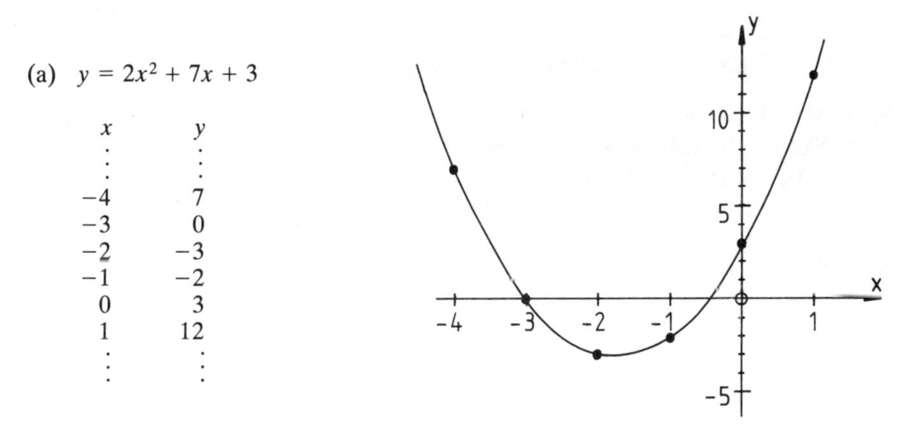

(a) $y = 2x^2 + 7x + 3$

x	y
⋮	⋮
−4	7
−3	0
−2	−3
−1	−2
0	3
1	12
⋮	⋮

Figure 3.13 A parabola $y = 2x^2 + 7x + 3$

Although integer values of x have been chosen (for convenience) there is no reason why we should not include intermediate values if this proves appropriate.

(b) $y = \dfrac{2}{x-1}$

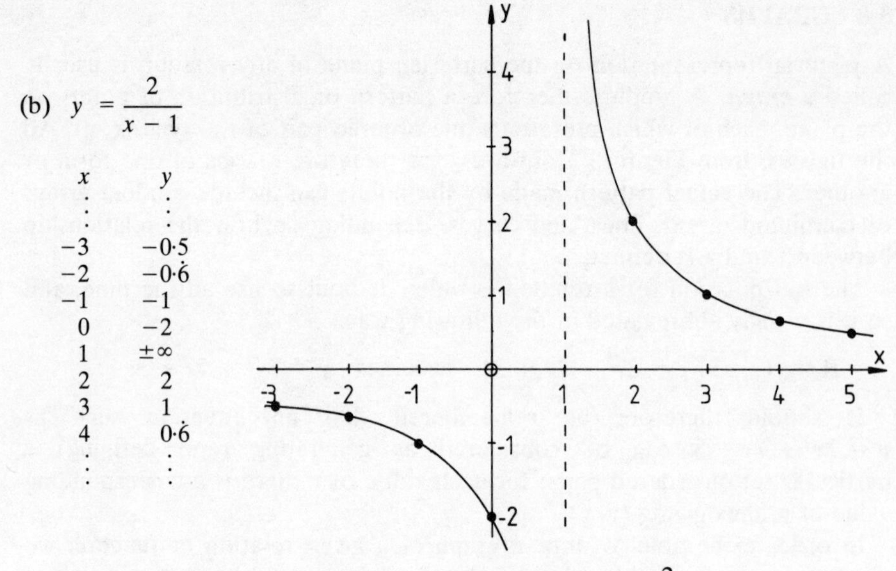

x	y
⋮	⋮
-3	$-0{\cdot}5$
-2	$-0{\cdot}\dot{6}$
-1	-1
0	-2
1	$\pm\infty$
2	2
3	1
4	$0{\cdot}\dot{6}$
⋮	⋮

Figure 3.14 A rectangular hyperbola $y = \dfrac{2}{x-1}$

This technique of *plotting* a curve is most useful for simple functions, over small regions of x and y. Later in the book, however, techniques of *curve sketching* will be introduced. These allow a better idea of the *overall* shape of a curve to be deduced quickly, and indicate which parts may then be usefully plotted.

3.9 BASIC EXAMPLES FOR THE STUDENT

B3.1 Sketch the set of points $\mathbf{A} = \{(0, 0), (1, 1), (2, 0), (3, 1), (-1, 1)\}$.

B3.2 Sketch the following relations

(a) $\mathbf{R} = \{(x, y) \mid -1 < x \leqslant 2, 0 < y \leqslant 1\}$

(b) $\mathbf{R} = \{(x, y) \mid x \in \mathbf{N}, -1 < y \leqslant 2\}$

(c) $\mathbf{R}\ \{(x, y) \mid -1 < x \leqslant 2, y \in \mathbf{N}\}$.

B3.3 If $\mathbf{R}_1 = \{(x, y) \mid y \leqslant 2 - x\}$

$\mathbf{R}_2 = \{(x, y) \mid y > 2x\}$

$\mathbf{R}_3 = \{(x, y) \mid x^2 + y^2 \leqslant 1\}$

sketch the following and state which, if any, are functions

(a) $\mathbf{R}_1 \cup \mathbf{R}_2$

(b) $\mathbf{R}_1 \cap \mathbf{R}_2$

(c) $\mathbf{R}_2 \cap \mathbf{R}_3$.

B3.4 Plot the following curves on the cartesian plane

(a) $3y + x + 1 = 0$

(b) $y = x^2 - 4x - 5$

(c) $y = 1/(x-2)^2$.

3.10 INTERMEDIATE EXAMPLES FOR THE STUDENT

I3.1 Sketch the intervals

(a) $\mathbf{I} = \{x \mid -8 < x \leqslant -1\}$

(b) $\mathbf{I} = \{x \mid 0 > x\}$.

I3.2 Sketch the following set on the cartesian plane

$\mathbf{A} = \{(2, 2), (-2, -2), (2, -2), (-2, 2), (0, 0)\}$.

I3.3 Sketch the relations marked with an asterisk

$\mathbf{R}_1 = \{(1, 2), (3, 4), (-3, -1), (0, 0)\}$

$\mathbf{R}_2 = \{(x, y) \mid x + y = 3\}$

$\mathbf{R}_3 = \{(x, y) \mid y > 2\}$

$^*\mathbf{R}_4 = \mathbf{R}_1 \cap \mathbf{R}_2$

$^*\mathbf{R}_5 = \mathbf{R}_2 \cap \mathbf{R}_3$

$^*\mathbf{R}_6 = \mathbf{R}_1 \cap \mathbf{R}_3$.

Which, if any, of these last 3 relations are also functions?

I3.4 A building contractor is tendering for 3 contracts x, y, and z. The probability that he will be awarded x is 0·6 and this is worth £2000 in profits. The probability he will be awarded y is 0·25, but y is worth £5000 in profits. The probability he will be awarded z is 0·05 and z is worth £10 000 in profits.

(a) Tabulate the possible outcomes as subsets of $\{x, y, z\}$.

(b) Tabulate those outcomes with a profit of at least £5000.

(c) What is the probability of this event?

I3.5 Plot a graph of $y = x^3 - 9x$.

More Advanced Examples for Part Two

A2.1 **A**, **B**, and **C** are three overlapping sets. Which two of the following five expressions are in fact the same?

(i) $\mathbf{C} \cap (\mathbf{A} \cup \mathbf{B})'$
(ii) $(\mathbf{C} \cap \mathbf{A}) \cup \mathbf{B}'$
(iii) $\mathbf{C} \cap (\mathbf{A}' \cup \mathbf{B}')$
(iv) $(\mathbf{A} \cup \mathbf{B} \cup \mathbf{C}')'$
(v) $(\mathbf{A} \cap \mathbf{C})' \cup \mathbf{B}$.

A2.2 Some sociological research in Wonderland produced the following evidence:

Aristocrats are always wealthy
No mathematician ever writes good poetry
People from non-academic professions are inevitably unintelligent
Academics are always poor
All poets and mathematicians are intelligent people.

What relationship exists between poets, mathematicians, and aristocrats?

Further evidence shows that:

30% of the population are wealthy
30% of the population are academics
10% of the population are intelligent
5% of the population are aristocrats
4% of the population are mathematicians
2% of the population are poets.

What proportion of the population are unintelligent academics? (According to the Common Sense Act, 1832, of Wonderland, the following points must be observed:

(i) All sociological evidence cannot be challenged
(ii) Facetious answers are a capital offence.)

A2.3 In the state of Vatopia, the tax authorities are asked to undertake a national survey on the incidence of tax on its citizens. They find that for the sake of convenience each member of the population can be classified according to whether or not he or she is

(i) Married (M)
(ii) Employed (E)
(iii) Subject to a higher tax (or surtax) rate (S).

The following table gives the percentages of the population which fall into each group identified, and also the percentage of the total tax revenue raised from each group. The group ME, for example, includes *all* married employed people, regardless of whether they are subject to the higher tax rate or not.

From this table of figures, calculate the percentages of people in each of the eight groups which can be logically identified, and likewise the equivalent contributions (%) to the total tax levy. Show

	% of population	% contribution to total levy
M	60	64
E	70	87
S	10	13
ME	51	56
MS	7	8
SE	2	4
MES	1	2

in principle (with an example) how a simple index for the *mean rate of taxation per head* could be devised for any group.

Using this index, answer the following, as far as your taxation is concerned:

(i) If you were married, would you be better-off employed or unemployed?

(ii) If you were employed, would you be better-off married or single?

A2.4 (a) Using Venn diagrams, demonstrate the basic laws of probability:

(i) $P(\text{A or B}) = P(\text{A}) + P(\text{B}) - P(\text{A and B})$

(ii) $P(\text{A and B}) = P(\text{A}) \times P(\text{B} \mid \text{A})$.

(b) From a survey of a group of managers, it is found that

40% have only a professional qualification

10% have only a university degree

30% have neither.

What proportion have both qualifications? On the basis of this survey, what is the probability of a manager having a professional qualification?

PART THREE

Introduction

We proceed in this section to the important topic of linear relationships. The algebraic and geometric (graphical) representations are developed in parallel to emphasize how particular features of an algebraic equation can be interpreted directly in terms of features on a graph. Linear functions, because of their relative simplicity, provide a useful basis for familiarizing the reader with the fundamental ideas involved in graphical representation.

The relative simplicity of linear relationships has also meant, however, that they are employed in mathematical models more often than might otherwise be justified. If one considers non-linear (or *curvilinear*) relationships over a small enough range, they can usually be considered as linear to a good approximation. The argument is similar to that of treating the Earth's surface as flat, an approximation which is good enough if one is considering an area the size of a field, rather than the size of Asia. Thus many systems, which may be described by a series of mathematical equations (which can then be termed a *model* of the real system), may be described by a series of *linear* equations if appropriate restrictions are imposed on their range of validity. A system of linear equations is not always appropriate, but if it is, it has the advantage of descriptive and analytical simplicity.

Chapter 5, introducing the basic ideas of linear programming, is concerned with an important class of models, in which *linear* relationships are employed, and which concern themselves with the problems of finding an *optimum* solution subject to constraints. Taking a broad group of problems common in linear programming, as an example, one might be seeking an appropriate mix of products to achieve maximum profits, but with limited production resources. Chapter 5 thus enables us to consolidate the section on linear relationships, as well as to introduce a valuable technique from the repertoire of quantitative methods for assisting decision-making.

CHAPTER 4

Graphical Representation: Straight Lines

4.1 A NOTE ON NOTATION

In the previous chapter we introduced the convention of the *cartesian plane*, and considered the graphical representation of relations and functions on this plane. As an example we describe a parabola in three ways:

(i) Formally, as a set of ordered pairs (x, y), thus

$$\mathbf{R} = \{(x, y) \mid y = 2x^2 + 7x + 3\}. \tag{4.1}$$

(ii) In less formal (though rigorous) terms as an equation, thus

$$y = 2x^2 + 7x + 3. \tag{4.2}$$

(iii) As a graph on the x–y plane, by generating a small number of ordered pairs (x, y pairs) directly from equation (4.2) and plotting (Figure 3.13).

From now on we shall almost invariably use the simplified form similar to that above in equation (4.2) to represent a particular relation, rather than the full set notation of equation (4.1).

4.2 INTRODUCTION

We shall concentrate in this chapter on a particular class of functions which are both simple in form, and very important, viz. straight lines. It should already be appreciated that most useful relations can be represented *both* as graphical curves *and* as algebraic equations. One of the purposes of this chapter, and some succeeding work, will be to emphasize this equivalence, and to show how specific features of the algebraic form correspond to specific features of the geometrical form. The approach here will be to begin with a discussion of the principal features of a straight line on a graph, and how these correspond to the algebraic form, or to *the equation of a straight line*.

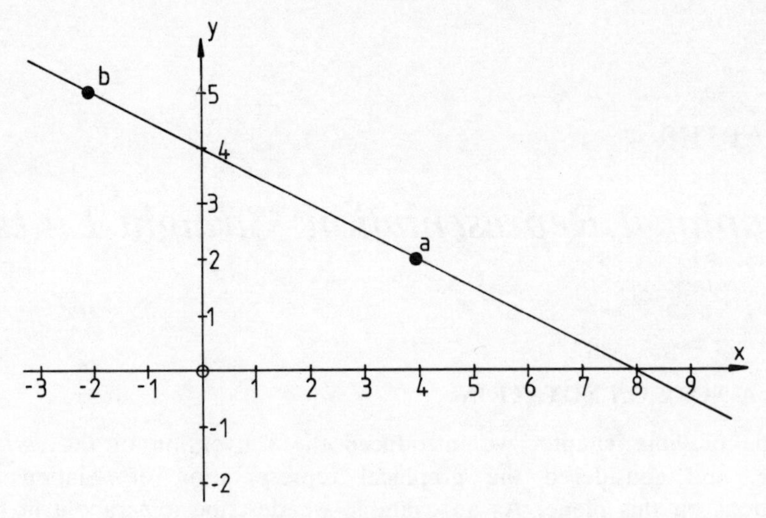

Figure 4.1 A straight line: $2y + x = 8$

4.3 BASIC CHARACTERISTICS OF A STRAIGHT LINE

If we draw a single line on a graph, such as in Figure 4.1, then we can list the most significant features which would distinguish it from any other line.

(a) The line cuts each axis once and has an x-intercept of $+8$ and a y-intercept of $+4$. One intercept of each type will always exist, except when the line lies parallel to one or other axis and so fails to intersect one axis (when it can be considered to be a line with an intercept at ∞). In particular it is worth noting that if the two intercepts are defined, then the line itself is defined uniquely, since *one* and *only one* line can be drawn through the two intercept values.

(b) If we imagine a point on the line in Figure 4.1 moving towards the right, then the value of the x-coordinate of the point *increases* as the value of the y-coordinate *decreases*. Moving to the left, the reverse occurs and x decreases as y increases. This is characteristic of a *negative slope* or *gradient* to the line. Roughly speaking, therefore, a negative slope means a line (or part of a curve) which slopes downwards towards the right.

A *positive slope* is characteristic of a line which slopes upwards towards the right. In this case the values of both the x-coordinate and the y-coordinate of a point on the line increase *together* (to the right) or decrease *together* (to the left); an example of such a line is given in Figure 4.2.

(c) We can ascribe a magnitude as well as a sign to the slope of a straight line. If we imagine the lines in Figures 4.1 and 4.2 as being the profile

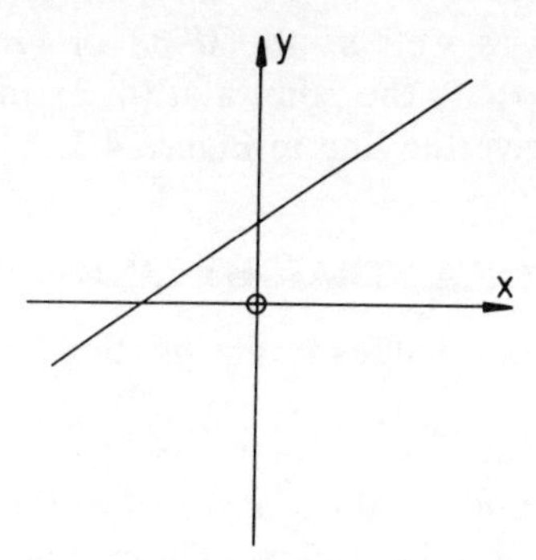

Figure 4.2 A straight line with positive slope or gradient

of a cross-section through a hill with the y-direction being vertical and the x-direction horizontal then a steep hill will have a large slope or gradient of either sign, as in Figure 4.3.

(d) Referring back to Figure 4.1 we can see that the line passes through two points at a(4, 2) and at b(−2, 5). Alternatively we would say that given the two points at a and b, only one line will pass through both points, which is of course the line that we have already drawn in. It follows therefore that one way to define a line uniquely is to give the coordinates of two points through which it must pass.

(e) Another way of defining a line would be to specify *one* point through

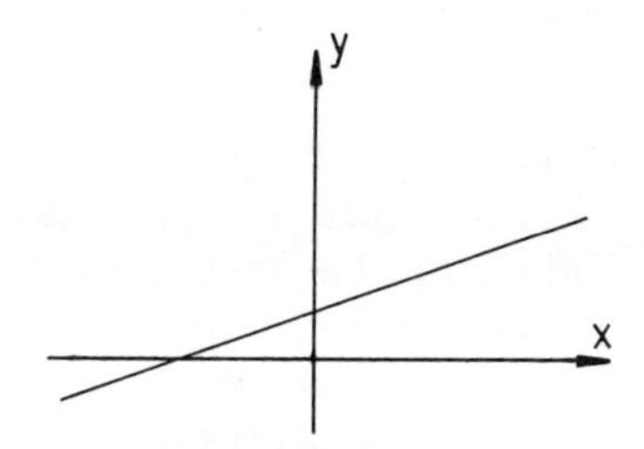

Figure 4.3(a) A line with positive slope of small magnitude

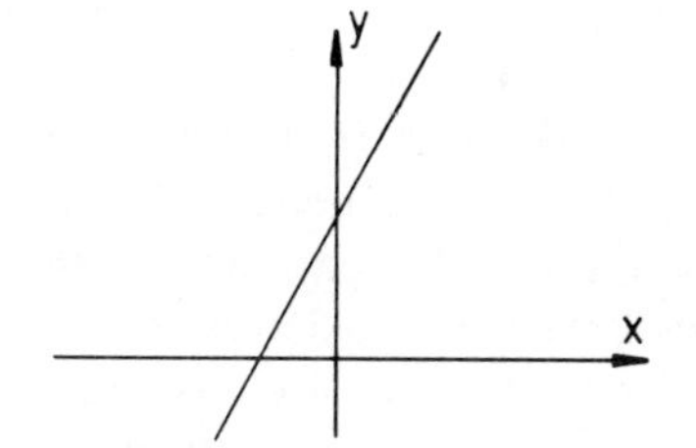

Figure 4.3(b) A line with positive slope of large magnitude

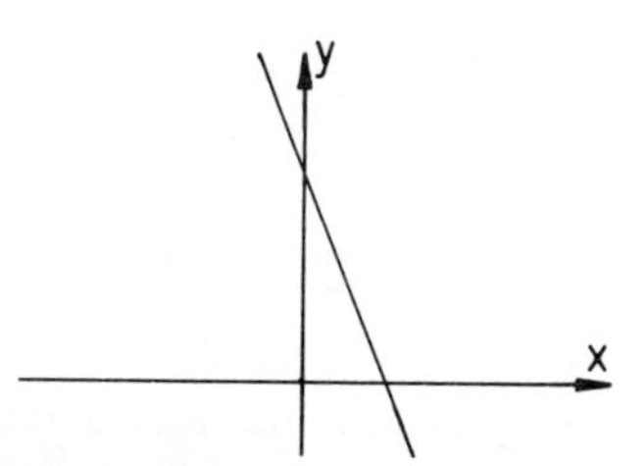

Figure 4.3(c) A line with negative slope of large magnitude

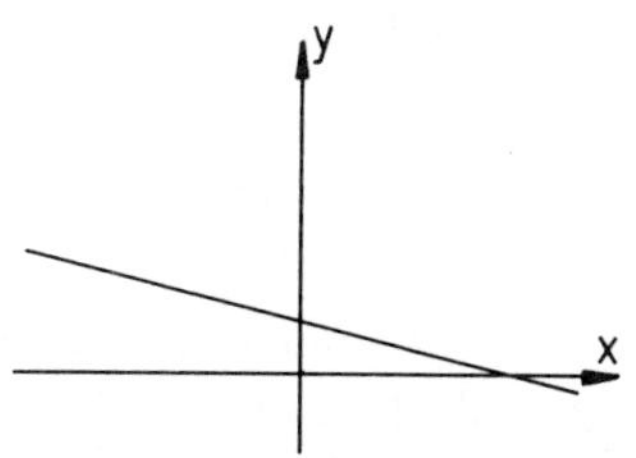

Figure 4.3(d) A line with negative slope of small magnitude

which it must pass as well as the *slope* or *gradient* of the line. For example we could specify the point a at (4, 2) and a slope (of $-\frac{1}{2}$ in this case) which would give the line in Figure 4.1.

4.4 THE EQUATION OF A STRAIGHT LINE

By substituting a range of x-values in the relation

$$y = -0{\cdot}5x + 4$$

we generate corresponding y-values and can build up a table of ordered pairs such as those in the table below. These points have been illustrated on the graph, and a line drawn to join all points.

x	y
-2	5
0	4
2	3
4	2
6	1
8	0
10	-1

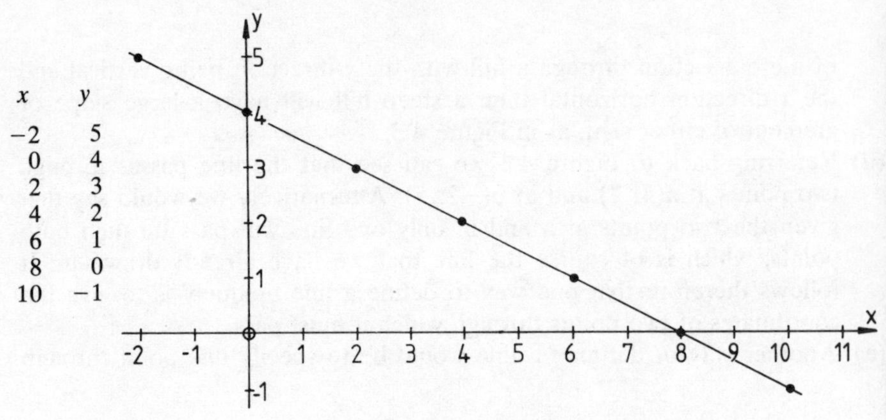

Figure 4.4 The line $y = -0{\cdot}5x + 4$

Clearly the line is straight and is identical to that in Figure 4.1. Consider each part of the linear relation $y = -0{\cdot}5x + 4$. The figure 4 is the y-value when x equals zero, and is therefore the y-intercept. The figure $-0{\cdot}5$ is the slope since it determines how rapidly y changes as x changes; the sign indicates that the slope is negative.

If we plot the graph of any equation which takes the general form

$$y = mx + c \qquad m, c \text{ are constants}$$

then we find it takes the form of a stright line with

y-intercept	c
x-intercept	$-c/m$
slope	m

There are of course alternative ways of writing the same relation. For example, all the following are equivalent, and all give the line illustrated in Figure 4.4, but if the form can be reduced to the standard $y = mx + c$ then the geometric interpretation is a straight line:

$y = -0{\cdot}5x + 4 \qquad y + 0{\cdot}5x = 4$

$2y + x = 8 \qquad 2y + x - 8 = 0$

$$\frac{1}{y} = \frac{1}{4 - 0{\cdot}5x} \qquad x = 8 - 2y$$

$$\frac{1}{x} = \frac{1}{8 - 2y}.$$

Example

A straight line has an x-intercept of 5, and passes through the point (7, 2). Determine its equation.

Since we are told the line is straight then we can assume its equation takes the form

$$y = mx + c$$

and we also know that the ordered pair (7, 2) satisfies the relation. Also, since the x-intercept is 5, the line also passes through (5, 0)

$$\therefore 2 = 7m + c$$

and

$$0 = 5m + c$$

This is a pair of *simultaneous equations* in m and c, for which only one solution exists.

Subtraction gives

$$2 = 2m + 0$$
$$\therefore m = 1.$$

But

$$c = -5m$$
$$\therefore c = -5.$$

Hence the line is $y = x - 5$.

In general we need two pieces of information to define the position of a line exactly. Examples of such pieces include:

(i) An intercept on the x or y-axis.
(ii) The coordinates of a point on the line.
(iii) The slope (gradient) of the line.

The example given used one intercept and one point. Nevertheless the general method of finding the equation of the line follows the same principles whatever the initial information. The form $y = mx + c$ is assumed at the outset, and the problem is therefore to find values for m and c using the information given. The x and y-coordinates of points on the line can be substituted in this form of the equation, while the gradient gives m directly.

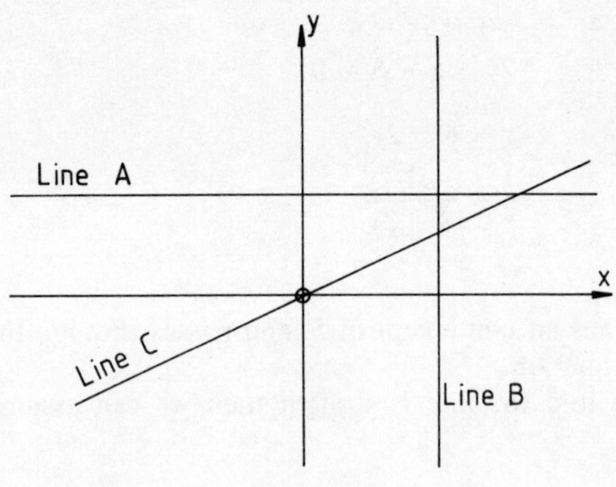

Figure 4.5

4.5 SPECIAL CASES

In Figure 4.5 the following lines are included:

(a) Line A, parallel to the x-axis. Gradient $m = 0$

$\therefore\ y = c$ (a constant)
e.g. $y = 4$.

(b) Line B, parallel to the y-axis. This is the only line which does not fit the form $y = mx + c$, unless we consider it to have an infinite slope, i.e. $m = \infty$. The equation is

$x =$ constant
e.g. $x = 5$.

(c) Line C, through the origin. Both intercepts are zero $\therefore\ c = 0$

$\therefore\ y = mx$
e.g. $y = 3x$.

This is the case where not only is y *linearly related* to x, but also y is proportional to x

$\therefore\ y \propto x$

(d) The axes.

The equation of the x-axis is $y = 0$
The equation of the y-axis is $x = 0$.

4.6 SIMULTANEOUS LINEAR EQUATIONS

Suppose that two variables, x and y, are linearly related in two different ways so that they must satisfy two equations simultaneously; for example,

$$y = 0{\cdot}5x + 1 \qquad \text{(Line A, } y\text{-intercept } + 1\text{, gradient } +0{\cdot}5)$$

and

$$y = -x - 1 \qquad \text{(Line B, } y\text{-intercept } -1\text{, gradient } -1).$$

These can be represented geometrically by two lines on a graph which intersect at *one* and *only one* point; this is illustrated in Figure 4.6.

We suppose that the two lines intersect at the point (x_1, y_1). The geometry of the diagram tells us that only one point, or ordered pair (x, y), will satisfy *both* equations simultaneously, and so such groups of equations are usually referred to as *simultaneous equations*. We can find the values of (x_1, y_1) either *geometrically* from a graph or *algebraically* by solving the two equations, thus:

$$y = 0{\cdot}5x + 1$$
$$y = -x - 1.$$

Subtract,

$$\therefore\ 0 = 1{\cdot}5x + 2$$
$$\therefore\ x = -2/1{\cdot}5 = -4/3 = -1{\cdot}\dot{3}$$

Also

$$y = -x - 1 = 4/3 - 1 = 1/3 = 0{\cdot}\dot{3}$$

or

$$y = 0{\cdot}5x + 1 = -2/3 + 1 = 1/3 = 0{\cdot}\dot{3}.$$

Hence the intersection of the two lines occurs at $(-4/3, 1/3)$.

If we are dealing with just two dimensions (the x–y plane) then the following difficulties can arise:

(a) *Parallel Lines* If the gradients of the two lines are the same then the

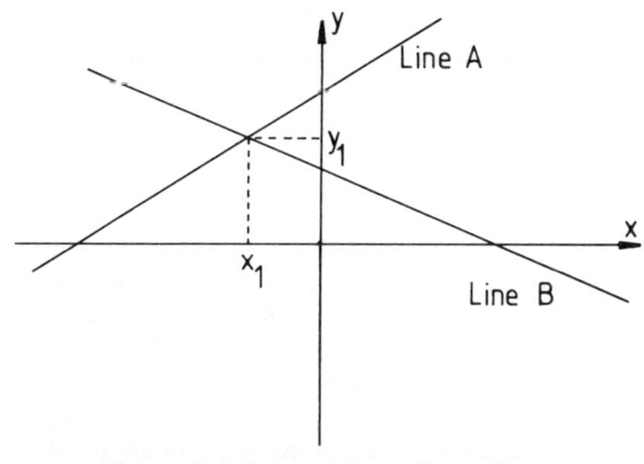

Figure 4.6 Intersection of two lines

two lines are parallel, and there is no intersection on a graph. This may be interpreted as meaning that there is no solution for x and y. The algebraic equations appear to be *inconsistent* in fact, because they show apparently contradictory information; for example,

$$x + y = 4$$
$$x + y = 6.$$

(b) *Coincident Lines* If both gradients and the intercepts of the two lines are the same, then both equations convey the same information and the lines are coincident on the graph, i.e. they are really identical lines and intersect along their entire length. No unique solution for x and y is then possible.

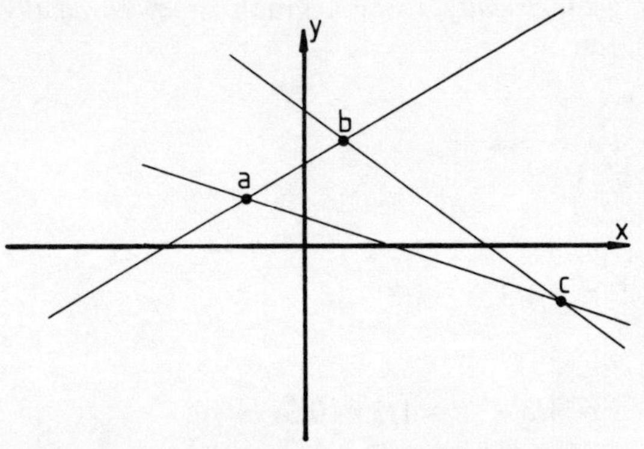

Figure 4.7 The general case of 3 lines. Any two lines or equations will provide a solution for (x, y) but there are different solutions for each pair of equations at a, b, and c

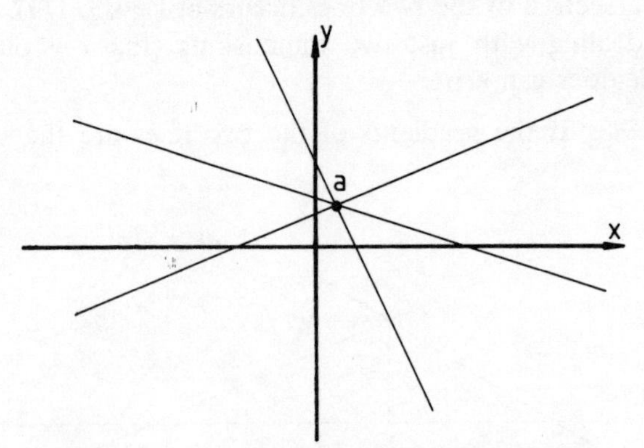

Figure 4.8 A special case of 3 lines. Any one equation is in fact a linear combination of the other two. This means that one equation is redundant, and gives no new information beyond that of the first two. To find the point a, therefore, one equation can in fact be ignored

(c) *More than 2 equations* If more than two simultaneous linear equations relate x and y then the question arises as to whether a single solution for x and y can occur. The answer is most easily seen from the geometry of a graphical representation (Figures 4.7 and 4.8).

Other cases can arise, such as when two or more of the lines are parallel, or are coincident. Again, a graph can show how many, and what sort of, solutions arise.

4.7 THREE VARIABLES, THREE DIMENSIONS

If three variables are introduced, x, y, and z say, then we have a cartesian volume rather than a plane. Three axes can be envisaged at right angles to each other as along the edges of a cube. Any point in this space can then be specified by an *ordered triple* (x, y, z).

Now in three dimensions a linear equation such as

$$z = ax + by + c$$

or

$$px + qy + rz = s$$

$(a, b, c, p, q, r, s,$ are constants) corresponds to a plane surface, and not to a straight line. Again there are intercepts on each of the three axes, but the concept of a gradient is not so easy to define and will not be considered here (see Chapter 8).

If we have *two* simultaneous equations in *three* dimensions then we have generally defined two planes. These will intersect along a *straight line* (unless they are parallel) and this line will not generally lie parallel to, or perpendicular to, any of the three axes.

With *three* simultaneous equations in *three* dimensions we have generally defined three planes, and there will be *one* and *only one* point in space where all three planes will intersect. Hence generally we need three equations if we wish to find unique solutions for three variables. Similar complications can occur as in the two-variable case. For example, parallel planes never intersect and in the corresponding *algebraic* solution an apparent contradiction will arise: for example,

$$x + y + z = 1$$

and

$$x + y + z = 2$$

can be represented by parallel planes, and the two pieces of information seem to contradict in algebraic forms. The two equations are again said to be algebraically *inconsistent*.

4.8 MORE THAN THREE DIMENSIONS AND VARIABLES

Geometric representation breaks down here since we cannot represent a four-dimensional surface. However, there is one important result which we can carry over into the case of 4 or more variables. If we have (linear) equations in n variables, then we generally need n independent equations to achieve unique values for each of the variables. The description *independent* can be interpreted here to mean that none of the equations are redundant.

4.9 LINEAR INEQUALITIES

We can consider linear relationships which involve the various inequality signs, rather than the equality sign. The notation for defining areas in such intervals has been described in the previous chapter (Figures 4.9 and 4.10).

4.10 LINEAR MODELS

The usual purpose of a mathematical model is to relate two or more variables in a rigorous way, in an attempt to imitate or simulate the relationships which arise in the real world. Linear models have the attraction of simplicity since if a large number of variables are being used, too much mathematical complexity can often prove self-defeating: the equations cannot actually be solved.

However, it is most important to realize that the quality of the model is only as good as the assumptions which go into it. If the assumptions are wrong or dubious over some range of values of the variables, then so must be the output or results of the model.

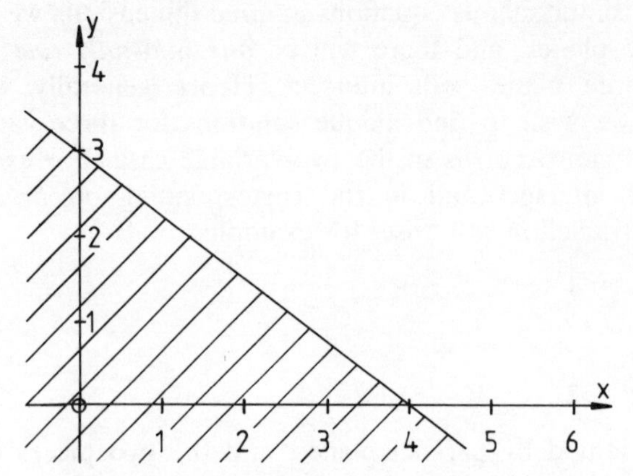

Figure 4.9 The shaded area defines all points (x, y) which satisfy $3x + 4y \leqslant 12$. The *solid* line indicates that it is *included* in the shaded region because of the equality option

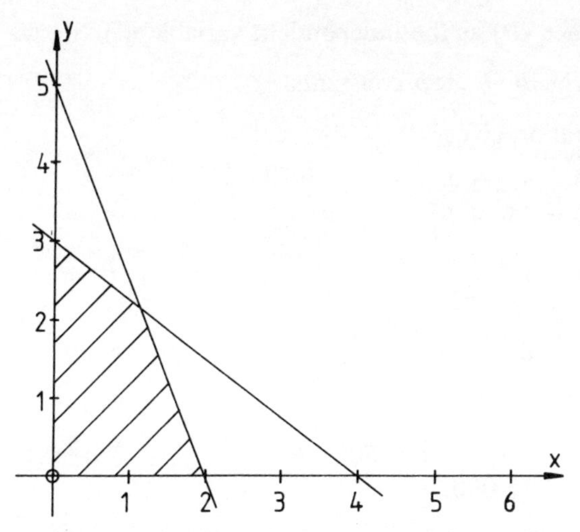

Figure 4.10 The shaded area defines all points (x, y) which satisfy all four inequalities:

$$x \geqslant 0$$
$$y \geqslant 0$$
$$3x + 4y \leqslant 12$$
$$2y + 5x \leqslant 10$$

Points outside this area may satisfy some of the inequalities, but not all four

The choice of relationships within a model may well be affected by an attempt to incorporate the results of a rigorous theoretical analysis. Such relationships then go beyond mimicry of reality and attempt to take account of structural relationships between some of the variables under consideration. This may well lead to a model which is more rigorous, but which is too complicated to solve analytically. We are thus faced with a common dilemma of choosing between an approximate solution to a rigorous statement of a problem, and a rigorous solution to an approximate (usually simplified) statement of the problem. Generally the former is to be preferred, since it is not always clear how serious the errors arising from the simplification of the model itself will be.

Examples

(i) 10 000 Easter eggs are sold at 50p each; 9500 Easter eggs are sold when the price is increased to 55p each. How many might we expect to sell at 60p each?

Assume a linear relationship between price and number sold over a small range.

We treat the number sold (N) as the dependent variable (y) and the price in pence (P) as the independent variable (x). So, we assume

$$N = aP + b \qquad a, b \text{ constants.}$$

Our information gives

$$10\,000 = 50a + b$$
$$9500 = 55a + b.$$

Subtract,

$$\therefore \quad 500 = -5a$$
$$\therefore \quad a = -100$$

Also

$$10\,000 = 50a + b = -5000 + b$$
$$\therefore \quad b = 15\,000.$$

Therefore

$$N = -100P + 15\,000.$$

This is our model for the number sold–price relationship. At a price of 60p, we would therefore expect the number sold to be given by

$$N = -6000 + 15\,000 = 9000$$

i.e. 9000 eggs should be sold.

The range of validity. Our original data was given for values of P between 50p and 55p. We must face the question as to how far one model can be treated as valid beyond this range of prices. The choice is, strictly speaking, arbitrary, but this does not prevent us from making some judgement based on an intuitive feel for the particular problem in hand. In this example, we might reasonably expect to be able to extend the range of validity to between 45p and 60p, a further 5p beyond our data, each way. If we knew that considerations of market competition, for example, operated very severely beyond our original data (50p to 55p), then we might well feel that the range of validity was more restricted. On the other hand, experimental data beyond the 50p to 55p range might show that the model held good well beyond this range. The most important point is perhaps that the range of validity must be viewed critically and not be assumed to be as great as mathematical convenience would permit.

(ii) Car hire costs £10 per week, plus 3p per mile travelled. Represent the total weekly cost as a function of the mileage travelled.

Let the weekly mileage be x, and let the total weekly cost be C in £.

$$\therefore \quad C = \text{hire cost} + \text{mileage cost}$$
$$= 10 + 0{\cdot}03(x)$$
$$\therefore \quad C = 0{\cdot}03x + 10.$$

The range of validity. In this case the hire cost is *defined* in terms of weeks and mileage covered. So the above relationship holds for the entire range of x unless different hire charges are specified for particularly high mileages (say). Note of course that x cannot be negative.

4.11 FAMILIES OF LINES

The standard form of the equation of a straight line is

$$y = mx + c.$$

We now consider the effect of varying the values of m or c while keeping the other fixed.

Varying m; c fixed

Since m is the gradient of the line and c is the y-intercept of the line, then varying m while fixing c implies that the line pivots about the y-intercept. We generate a whole series or *family* of lines which have the same y-intercept but different gradients (Figure 4.11).

Varying c; m fixed

In this case, varying c while fixing m implies that the line retains its gradient, or slope, but shifts its position. We generate a *family* of lines which are all parallel to one another (same slope), but have different y-intercepts (Figure 4.12).

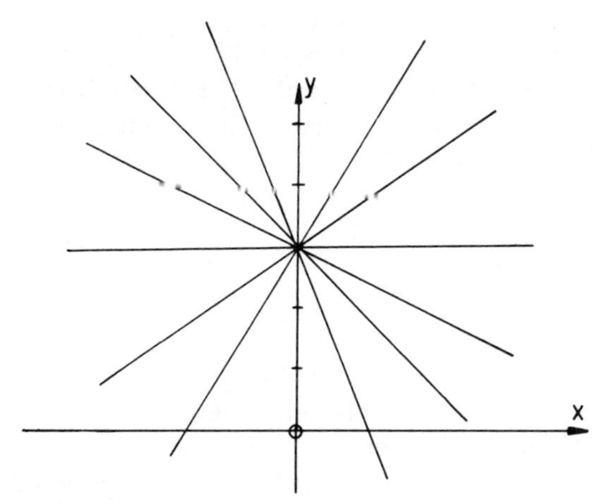

Figure 4.11 $y = mx + 3$. c fixed (at 3), m varied

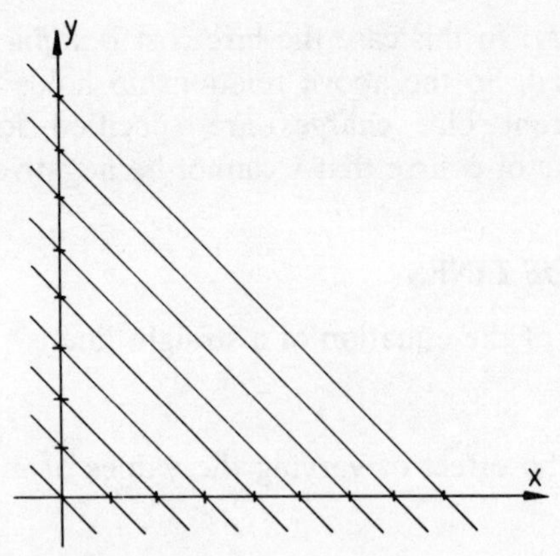

Figure 4.12 $y = -x + c$. m fixed (at -1), c varied

4.12 BASIC EXAMPLES FOR THE STUDENT

B4.1 Sketch the following relations, showing the values of the intercepts and the gradients.

(a) $y + 3 = 2x$
(b) $5y + x \leqslant 30$
(c) $3y + 5x < 45$
(d) $2y + 3x + 8 = 0$
(e) $3y + 4x \geqslant 18$.

B4.2 Shade the area which satisfies all of the relations (a) to (d) below simultaneously; and then the area which satisfies (e) to (h) simultaneously.

(a) $3x + 4y \leqslant 12$
(b) $7x + 2y \leqslant 14$
(c) $x \geqslant 0$
(d) $y \geqslant 0$

(e) $6y + x \geqslant 15$
(f) $4y + 7x \geqslant 28$
(g) $x \geqslant 0$
(h) $y \geqslant 0$.

B4.3 Solve the simultaneous equations below. How can they be interpreted geometrically? What is the significance of your solution in this geometric model?

(a)
$$x + y + z = 0$$
$$2x - y - 3z = -4$$
$$-3x - 2y + 4z = -4$$

(b)
$$y + 3x = 2$$
$$2y = 7 - 6x.$$

B4.4 Derive the equations of the straight lines which satisfy the following pairs of conditions:

(a) Passes through $(-1, 3)$ and $(1, -3)$.
(b) Has a y-intercept 4 and an x-intercept of 3.
(c) Has intercepts on both axes of zero.
(d) Has a gradient of $-1/7$ and an x-intercept of $+2$.

B4.5 The State of Evenlazia has a novel income tax scheme, whereby the taxpayer has 2 options for the way in which he pays tax.

Option A allows him to pay 30p in the pound on the first £2000 of income, and 50p in the pound on all income earned over and above £2000.

Option B allows him to pay 40p in the pound on all income earned, whatever the income.

Over what income ranges would you advise him to take option A and option B respectively? At what income would you advise him to emigrate to England with a rate of 33p in the pound (for all income earned—ignore allowances and surtax, etc.).

4.13 INTERMEDIATE EXAMPLES FOR THE STUDENT

I4.1 Write down the equations of the straight lines with the following properties.

(a) Gradient 2 and intercept 4.
(b) Gradient -2, and through the point $(1, 1)$.
(c) Intercept 4, and through the point $(-1, 2)$.
(d) y-intercept -1, x-intercept -3.

I4.2 A straight line passes through the points $(4, 5)$ and $(-1, 6)$. Use an algebraic method to find:

(i) the x-intercept
(ii) the y-intercept
(iii) the gradient.

I4.3 Solve the following simultaneous equations if possible (i) by algebra, (ii) by graphs.

(a) $x + 2y = 4$ $\quad$ $x - y = 7$
(b) $2x - y = 7$ $\quad$ $4x - 2y = 3$.

I4.4 The manager of a small haulage business must decide whether to hire a new van or to buy one. If he hires a van from Company A he will have to pay a fixed cost of £30 per week, but nothing for mileage. Company B charges purely on a mileage basis at a rate of $7\frac{1}{2}$p per mile. If he buys a van he estimates that he will incur a fixed cost of £10 per week plus a mileage cost of $2\frac{1}{2}$p per mile.

By expressing the cost of each of three possible courses of action as functions of x (the weekly mileage), and by sketching these functions in one diagram, find the most economic decision for different regions of x.

I4.5 When a company spends £20 on advertising annually it finds that it sells 11 000 items. When it spends £50 annually, its sales rise to 14 000 items. Estimate the number of items it would sell with annual expenditures of

(a) £35 (b) £200 (c) 0

on advertising making clear any assumptions you make in the calculation. If the company makes the following gross profit on each item (before advertising costs are deducted), what advertising policy would you advise it to follow in each case

(i) 5p (ii) 1p (iii) 0·5p.

CHAPTER 5

Linear Programming: An Introduction

5.1 INTRODUCTION

At various stages in this book, several techniques will be introduced which have practical application in the general area of *resource allocation*. For example, the management scientist might typically be asked how costs may be minimized or profits maximized by varying such things as the scale of production, the size of delivery batches, the price charged per unit, the mix of products in the output of some manufacturing process, etc. We shall be giving a brief introduction to the most useful of some of these methods, of which *linear programming* is the first.

Linear because the equations and relationships introduced are usually linear. Linearity is not in fact essential, but the most common and the most tractable problems are linear.

Programming is used in the sense of *method*, rather than in the computing sense. Nevertheless, the more realistic examples of linear programming need to be solved by computer methods because of the large number of variables which must be handled.

The examples which we shall use are in fact unrealistic in that we shall be concerned with a few equations involving at most two variables. This lends itself to geometric as well as algebraic interpretation, which considerably aids an understanding of the basic idea. However, most practical examples involve large numbers of both equations and variables, perhaps measured in thousands. The examples have therefore been chosen to demonstrate the *basic principles* rather than the problems of actual case studies.

5.2 THE BASIC IDEA

There are many variations on the basic pattern, but the standard problem amenable to solution by linear programming methods has the following features:

(a) We are concerned with the use of several *resources*, such as plant, labour, finance, or of *demands* such as minimum requirements or supply levels. These resources or demands impose *constraints* on levels of production or consumption because resources impose a *maximum* level of production, demands require a *minimum* level of production or consumption.

(b) Because of these constraints, the level of production is restricted in some way, and we do not have completely free choice in determining output.

(c) We are concerned with more than one product or good and the basic problem is to determine the level of production or consumption of each product or good so that the *mix* chosen is the *most advantageous*.

(d) The term *most advantageous* must be definable in explicit terms, such as the maximum profit or minimum cost mix among all feasible mixes. The function which defines our advantage—for example, the function defining *profits*, or *costs* in terms of the overall production/consumption level—is called the *objective function*.

(e) The end product of the calculation is therefore a recipe for the mixture of goods which when produced will give rise to the highest profits, or lowest costs, within the constraints imposed by resource availability.

The basic technique, and variations, are best illustrated through examples.

5.3 EXAMPLES (TWO CONSTRAINTS)

Example 1

Suppose a farmer has 100 acres of land, and £5000 capital available, and wishes to choose between two crops—wheat and barley. He estimates that wheat costs £40 per acre and barley £60 per acre to seed, and the respective profits per acre are £30 (wheat), and £40 (barley). Purely from a financial viewpoint (i.e. ignoring agricultural considerations) what mixture of wheat and barley should he grow?

Resources: Cash, limited to £5000.
Land, limited to 100 acres.
Questions: How many acres of wheat?
How many acres of barley?

Let him plant x acres of wheat and y acres of barley. Let the profit be P (£) in total.

We can now tabulate the constraints and the profitability, thus:

	Wheat (x acres)	Barley (y acres)	Constraint
Land	1 (acre per acre)	1 (acre per acre)	$\leqslant$100 acres
Cash	40 (£ per acre)	60 (£ per acre)	$\leqslant$5000 £
Profit	30 (£ per acre)	40 (£ per acre)	maximize P

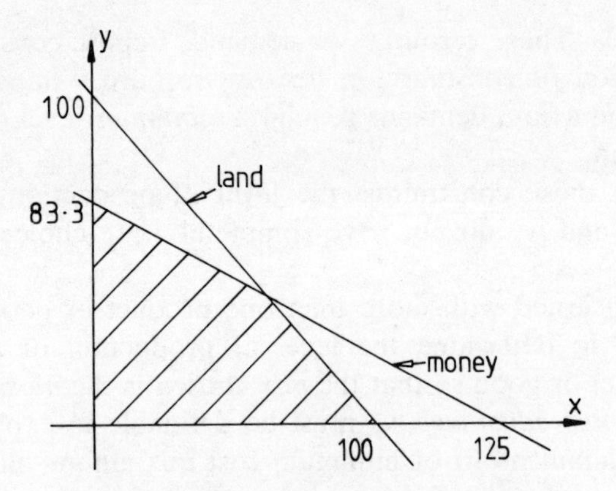

Figure 5.1 The shaded area is the *feasible region* defined by the four inequalities. (Not to scale)

So that, with x acres of wheat production and y acres of barley production

$$1x + 1y \leqslant 100 \qquad \text{(gradient } m_1 = -1\text{)}$$
$$40x + 60y \leqslant 5000 \qquad \text{(gradient } m_2 = -2/3\text{).}$$

Also $x \geqslant 0$, $y \geqslant 0$

$$P = 30x + 40y \qquad \text{(gradient } m_3 = -3/4\text{).}$$

From the four inequalities above, we can define the *feasible region* (shaded in Figure 5.1).

Also, for any particular value of P, we can draw in the *objective function* as a straight line with gradient $(-3/4)$. This is because

$$P = 30x + 40y$$

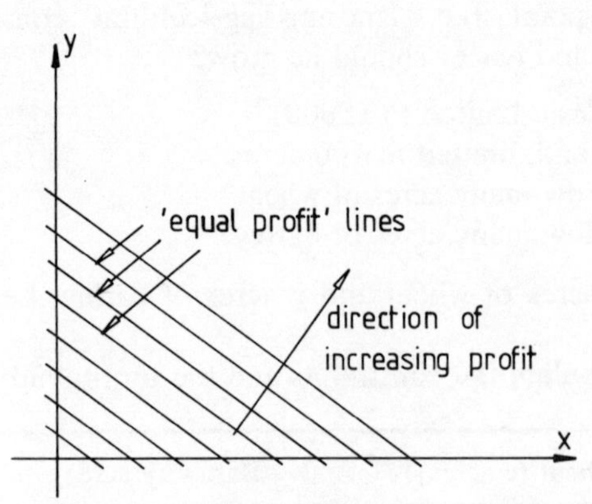

Figure 5.2 The lines show the *objective function* with gradient $-3/4$. Each of the points (x, y) on each line generates an equal profit

is a straight line, which can be written

$$y = (-3/4)x + P/40$$

and this generates a *family* of lines, all with gradient $-3/4$ but with different intercepts (Figure 5.2). On any particular line, the different combinations of x and y (wheat and barley) all yield the same profit (P).

However, as P increases so does $P/40$ (the y-intercept), so that lines with the higher y-intercepts yield higher profits.

The problem is therefore to maximize the profit, while at the same time remaining within the constraints (or feasible region). The particular combination of x and y which achieves this can be found by superimposing Figures 5.1 and 5.2, as in Figure 5.3.

Only those parts of the profit line which fall *within* the feasible region have been heavily drawn. Profit lines drawn furthest away from the origin (0, 0) yield the highest profits. Therefore the highest profit yielded *within* the feasible region must be at point A on the figure, which can be found from the simultaneous equations:

$$\begin{aligned} x + y &= 100 \qquad \text{or} \qquad & 40x + 40y &= 4000 \\ 40x + 60y &= 5000 & 40x + 60y &= 5000. \end{aligned}$$

Subtract,

$$20y = 1000, \qquad \therefore y = 50.$$

Also

$$x = 100 - y = 100 - 50 \qquad \therefore x = 50.$$

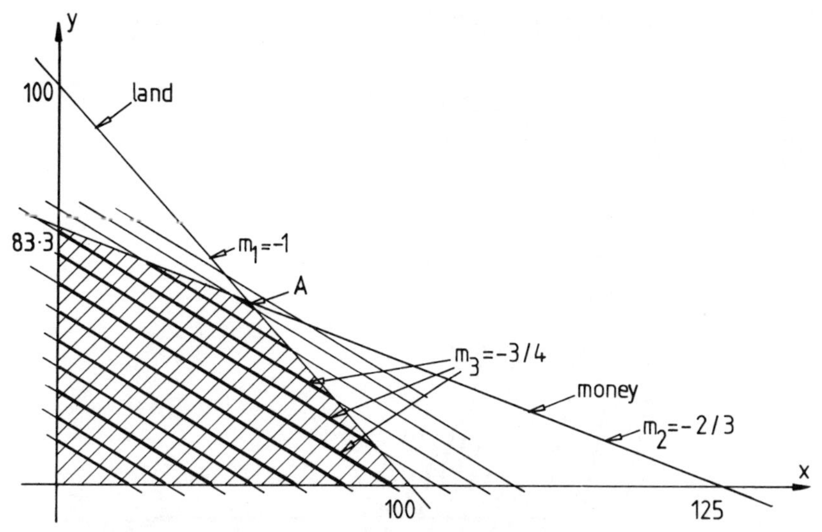

Figure 5.3 Figures 5.1 and 5.2 are superimposed. (Not to scale)

Also

$$P = 30x + 40y = 30(50) + 40(50) = 3500.$$

Hence, for maximum profit, the farmer should grow 50 acres of wheat and 50 acres of barley, which would yield £3500 profit.

It is very important to realize why the point of maximum profit yield should be at A. Because of the relative gradients of the three lines

$$\begin{aligned} x + y &= 100 \qquad & m_1 &= -1 \\ 40x + 60y &= 5000 \qquad & m_2 &= -2/3 \\ 30x + 40y &= P \qquad & m_3 &= -3/4 \end{aligned}$$

we have

$$m_1 < m_3 < m_2$$

and the gradient of the profit line falls *between* that of the two constraints. Suppose that the profits on wheat and barley change. The following examples indicate how the *most advantageous mix* also varies.

Example 2

As example 1 except that the profit on wheat becomes £40 per acre and the profit on barley becomes £35 per acre.

$$P = 40x + 35y \qquad \therefore m_3 = -\frac{40}{35} = -\frac{8}{7}$$

$$m_1 = -1, \qquad m_2 = -2/3 \quad \text{as before}$$

$$\therefore m_3 < m_1 < m_2$$

and the profit line falls as in Figure 5.4.

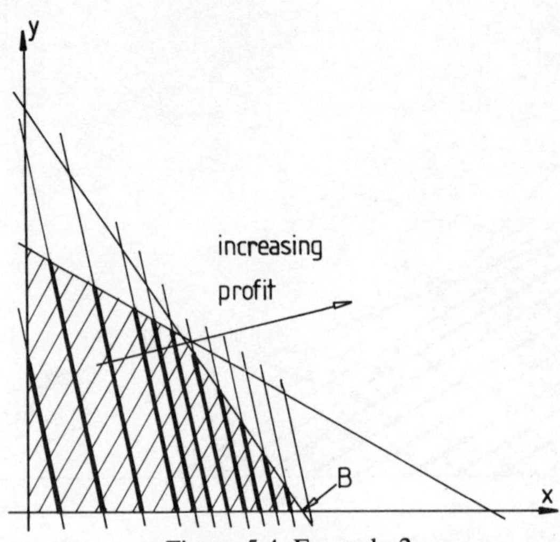

Figure 5.4 Example 2

The point which both lies in the feasible region and achieves the highest profit is at point B where

$$x + y = 100 \quad \text{and} \quad y = 0$$
$$\therefore x = 100, \qquad y = 0, \qquad P = 4000.$$

The farmer should grow 100 acres of wheat, and no barley, at a profit of £4000, and a cost of £4000.

Example 3

As example 1 except that the profit on wheat falls to £25 per acre and the profit on barley remains at £40 per acre.

$$P = 25x + 40y \qquad \therefore m_3 = -\frac{25}{40} = -\frac{5}{8}$$

$$m_1 = -1, \qquad m_2 = -2/3 \text{ as before}$$

$$\therefore m_1 < m_2 < m_3$$

and the profit line falls as in Figure 5.5.

The point which now falls in the feasible region and yields the highest profit is at point C where

$$40x + 60y = 5000, \qquad x = 0$$
$$\therefore y = £5000/60 = 83{\cdot}3, \qquad x = 0, P = £3333.$$

The farmer should now grow only 83·3 acres of barley and no wheat with a profit of £3333 and a cost of £5000.

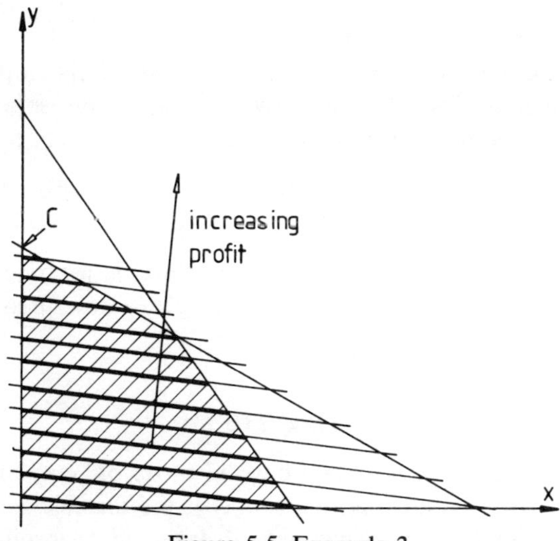

Figure 5.5 Example 3

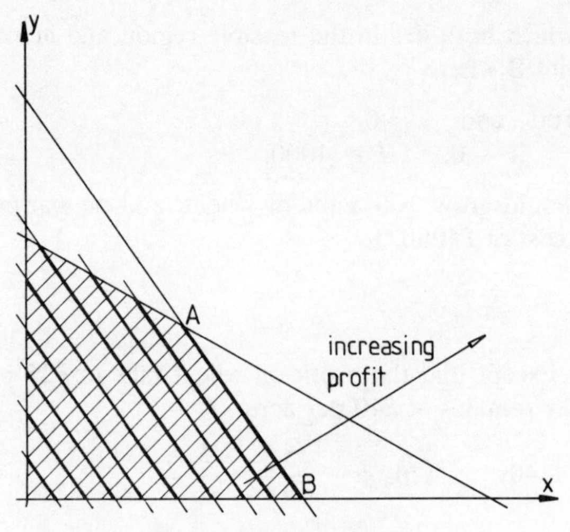

Figure 5.6 Example 4

N.B. Notice in example 1, all land and cash were used up; in example 2, all land was used, but not all the cash; and in example 3, all the cash was used, but not all the land.

Example 4

As example 1 except that the profit on wheat becomes £35 per acre and the profit on barley becomes £35 per acre.

$$P = 35x + 35y \qquad m_3 = -1$$
$$m_1 = -1, \qquad m_2 = -2/3 \text{ again}$$
$$\therefore m_3 = m_1 < m_2$$

and the profit line falls as in Figure 5.6. Here, *all* points along the line AB yield the highest profits, and we have a *degenerate* solution. We can calculate the profit yielded from *any* point on this line, so, taking the point B again

$$x = 100, \qquad y = 0, \qquad P = 3500 \qquad \text{i.e. a profit of £3500.}$$

Since any one of the infinite number of solutions for x and y along the line AB is optimal, this situation is also described as one of *non-uniqueness*.

5.4 MORE THAN TWO CONSTRAINTS

If further constraints are introduced beyond those of, say, land and cash in the example we have been using, the shape of the feasible region will alter.

Suppose that the farmer in example 2 takes into account the cost of fertilizer, and that he can only afford to use a total amount of 33 tons. Also suppose that it is desirable to apply it at a rate of 7 cwt per acre for wheat, and at 6 cwt per acre for barley. The new table of data becomes:

	Wheat	Barley	Constraint
Land	1 (acre per acre)	1 (acre per acre)	≤100 acres
Cash	40 (£ per acre)	60 (£ per acre)	≤5000 £
Fertilizer	7 (cwt per acre)	6 (cwt per acre)	≤660 cwt
Profit	40 (£ per acre)	35 (£ per acre)	maximize P

Hence the constraint equations become

Land:	$x + y \leqslant 100$	$m_1 = -1$
Cash:	$40x + 60y \leqslant 5000$	$m_2 = -2/3$
Fertilizer:	$7x + 6y \leqslant 660$	$m_4 = -7/6$

Also $x \geqslant 0$, $y \geqslant 0$, and

$$\text{Profit } P = 40x + 35y \qquad m_3 = -8/7.$$

The modified diagram is shown in Figure 5.7. In this case, point D becomes the maximum profit point. Note that

$$m_4 < m_3 < m_1 < m_2$$

which partly explains why this is so.

However, it is important to realize that the gradients of the line alone

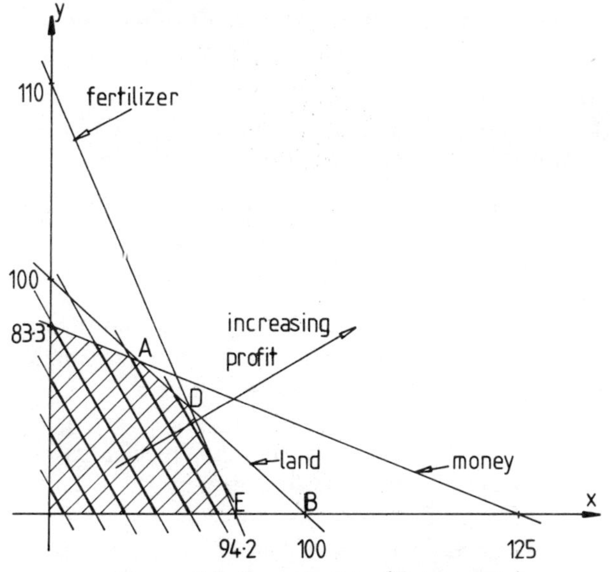

Figure 5.7 3 constraints. (Not to scale)

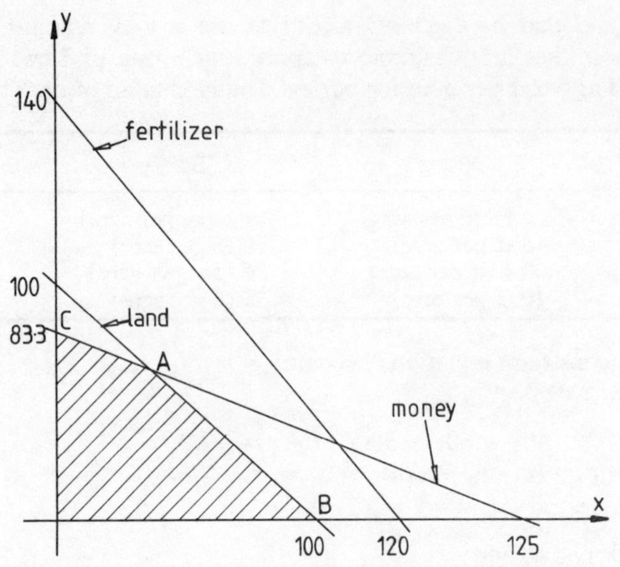

Figure 5.8 3 constraints, 1 redundant. (Not to scale)

are not enough justification for drawing a diagram such as Figure 5.7. If the farmer had been able to increase the fertilizer application to more than 35 tons (700 cwt) then the fertilizer constraint becomes *redundant*, and we are back to the original problem. We show the case where 42 tons (840 cwt) of fertilizer are available in Figure 5.8. It should make clear the point that if the land and the cash constraints are satisfied then the fertilizer constraint is automatically satisfied. In other words, the fertilizer constraint itself does not affect the shape of the feasible region.

$$x + y \leqslant 100$$
$$40x + 60y \leqslant 5000$$
$$7x + 6y \leqslant 840.$$

Depending on the relative profitabilities, the solution lies at A, B, or C again (examples 1, 2, or 3) or is degenerate (example 4).

So the major conclusions to be drawn from these examples are

(a) We need to determine the feasible region, and which constraints determine it. Are any constraints redundant, for example?
(b) If the solution is not degenerate, then it will lie at a corner or vertex of the feasible region (see points A, B, C, D, and E in Figure 5.7).
(c) The solution is degenerate only for the special cases, where two gradients happen to be exactly equal (fairly rare).
(d) Having decided which constraints are not redundant, the relative gradients of the constraints and the objective function help us to determine which vertex is the appropriate one.

5.5 MORE THAN TWO PRODUCTS

If the farmer in the previous examples had chosen to include another cereal as a possible crop (maize say) then we would have to introduce another variable for the acreage of maize (z say). The problem is then three-dimensional with 3 axes x, y, and z and a simple solution using a graph is no longer possible. With further variables (or crops, or products) we have even more dimensions to cope with, and there are computer methods for dealing with these cases. However the points (a) to (c) above still apply, in particular that the solution will lie at a vertex in the appropriate multi-dimensional space.

Methods for dealing with very complicated problems are discussed in Chapter 19.

5.6 INTEGER PROGRAMMING

Often the product or the resources can only occur as whole objects, so that the solution to the problems must only include integral values. For example, if we had been considering the building of houses and bungalows from several resources (land, finance, labour, etc.) then only whole houses and whole bungalows are useful: 5·3 houses and 0·9 bungalows would be a meaningless solution. The implication of this integral requirement for the product is that the equivalents of x and y in our previous example must be whole numbers. The feasible region is no longer an area, but consists of a finite number of cartesian pairs (x, y) spread over a region as a regular array (Figure 5.9).

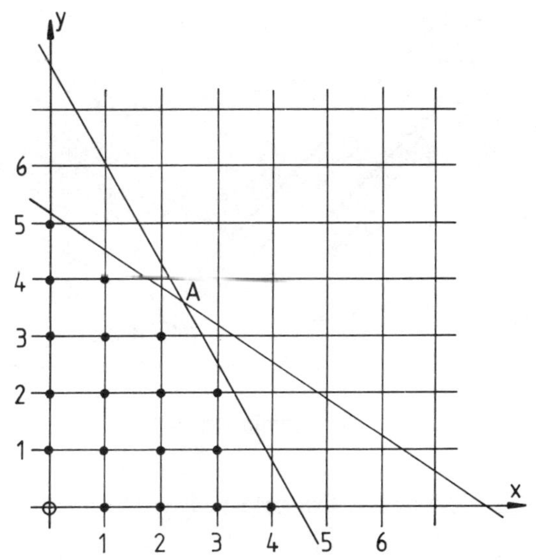

Figure 5.9 Integer requirement reduces feasible region to an array of points

In solving this type of problem, some search procedure must be established to determine which feasible point represents the optimal solution (highest profit say). In Figure 5.9 for example, the integer requirements do not allow a solution at point A. Although the search may *commence* from the vertex which would be optimal if the example were continuous, the feasible point which is eventually found to be optimal in the integer case may well be quite remote from this vertex, or indeed from any other vertex of the (continuous) feasible region.

5.7 MINIMUM COST PROBLEMS

We have considered problems which all involve the *maximization of profit*. However, the *minimization of costs* may be a valid aim in some cases. What usually happens in these cases is that the constraints involved will be of the type

$$ax + by \geqslant c$$

(e.g. nutritional value must exceed a stated level or dose) and we must determine appropriate values for x and y (e.g. food mixtures of 2 types) to satisfy the constraints at the lowest cost. The diagrams then take the form

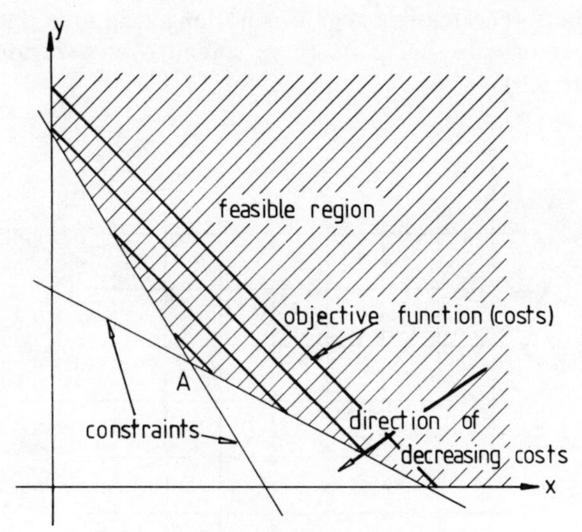

Figure 5.10 A minimum cost problem

shown in Figure 5.10. For the example shown the minimum cost solution will be at point A.

5.8 BASIC EXAMPLES FOR THE STUDENT

B5.1 Shade the regions defined by the following relations on a graph.

(a) $x \geqslant 0$
$y \geqslant 0$
$y - x \leqslant 0$
$x + y \leqslant 3$

(b) $x \geqslant 0$
$y \geqslant 0$
$2x + 3y \leqslant 6$
$3x + 2y \leqslant 6.$

B5.2 Chardust, tea-importers, produce two types of blended tea. Their Afternoon tea (type A) requires 2 tons of Indian and 3 tons of Chinese tea for each unit of final product of A. Their Breakfast tea (type B) requires 1 ton of Indian and 4 tons of Chinese tea. Each type of tea generates the same profit per unit of £5000. 200 tons of Indian and 600 tons of Chinese tea are available in a particular season, so what combination of type A and type B teas should the firm produce on the basis of overall profit?

B5.3 In another year the data given in question B5.2 still hold, except that the profit of Afternoon tea drops to £4000 per unit and the profit on Breakfast tea rises to £6000 per unit. What strategy would you now advise for Chardust?

B5.4 A builder acquires a building site and plans an estate of houses and bungalows. The amounts of land and building materials needed for a house and a bungalow, expressed in arbitrary units, are

	Land	Materials
House	10	25
Bungalow	15	20

The builder calculates that the profit he will make per house is £400, and per bungalow £500. If the site area is 300 units, and the total amount of materials available is 500 units, how many houses and/or bungalows should he build, and what is his expected profit, assuming that all unused resources are written off?

What would you advise the builder to do if he were offered another 5 units of land on the same site at a cost of £40 per unit?

5.9 INTERMEDIATE EXAMPLES FOR THE STUDENT

I5.1 In a sketch graph shade the regions defined by the following sets of inequalities:

(a) $x > 0$
$y > 2$
$x + y < 3$

(b) $x \geqslant 0$
$y \geqslant 0$
$3x + y \leqslant 6$

(c) $x \geqslant 0$
$y \geqslant 0$
$3x + y \leqslant 6$
$x + y \leqslant 3.$

I5.2 A person trying to devise a simple perfect diet decides that ideally he should consume at least 70 g of protein, 23 mg of vitamin C and

3300 kcal of carbohydrate per day. On the market there are two brands of balanced food, Ambrosia (A) and Nectar (N), both of which cost 30p per tin. Each tin of food A contains 10 g of protein, 3 mg of vitamin C, and 900 kcal of carbohydrate, while each tin of N contains 15 g protein, 6 mg vitamin C and 600 kcal of carbohydrate. What daily mixture of foods would you recommend to minimize the total cost and to satisfy the nutritional requirements (part tins may be used). What daily cost would be incurred?

How would your recommendation be affected if

(a) the price of N rose to 40p per tin and A remained at 30p?

(b) the price of A rose to 40p per tin and N remained at 30p?

kcal means kilocalories
g means grams
mg means milligrams

} These are the usual units used for the three food components mentioned in the question.

15.3 A manufacturer can make two products from three resources and he wishes to *maximize* his profits by producing the correct combination of products. Show on a graph the form that the feasible region will take, in general, and indicate the possible points which could prove to be the best combination (ignoring degenerate solutions). Include the possibility of redundant constraints in your answer.

More Advanced Examples for Part Three

A3.1 Solve algebraically the 3 sets of simultaneous equations given below, and also give a graphical solution for two of them.

(a) $4x + 3y = 12$
$x + 2y = 4$

(b) $x = 3 + 2y$
$10y - 5x = 8$

(c) $x + y = 6$
$3x + y = 8$
$x + y + z = 0.$

A3.2 A builder wishes to erect a small estate of houses and bungalows on a 30 acre site, within a total budget of £1 040 000. Unfortunately there is a world shortage of roofing tiles, so that he has to limit himself to a maximum of 180 000 tiles. Houses make a profit of £1500 each, on a cash outlay of £8000, use 2000 roofing tiles and take up 1/4 acre each. Bungalows make him a profit of £2000 each on an outlay of £13 000, use 2500 tiles and take up 1/2 acre of land each. How many houses and bungalows should he build to maximize his profit, and what profit does he make? To what level would the profit on each house have to rise in order to alter your suggested mix of houses and bungalows, all other quantities remaining the same? (A total recalculation is not required.)

A3.3 Grease, an oil-producing country, is planning an investment in the construction of a series of oil distillation plants within its own territory. It has to import the plants and, after initial discussions with representatives of appropriate manufacturers, decided that the choice lies between two types of plant (or a mix of both):

Plant A, manufactured by the Asso Oil Company, capable of producing 25 000 tons of high grade oil and 30 000 tons of low grade oil per year.

Plant S, manufactured by the Smell Oil Company, capable of producing 20 000 tons of high grade oil and 40 000 tons of low grade oil per year.

The cost of the two types of plant is the same, $5 000 000 each. In the short term the country hopes to be able to produce at least 200 000 tons of high grade oil per year, and 320 000 tons of low grade oil per year.

The Government of Grease is trying to decide how many plants of type A and S it needs and asks your advice on what it should buy, if its only criterion were that of minimizing the total cost. What recommendation would you make?

The longer term oil requirements are likely to show a higher demand for high grade oil, and less demand for low grade oil. How would this tend to affect your recommendation if these long term forecasts are to be taken into account? (A *qualitative* description and justification of how your recommendation would be changed is required here.)

A3.4 A student is trying to decide the best strategy for sharing his time in an examination between *thinking-time* and *writing-time*. Altogether, the examination lasts three hours, and he reckons that he could not spend more than $1\frac{1}{2}$ hours in thought, or more than $2\frac{1}{2}$ hours just writing. Also he decides that, since quantity alone has some merit, he must write for at least $1\frac{3}{4}$ hours, but that it would also be feasible to pass without any thought at all.

His problem is that he is not certain how many marks would be awarded on average for each minute of thought, and how many for each minute of writing, so he decides to compare two assumptions:

(a) That he would gain 30 marks on average for each hour of thought, and 20 marks on average for each hour of writing.

(b) That he would gain 20 marks on average for each hour of thought, and 30 marks on average for each hour of writing.

What conclusions does he draw on

(i) The total time he should spend on each activity, and

(ii) The maximum number of marks he can expect to gain for each of the two assumptions (a) and (b) mentioned above?

PART FOUR

Introduction

This fourth part, the most substantial in terms of length, tackles the problem of nonlinear relationships, and methods particularly suited to their analysis. Chapter 6 includes some necessary preliminary remarks, and discusses the idea of a *mathematical limit*, which is essential to an understanding of several subsequent sections. The main substance of the pure mathematics introduced is contained in the four chapters on the calculus: 7, 8, 12, and 13. The differential calculus is introduced from first principles, taking the gradient or slope of a curve as the starting point. While Chapters 7 and 8 explain the technique of differentiation as a branch of pure mathematics, Chapters 9 and 10 demonstrate two rather different applications. The economic model used in Chapter 9 is aimed not so much at explaining a topic in economics, but at illustrating how a few basic assumptions and definitions can provide the basis for an extended mathematical development. The reader is reminded of the importance of the basic inputs and assumptions to any such quantitative model, and the dangers of accepting uncritically the output from such calculations. Chapter 10, on curve sketching, demonstrates the power of differential calculus in the analysis of any given equation, and shows how a surprising amount of information can be gained (in graphical form) from an intelligent use of mathematical arguments. A facility in analysing any equation, not just a standard type, is an immense asset in model building. Chapter 11 then goes on to consider three important types of curve useful in models, and which can be more efficiently sketched and analysed with the use of differential calculus. Finally, in Chapter 15, another technique concerned, like linear programming, with problems of optimization under constraints is introduced. The method of Lagrange multipliers, which involves partial differentiation methods, is explained and the similarity of some of the problems it can tackle to those of linear programming is illustrated.

Integration is introduced, in the first instance, as the process of differentiation in reverse, though its value in the calculation of areas and volumes is clearly explained (Chapter 12). Besides tackling more complicated functions, Chapter 13 goes on to illustrate the value of integration in connection with probability distribution functions from statistical theory. Chapter 14 introduces two more models explicitly (stock control and decay rates), besides considering some of the more general problems involved in building such models. Again, the importance of the basic assumptions underlying any model is emphasized, and the two models chosen allow a further demonstration of the differential and integral techniques recently introduced.

CHAPTER 6

Graphical Representation: Curves, Limits, Continuity

6.1 INTRODUCTION

We have already considered (Section 3.8) the translation of a relation into a graph by building up a table of ordered pairs (x, y) and plotting them on the cartesian $(x–y)$ plane. It was also mentioned that there are some difficulties associated with the technique of plotting a few points on a graph, and linking these with a smooth curve. Unless great care is taken in the selection of these points to be plotted it is possible to omit significant features of the curve altogether by making incorrect assumptions of *smoothness* or *continuity*.

For example, consider the relation

$$y = +\sqrt{\frac{4}{(2x-3)^2}}$$

from which we can generate the cartesian pairs: $(-2, 2/7)$, $(-1, 2/5)$, $(0, 2/3)$, $(1, 2)$, $(2, 2)$, $(3, 2/3)$, $(4, 2/5)$, $(5, 2/7)$ etc. These have been plotted in Figure 6.1 and joined by the broken line. The question clearly arises as to whether the two points $(1, 2)$ and $(2, 2)$ should be linked, to produce a smooth peak. In fact, they should not, as the curve approaches an infinite peak at $x = 1{\cdot}5$. In more formal language, the curve is *discontinuous* at $x = 1{\cdot}5$ as there is a break or gap in the curve at this point. To detect this one needs to generate some points in the interval

$$1 < x < 2$$

or to notice that as x approaches the value $3/2$, then y approaches $+\infty$.

It is also possible to devise a function with *bad behaviour* artificially such as in the following example (Figure 6.2).

$$\mathbf{R} = \{(x, y) \mid y = f(x) \quad \text{where} \quad f(x) = x + 3 \quad \text{for} \quad x < 1,$$
$$\text{and} \quad f(x) = 0{\cdot}5x + 1{\cdot}5 \quad \text{for} \quad x \geqslant 1\}.$$

Here a deliberate break or discontinuity is introduced by defining the relation in two parts.

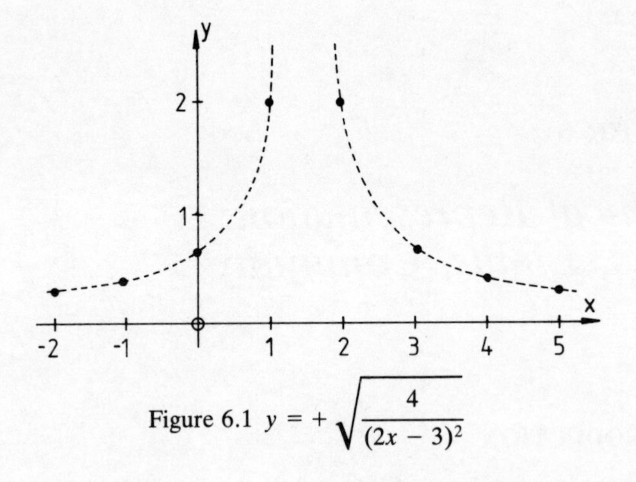

Figure 6.1 $y = +\sqrt{\dfrac{4}{(2x-3)^2}}$

However the relation

$$\mathbf{R} = \{(x, y) \mid y = f(x) \quad \text{where} \quad f(x) = x + 2 \quad \text{for} \quad x \leqslant 2$$
$$\text{and} \quad f(x) = -2x + 8 \quad \text{for} \quad x > 2\}.$$

also defined in two parts, does retain continuity since the two sections meet at the point (2, 4) which is *included* in at least one of the parts (Figure 6.3). Here no break or discontinuity occurs since the two parts meet at the point A at (2, 4). Notice that if the first part of the relation had been defined by

$$f(x) = x + 2 \quad \text{for} \quad x < 2$$

then the point (2, 4) would not have been included anywhere in the relation, and the curve would have been *discontinuous* at (2, 4).

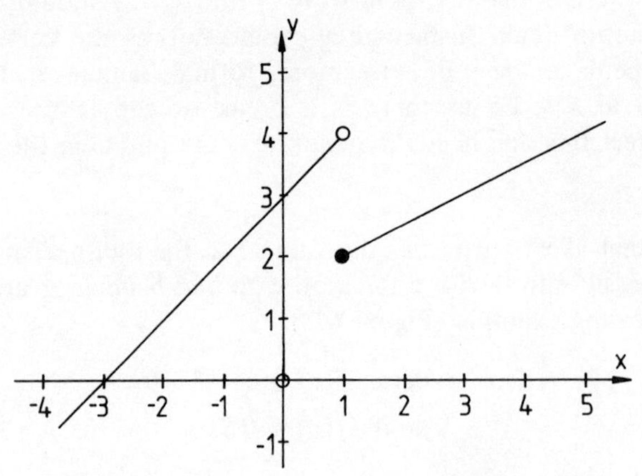

Figure 6.2 $y = x + 3$ for $x < 1$, $y = 0{\cdot}5x + 1{\cdot}5$ for $x \geqslant 1$

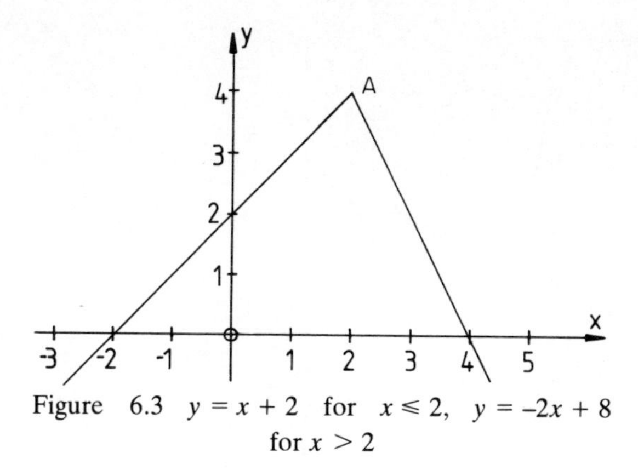

Figure 6.3 $y = x + 2$ for $x \leqslant 2$, $y = -2x + 8$ for $x > 2$

So, *discontinuity* in a curve is associated with any gap or break within the finite region of the curve. We now go on to consider the idea of a *limit* or *limiting value* informally, before introducing formal notation and definitions.

6.2 LIMITS: AN INTRODUCTION

In considering continuity we were principally concerned with the behaviour of a curve at a point (or points) and were asking whether the curve actually continues at each point. In the case of limits we are concerned with the way in which a curve appears to be behaving over some range and what would happen if the curve were to continue behaving in the same way. So, as an introductory example, consider the curve (parabola):

$$y = x^2$$

near the origin. We wish to consider the value which y approaches or eventually appears to reach as x moves towards the value zero. In other words we are asking the question: *what is the limiting value of y as x tends towards zero?*

To answer this we consider the value of y for various values of x near zero, and on *both sides* of zero (Figure 6.4). The nearer x approaches zero from the positive side (right-hand side of the figure) the closer y approaches the value zero as well. Also, the nearer x approaches zero from the negative side (left-hand side of the figure) the closer y approaches the value zero.

In *both* cases therefore the limiting value of y, as x approaches zero, is zero. We can then say that *y has a limiting value of zero* as x approaches zero. One important, and slightly curious, point to note is that we do not actually have to consider the exact point $x = 0$, but only have to be able to move as close to it as we like. Nevertheless it will be noted that in the case of a well-behaved function like the one chosen, it is the case that the

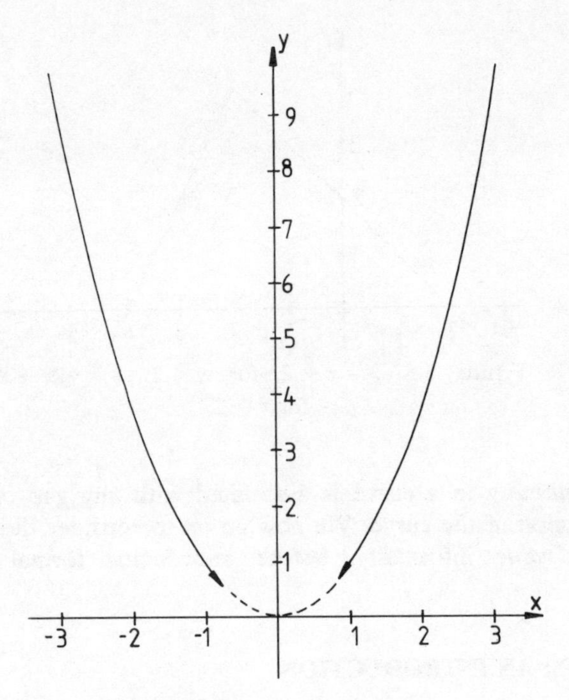

Figure 6.4 $y = x^2$ as $x \to 0$

limiting value of y as x approaches zero is the same as *the value of y at x equals zero*. (In fact this will always be the case for continuous functions from the definitions which follow below.)

6.3 FORMAL DEFINITIONS

Limits The limiting value of a function of x, $f(x)$, as x approaches the value c (constant) means the value which $f(x)$ approaches as x tends towards c from either direction, but without x ever actually taking the value c itself. It is written as

$$\lim_{x \to c} f(x).$$

So, if $\lim_{x \to c} f(x) = L$, then the closer x approaches c the closer the value of $f(x)$ approaches L.

Continuity If the limiting value of a function of x, $f(x)$, as x approaches the value c is actually $f(c)$, then the function f is continuous at the point $x = c$. That is, a function is continuous at the point $x = c$ if

$$\lim_{x \to c} f(x) = f(c).$$

6.4 EXAMPLES (GEOMETRIC REPRESENTATION)

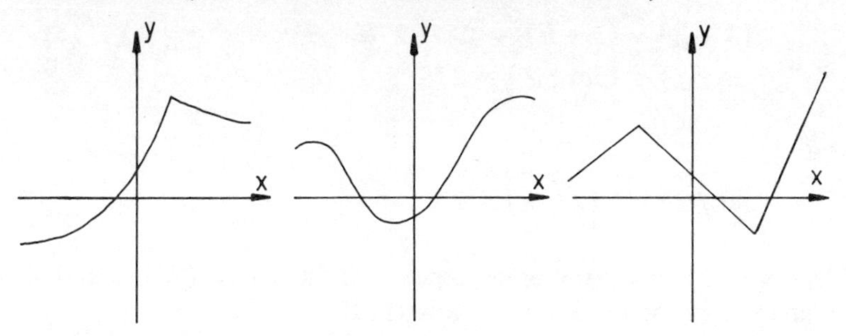

Figure 6.5 Some continuous functions

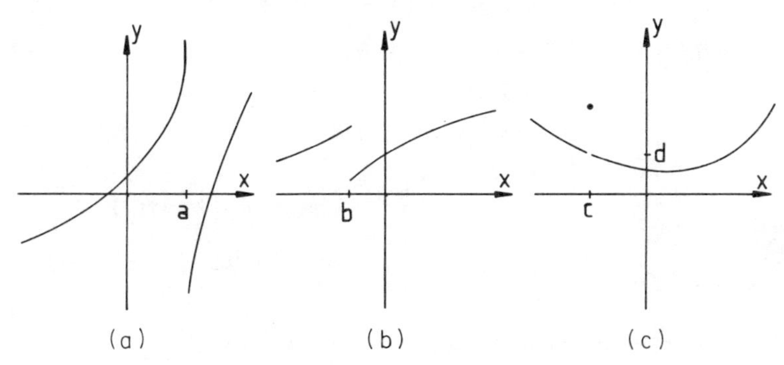

Figure 6.6 Some discontinuous functions

For all cases in Figure 6.5, the limit of each function at any point is the value of the function at that point. That is,

$$\operatorname*{Lim}_{x \to c} f(x) = f(c) \quad \text{for all possible } c.$$

For case 6.6(a) this relation also holds except where $x = a$ since here the curve approaches $\pm\infty$ from different sides, and no limit is defined. For case 6.6(b) a limit exists except for $x = b$ where the curve is also discontinuous. For case 6.6(c) the curve is discontinuous at $x = c$, but a limit is defined:

$$\operatorname*{Lim}_{x \to c} f(x) = d \neq f(c).$$

6.5 EXAMPLES (ALGEBRAIC REPRESENTATION)

Find $\operatorname*{Lim}_{x \to c} f(x)$ for the cases

$f(x) =$ (a) x^3

(b) $2x^2 - x - 14$

(c) $1/(x - 3)$.

(a) $f(x) = x^3$, and putting $x = 3 + h$ we have

$$f(3+h) = (3+h)^3 = 27 + 27h + 9h^2 + h^3 = g(h)$$

$$\therefore \lim_{x\to 3+} f(x) = \lim_{h\to 0+} g(h) = 27$$

and

$$\lim_{x\to 3-} f(x) = \lim_{h\to 0-} g(h) = 27.$$

Here $x \to 3+$ means x approaches 3 from above (RHS) and $x \to 3-$ means x approaches 3 from below (LHS)

so that $\lim_{x\to 3} f(x) = 27 = f(3)$ (continuous at this point).

(b) $f(x) = 2x^2 - x - 14$, and putting $x = 3 + h$ we have

$$\begin{aligned} f(3+h) &= 2(3+h)^2 - (3+h) - 14 \\ &= 18 + 12h + 2h^2 - 3 - h - 14 \\ &= 1 + 11h + 2h^2 = g(h) \end{aligned}$$

$$\left.\begin{aligned} &\therefore \lim_{x\to 3+} f(x) = \lim_{h\to 0+} g(h) = 1 \\ &\text{and} \\ &\lim_{x\to 3-} f(x) = \lim_{h\to 0-} g(h) = 1 \end{aligned}\right\} \therefore \lim_{x\to 3} f(x) = 1 = f(3)$$

(continuous at this point).

(c) $f(x) = 1/(x-3)$, and putting $x = 3 + h$ we have

$$\begin{aligned} f(3+h) &= 1/(3+h-3) \\ &= 1/h = g(h) \end{aligned}$$

$$\left.\begin{aligned} &\therefore \lim_{x\to 3+} f(x) = \lim_{h\to 0+} g(h) = \lim_{h\to 0+} (1/h) = +\infty \\ &\text{and} \\ &\lim_{x\to 3-} f(x) = \lim_{h\to 0-} g(h) = \lim_{h\to 0-} (1/h) = -\infty \end{aligned}\right\} \therefore \text{No single limit is defined.}$$

Hence no limit is defined at the point $x = 3$, where the curve is discontinuous.

It will be noted that where continuity does exist (examples (a) and (b)) then we can write

$$\lim_{x\to c} f(x) = f(c) \quad \text{or} \quad \lim_{x\to 3} f(x) = f(3)$$

so that the quickest way to find the limit is simply to substitute the value 3 instead of x, in each function. In other words, if we can be sure that the curve is continuous at the point in question, then the rather tedious procedure which has been demonstrated in full can be omitted and the appropriate value (c) substituted for x immediately; for example,

$$\lim_{x \to 2} (x^3 + 4x - 1) = (2^3 + 4.2 - 1) = (8 + 8 - 1) = 15$$

since all cubic functions are continuous.

6.6 RULES FOR LIMITS

(a) $\lim_{x \to c} [Kf(x)] = K[\lim_{x \to c} f(x)]$, K is a constant. For example,

$$\lim_{x \to 2} [4x^3] = 4[\lim_{x \to 2} x^3] = 4[8] = 32.$$

(b) $\lim_{x \to c} [x^n] = c^n$. For example,

$$\lim_{x \to 3} [x^4] = 3^4 = 81.$$

(c) If $\lim_{x \to c} f(x) = A$ and $\lim_{x \to c} g(x) = B$, then

$$\lim_{x \to c} [f(x) \pm g(x)] = A \pm B \quad \text{(both +, or both −)}$$

and

$$\lim_{x \to c} [f(x)g(x)] = AB.$$

(d) If $\lim_{x \to c} f(x) = A$ and $\lim_{x \to c} g(x) = B$, then

$$\lim_{x \to c} [f(x)/g(x)] = \frac{A}{B} \quad \text{if } A \neq 0, \text{ and } B \neq 0.$$

If, however,

$A = 0, B \neq 0$ then limit is zero
$A \neq 0, B = 0$ then no limit is defined (infinity involved)
$A = 0, B = 0$ there may or may not be a limit.

This last result shows that, if a limit is defined in both cases, we can write

$$\lim_{x \to c} [f(x)/g(x)] = \frac{1}{\lim_{x \to c} [g(x)/f(x)]}$$

since both sides of the equation equal A/B.

The case where both A and B equal zero is in practice very important, since it underlies the development of differential calculus (Section 7.3).

(e) From the above it can be shown that

$$\lim_{x \to c} [a + bx + dx^2 + ex^3 + \ldots] = a + bc + dc^2 + ec^3 + \ldots$$

so that all functions of the form $y = a + bx + dx^2 + ex^3 + \ldots$ are continuous. (Such functions are called *polynomials*.)

6.7 BASIC EXAMPLES FOR THE STUDENT

B6.1 Draw a sketch graph of each of the following functions and indicate where any discontinuities occur.

(a) $y = 2x^2 + x + 3$
(b) $y = 3x/(x - 1)$
(c) $y = (x^2 + x - 2)/(x - 1)$
(d) $y = f(x)$ where $f(x) = x^2$ for $-1 < x < 1$
and $f(x) = -x^2 + 2$ for $x < -1$ and $x > 1$
(e) $y = f(x)$ where $f(x) = x^2$ for $-1 < x < 1$
and $f(x) = -x^2 + 2$ for $x \leqslant -1$ and $x \geqslant 1$.

B6.2 Evaluate the following limits:

(a) $\lim_{x\to 3}\left(\dfrac{x^2 - 1}{x - 1}\right)$ (d) $\lim_{x\to 1}\left(\dfrac{3}{x - 1}\right)$

(b) $\lim_{x\to 1}\left(\dfrac{x^2 - 1}{x - 1}\right)$ (e) $\lim_{x\to 3}(x + 2)(x - 3)$

(c) $\lim_{x\to 2}(x + 2)(x - 3)$ (f) $\lim_{x\to 3}\left(\dfrac{x^2 - 6x + 9}{x^2 - 2x - 3}\right)$

(g) $\lim_{x\to 2} f(x)$ where $f(x) = x^2$ for $x \neq 2$
$f(x) = 1$ for $x = 2$

(h) $\lim_{x\to 0} f(x)$ where $f(x) = x + 1$ for $x < 0$
$f(x) = 2 - x$ for $x \geqslant 0$

(i) $\lim_{x\to 1} f(x)$ where $f(x) = 2 - x$ for $x < 1$
$f(x) = x$ for $x \geqslant 1$.

6.8 INTERMEDIATE EXAMPLES FOR THE STUDENT

I6.1 Draw a sketch graph of each of the following functions; indicate any discontinuities; evaluate the limits indicated.

(a) $f(x) = (x + 1)/(x - 2)$; $\lim_{x\to 1} f(x)$

(b) $f(x) = (x + 2)/(x - 1)$; $\lim_{x\to 1} f(x)$

(c) $f(x) = \left(\dfrac{x^2 - 1}{x^2 - 3x + 2}\right)$; $\lim_{x\to 1} f(x)$ (Hint: factorize numerator and denominator)

(d) $y = f(x)$ where $f(x) = x + 2,\ x < 1$
$f(x) = 4 - x,\ x \geqslant 1$; $\lim_{x\to 1} y$

(e) $y = f(x)$ where $f(x) = x + 2,\ x < 1$
$f(x) = x,\ \ x \geqslant 1$; $\lim_{x\to 1} y$.

CHAPTER 7

Calculus: Differentiation I

7.1 GRADIENTS

In this chapter we shall be extending the theory of graphs which we have been developing by considering the idea of a gradient or slope for non-linear functions.

In the case of straight lines, the gradient is the constant m in the equation

$$y = mx + c$$

and if the line passes through the two points $A(x_1, y_1)$ and $B(x_2, y_2)$ then $y_1 = mx_1 + c$ and $y_2 = mx_2 + c$. Hence

$$y_2 - y_1 = mx_2 + c - mx_1 - c = mx_2 - mx_1 = m(x_2 - x_1)$$

so that

$$m = \frac{y_2 - y_1}{x_2 - x_1} = \frac{\text{change in the } y\text{-coordinate}}{\text{change in the } x\text{-coordinate}} = \frac{\Delta y}{\Delta x} \quad \text{(Figure 7.1).}$$

The change in the value of the y-coordinate is given the symbol Δy and the change in the value of the x-coordinate is given the symbol Δx. So, for a straight line, no matter how far apart or how close A and B are, the ratio $\Delta y/\Delta x$ is a constant and is the gradient of the line.

In the case of a curve, however, the gradient at each point is, by convention, taken to be the gradient of the tangent touching at that point. The value of the gradient therefore varies, and is itself a function of x. In Figure 7.2 for example:

the gradient at A is large and positive,
the gradient at B is small and positive,
the gradient at C is zero, and
the gradient at D is negative.

7.2 INTERPRETATIONS OF THE GRADIENT

If we consider the gradient of a particular point on a curve (such as at B in Figure 7.2), we can interpret it as giving information on how fast y changes

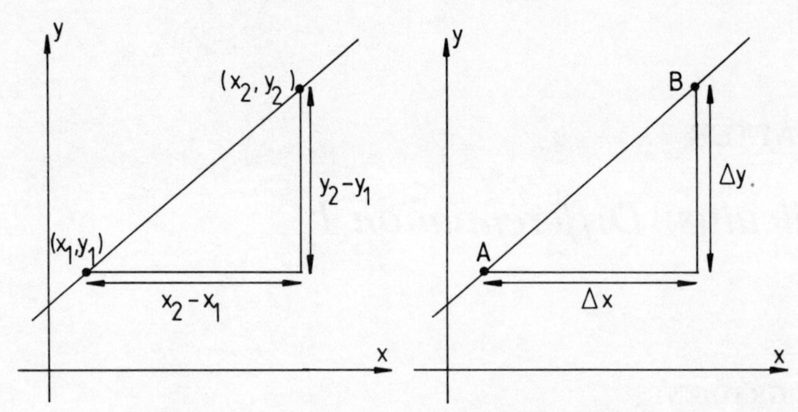

Figure 7.1 A line $y = mx + c$ through two points

with respect to changes in x. In other words, the gradient gives information on *rates of change*. For example,

(i) If y represents the price index, x represents time, then the gradient gives the rate of price changes with time (inflation).
(ii) If y represents the total manufacturing costs, x the total production level, then the gradient represents the rate of increase of production costs with respect to changes in production level (marginal costs).
(iii) At the point C in Figure 7.2 the value of y reaches a maximum value within the range of the graph. In addition the gradient at C is zero, so that local *maximum* and *minimum* values of y correspond to *zero rates of change*, and *zero gradients*. This property of zero gradients is invaluable in helping us to locate the positions of local maxima and minima in functions of a variable (say x).

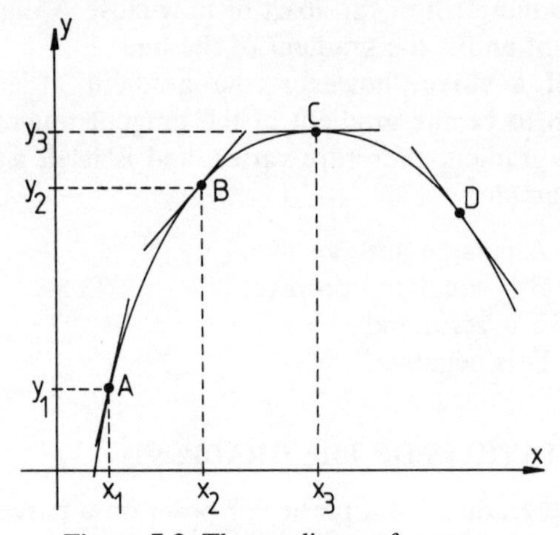

Figure 7.2 The gradients of a curve

It is worthwhile pointing out here the distinction between *instantaneous rates of change*, represented by gradients at a point and described above, and rates of change averaged over considerable finite intervals represented by gradients of chords of curves. Only when the relationship is linear over the entire interval are the two rates the same, in general.

So, referring to Figure 7.2 once again, we could imagine a chord between points A and C for example (i.e. the straight line AC). The gradient of AC would then measure the average rate of change over the interval x_1x_3, and would be given by

$$\frac{y_3 - y_1}{x_3 - x_1} = \frac{\text{total change in } y}{\text{total change in } x}.$$

Figures published for the annual rate of inflation, for example, usually use this procedure over one year so that

$$\text{Rate of inflation} = \frac{\text{change in retail prices index over one year}}{\text{original retail prices index}}.$$

Clearly, the disadvantage of averaging over such an interval is that it hides short-term changes.

7.3 CALCULATION OF THE GRADIENT: THE PRINCIPLE

We wish to generate a simple and straightforward methodology for evaluating the gradient at any point (x, y) of a function $y = f(x)$. The approach we shall take will be to find an approximate value of the gradient at a particular point, and then to improve on this by closer and closer approximation.

In Figure 7.3, we first define two points on a curve at $A(x_1, y_1)$ and

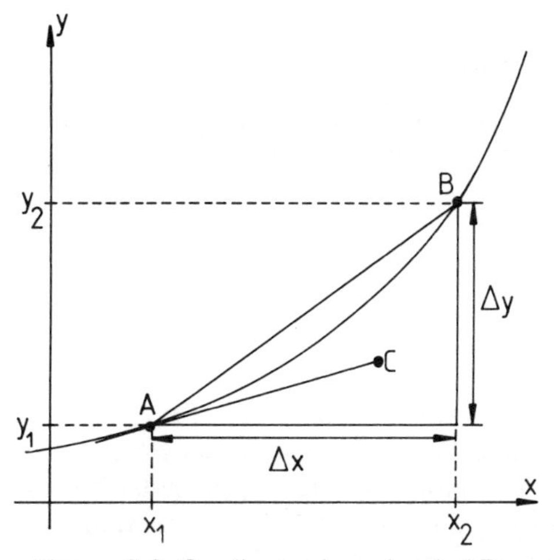

Figure 7.3 Gradient of a chord AB and tangent AC on a curve

$B(x_2, y_2)$ and link them by a line (or chord) AB. Also we define the distances

$$\Delta x = x_2 - x_1 \qquad \Delta y = y_2 - y_1.$$

Our aim is to find the gradient at the point $A(x_1, y_1)$, and this is the same as the gradient of the tangent, AC. We take the gradient of AB as the first approximation to the gradient of AC, and then imagine that the point B moves towards the point A. As this happens, the gradient of AB moves closer and closer towards the gradient of AC, while Δy and Δx simultaneously diminish in size. Eventually the gradients are identical, and Δy and Δx approach zero. Therefore,

$$\text{gradient of AB} = \frac{y_2 - y_1}{x_2 - x_1} = \frac{\Delta y}{\Delta x}$$

and

$$\text{gradient of AC} = \lim_{\Delta x \to 0} (\text{Gradient of AB}) = \lim_{\Delta x \to 0} \left(\frac{\Delta y}{\Delta x}\right).$$

The quantity $\lim_{\Delta x \to 0}(\Delta y/\Delta x)$ is given the simplified notation

$$\frac{dy}{dx}$$

where this is a single quantity, and should not be treated as though it were a fraction.

Now both the points (x_1, y_1) and (x_2, y_2) lie on the curve $y = f(x)$ so that $y_1 = f(x_1)$ and $y_2 = f(x_2)$. But also, we have defined Δy by

$$\Delta y = y_2 - y_1 = f(x_2) - f(x_1)$$

and $\Delta x = x_2 - x_1$, giving $x_2 = x_1 + \Delta x$, so that

$$\frac{\Delta y}{\Delta x} = \frac{f(x_2) - f(x_1)}{\Delta x} = \frac{f(x_1 + \Delta x) - f(x_1)}{\Delta x}$$

and

$$\frac{dy}{dx} = \lim_{\Delta x \to 0} \left(\frac{f(x_1 + \Delta x) - f(x_1)}{\Delta x}\right)$$

which is the gradient of $f(x)$ as a function of x_1.

This general formula for dy/dx is as far as we can go for a general curve $y = f(x)$. Further development can proceed, however, for particular equations, and this is undertaken in the next section.

One further detail can be seen at this stage. The gradient (with the axes reversed in effect) given by $\mathrm{d}x/\mathrm{d}y$ must likewise be defined as

$$\begin{aligned}\frac{\mathrm{d}x}{\mathrm{d}y} &= \operatorname*{Lim}_{\Delta x\to 0}\left(\frac{\Delta x}{\Delta y}\right)\\ &= \operatorname*{Lim}_{\Delta x\to 0}\left(\frac{\Delta x}{f(x_1+\Delta x)-f(x_1)}\right)\\ &= \frac{1}{\operatorname*{Lim}_{\Delta x\to 0}\left(\frac{f(x_1+\Delta x)-f(x_1)}{\Delta x}\right)} \quad \text{from Section 6.6(d)}\\ &= \frac{1}{\mathrm{d}y/\mathrm{d}x}\end{aligned}$$

so that

$$\frac{\mathrm{d}y}{\mathrm{d}x} = \frac{1}{\mathrm{d}x/\mathrm{d}y} \quad \text{or} \quad \frac{\mathrm{d}x}{\mathrm{d}y} = \frac{1}{\mathrm{d}y/\mathrm{d}x}.$$

7.4 EXAMPLES

We now consider the details of how $\mathrm{d}y/\mathrm{d}x$ is calculated for some specific curves.

(a) Find $\mathrm{d}y/\mathrm{d}x$ for the curve

$$y = x^2 - 4x + 5.$$

Point A is at (x, y), point B is at $(x + \Delta x, y + \Delta y)$. Hence, at A,

$$f(x) = x^2 - 4x + 5$$

and at B,

$$f(x + \Delta x) = (x + \Delta x)^2 - 4(x + \Delta x) + 5.$$

Subtract,

$$\begin{aligned}f(x+\Delta x) - f(x) &= (x+\Delta x)^2 - 4(x+\Delta x) + 5 - x^2 + 4x - 5\\ &= x^2 + 2x\Delta x + (\Delta x)^2 - 4x - 4\Delta x - x^2 + 4x\\ &= 2x\Delta x - 4\Delta x + (\Delta x)^2\end{aligned}$$

$$\therefore \frac{f(x+\Delta x) - f(x)}{\Delta x} = \frac{\Delta y}{\Delta x} = 2x - 4 + \Delta x$$

$$\therefore \frac{\mathrm{d}y}{\mathrm{d}x} = \operatorname*{Lim}_{\Delta x\to 0}\left(\frac{\Delta y}{\Delta x}\right) = 2x - 4,$$

since the Δx term becomes zero when the limit is taken.

Conclusion: if $y = x^2 - 4x + 5$, then $dy/dx = 2x - 4$.

Positive gradients occur when

$$2x - 4 > 0$$
$$\therefore 2x > 4$$
$$\therefore \underline{x > 2}.$$

Negative gradients occur when

$$2x - 4 < 0$$
$$\therefore 2x < 4$$
$$\therefore \underline{x < 2}.$$

Zero gradient occurs when

$$2x - 4 = 0$$
$$\therefore 2x = 4$$
$$\therefore \underline{x = 2}.$$

These results are consistent with the sketch of the curve in Figure 7.4.

(b) Find dy/dx for the curve

$$y = 1/(x - 1)^2.$$

Point A is at (x, y), point B is at $(x + \Delta x, y + \Delta y)$. Hence, at A,

$$f(x) = 1/(x - 1)^2$$

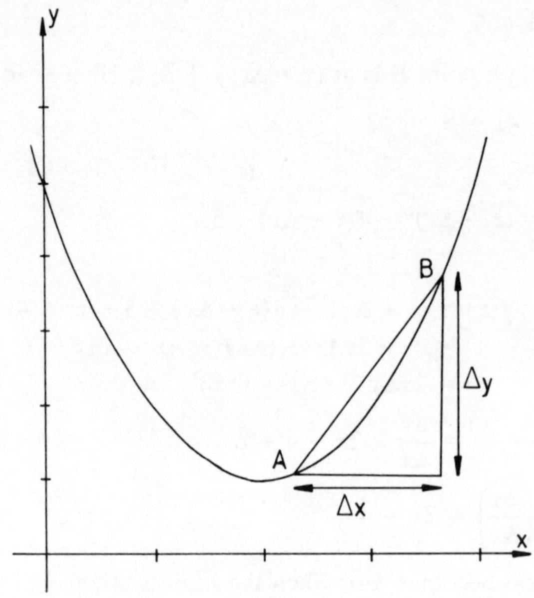

Figure 7.4 $y = x^2 - 4x + 5$

and at B,

$$f(x + \Delta x) = 1/(x + \Delta x - 1)^2.$$

Subtract,

$$\begin{aligned} f(x+\Delta x) - f(x) &= 1/(x + \Delta x - 1)^2 - 1/(x - 1)^2 \\ &= \frac{(x-1)^2 - (x+\Delta x - 1)^2}{(x + \Delta x - 1)^2(x - 1)^2} \\ &= \frac{(x^2 - 2x + 1) - (x^2 + (\Delta x)^2 + 1 - 2x - 2\Delta x + 2x\Delta x)}{(x + \Delta x - 1)^2(x - 1)^2} \\ &= \frac{-(\Delta x)^2 + 2\Delta x - 2x\Delta x}{(x + \Delta x - 1)^2(x - 1)^2}. \end{aligned}$$

Hence

$$\frac{f(x + \Delta x) - f(x)}{\Delta x} = \frac{\Delta y}{\Delta x} = \frac{2 - 2x - \Delta x}{(x + \Delta x - 1)^2(x - 1)^2}$$

$$\therefore \frac{\mathrm{d}y}{\mathrm{d}x} = \lim_{\Delta x \to 0}\left(\frac{\Delta y}{\Delta x}\right) = \frac{2 - 2x}{(x - 1)^2(x - 1)^2} = \frac{-2(x - 1)}{(x - 1)^4} = \frac{-2}{(x - 1)^3}.$$

Conclusion: if $y = 1/(x - 1)^2$, then $\mathrm{d}y/\mathrm{d}x = -2/(x - 1)^3$.

If $x - 1 < 0$, then $(x - 1)^3 < 0$ and $\mathrm{d}y/\mathrm{d}x > 0$

$\therefore$ Positive gradients occur when $\underline{x < 1}$.

If $x - 1 > 0$, then $(x - 1)^3 > 0$ and $\mathrm{d}y/\mathrm{d}x < 0$

$\therefore$ Negative gradients occur when $\underline{x > 1}$.

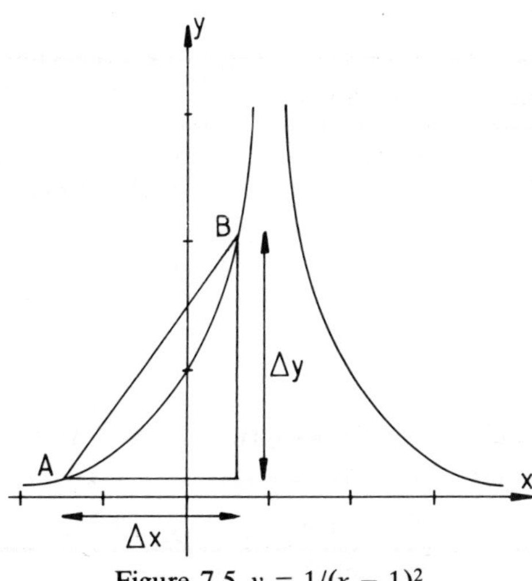

Figure 7.5 $y = 1/(x - 1)^2$

As x approaches $+1$, then $(x - 1)$ approaches zero and dy/dx approaches

$$+\infty \ (x < 1) \quad \text{or} \quad -\infty \ (x > 1).$$

These results are consistent with the sketch of the curve in Figure 7.5.

7.5 NOTATION

If $y = f(x)$, then the alternative notations and descriptions for dy/dx include:

$$\frac{dy}{dx} = \frac{df}{dx} = y'(x) = f'(x) = y' = f'$$

and dy/dx is
the gradient of y with respect to x
the first derivative of y (with respect to x)
the first differential coefficient of y (with respect to x).
When no confusion with the second and third derivatives is likely to occur (defined later), then the abbreviated form *the derivative of y* is acceptable.

7.6 FORMULAE FOR THE FIRST DERIVATIVE

The two examples in Section 7.4 both showed how the first derivative dy/dx can be calculated from first principles. However, this complete procedure, involving Δx and Δy, is actually carried out in full very rarely. Instead the first derivative is usually found by the application of a basic set of rules and formulae which can be used to cover most functions of importance. The more basic formulae and some examples of their application are given below.

Rule No.	Function	First derivative	Example
1	$y = \text{constant}$	$y' = 0$	$y = 3$, $y' = 0$ (graph of $y = 3$ is a horizontal line)
2	$y = x^n$	$y' = nx^{n-1}$	$y = x^3$, $y' = 3x^2$ (n +ve integer) $y = x^{-4}$, $y' = -4x^{-5}$ (n −ve integer) $y = x^{1/2}$, $y' = \frac{1}{2}x^{-1/2}$ (n fractional)
3	$y = e^x$	$y' = e^x$	
4	$y = \log_e x$	$y' = 1/x$	
5	$y = f(x) + g(x)$	$y' = f'(x) + g'(x)$	$y = x + x^2$, $y' = 1 + 2x$
6	$y = f(x) - g(x)$	$y' = f'(x) - g'(x)$	$y = x - 1/x$, $y' = 1 + 1/x^2$
7	$y = af(x)$	$y' = af'(x)$	$y = 3x^4$, $y' = 12x^3$ (a is a constant)

7.7 BASIC EXAMPLES FOR THE STUDENT

B7.1 From first principles (using Δx and Δy) find the first derivatives of the following functions and show that they agree with the general formulae given in the preceding section.

(a) $y = x^2$
(b) $y = x^3$
(c) $y = 1/x^2$
(d) $y = 1/x^3$.

B7.2 Using the formulae given in the preceding section, find the first derivatives of the following functions indicating which rule or rules were used for each one. Give the value of the gradient at $x = 1$.

(a) $y = 2x^2 + 3$
(b) $y = 3x + 4/x + 2/x^2$
(c) $y = ax^3 + bx^2 + cx + d$; a, b, c, d are constants
(d) $y = (x + a)^2$
(e) $y = 3 \log_e x$
(f) $y = 5e^x - \log_e x - 15/x$.

7.8 INTERMEDIATE EXAMPLES FOR THE STUDENT

I7.1 For the three equations below:

(a) $y = x^2 - 3x + 5$
(b) $y = 1/(x - 1)$
(c) $y = 1/x^2$

find

(i) the increment Δy in y corresponding to an increment Δx in x.
(ii) the ratio $\Delta y/\Delta x$
(iii) the derivative dy/dx
(iv) the gradient at $x = -1$.

I7.2 The total cost in pounds of producing x items is given by

$$TC = 0{\cdot}01x^2 + 4x + 100.$$

Find

(i) the extra cost ΔTC caused by an increase Δx in the number of items produced.
(ii) the extra cost to produce one more item if 150 are already being produced.
(iii) the instantaneous marginal cost at $x = 150$.
(iv) the average cost per item, for any x.

I7.3 Differentiate the functions:

$$y = x^3 - 3/x^2 + e^x - 2 \log_e x + 2 \log_e 3$$

$$y = (2x + 3)^2.$$

CHAPTER 8

Calculus: Differentiation II

8.1 INTRODUCTION

So far the technique of differentiation has been explained from first principles, using simple functions of a single independent variable. We now wish to extend the method to cover the cases of

(i) more complicated functions of a single variable,
(ii) functions of more than one variable (where 'partial differentiation' is introduced), and
(iii) higher order derivatives (where derivatives are themselves differentiated further).

In all cases new formulae will be introduced as given, rather than developed from first principles (using the Δx and Δy method). Nevertheless the original purpose of differentiation remains: it enables us to find the gradient or rate of change of a dependent variable with respect to independent variable(s).

8.2 PRODUCTS

We can, using the elementary formulae of Section 7.6, differentiate such functions as

$$y = x^2 \quad \text{and} \quad y = \mathrm{e}^x.$$

We now wish to consider the problem of differentiating functions which are *products* of simpler functions, such as

$$y = x^2\,\mathrm{e}^x.$$

The general problem can be written in the form: find $\mathrm{d}y/\mathrm{d}x$ if

$$y = uv$$

where $u = u(x)$, $v = v(x)$ (i.e. both parts of the product are themselves functions of x).

The solution is

$$\frac{\mathrm{d}y}{\mathrm{d}x} = v\,\frac{\mathrm{d}u}{\mathrm{d}x} + u\,\frac{\mathrm{d}v}{\mathrm{d}x} \quad \text{or} \quad y' = vu' + uv'.$$

For example, if $y = x^2\,\mathrm{e}^x$ then we put $u = x^2$ and $v = \mathrm{e}^x$ so that $u' = 2x$ and $v' = \mathrm{e}^x$. Then

$$y' = vu' + uv' = 2x\,\mathrm{e}^x + x^2\,\mathrm{e}^x = \mathrm{e}^x(2x + x^2).$$

Similar formulae hold for products of more than two functions:
If $y = uvw$ then $y' = uvw' + uwv' + vwu'$
If $y = tuvw$ then $y' = tuvw' + tuwv' + twvu' + uvwt'$.
Notice that effectively each factor in the product is differentiated in turn, so that the first derivative of $y = stuvw$ would be formed from the sum of five expressions ($stuvw'$, etc.). Also notice that we have assumed that all the functions involved are functions of x, i.e. $y = y(x)$, $t = t(x)$, $u = u(x)$, etc.

Example If $y = (1 + x^2)x^3\mathrm{e}^x \log_e x$, find $\mathrm{d}y/\mathrm{d}x$.
Let

$$t = (1 + x^2), \quad u = x^3, \quad v = \mathrm{e}^x, \quad w = \log_e x.$$

Then

$$t' = 2x, \quad u' = 3x^2, \quad v' = \mathrm{e}^x, \quad w' = 1/x$$

$$\begin{aligned}\therefore\ y' &= 2xx^3\mathrm{e}^x \log_e x + (1 + x^2)3x^2\mathrm{e}^x \log_e x \\ &\qquad + (1 + x^2)x^3\mathrm{e}^x \log_e x + (1 + x^2)x^3\mathrm{e}^x/x \\ &= \mathrm{e}^x(2x^4 \log_e x + 3x^2(1 + x^2)\log_e x + x^3(1 + x^2)\log_e x + x^2(1 + x^2)) \\ &= \mathrm{e}^x((3x^2 + x^3 + 5x^4 + x^5)\log_e x + x^2 + x^4) \quad \text{etc.}\end{aligned}$$

Derivation of the formulae

As we have pointed out earlier, we are more interested in the *use* than in the *derivation* of the formulae, but we show below how the formula can be obtained in the simplest case: $y = uv$.

$$y = uv \quad \text{where } y = y(x),\ u = u(x),\ v = v(x)$$

$$\begin{aligned}\therefore\ (y + \Delta y) &= (u + \Delta u)(v + \Delta v) \quad \text{at the point } (x + \Delta x, y + \Delta y) \\ &= uv + v\Delta u + u\Delta v + (\Delta u)(\Delta v)\end{aligned}$$

$$\therefore\ \Delta y = (y + \Delta y) - y = v\Delta u + u\Delta v + (\Delta u)(\Delta v)$$

$$\therefore\ \frac{\Delta y}{\Delta x} = v\,\frac{\Delta u}{\Delta x} + u\,\frac{\Delta v}{\Delta x} + \frac{\Delta u}{\Delta x}\,\Delta v$$

$$\therefore\ \frac{\mathrm{d}y}{\mathrm{d}x} = \lim_{\Delta x \to 0}\left(\frac{\Delta y}{\Delta x}\right) = \lim_{\Delta x \to 0}\left(v\,\frac{\Delta u}{\Delta x} + u\,\frac{\Delta v}{\Delta x} + \frac{\Delta u}{\Delta x}\cdot\Delta v\right)$$

$$\therefore\ \frac{\mathrm{d}y}{\mathrm{d}x} = v\,\frac{\mathrm{d}u}{\mathrm{d}x} + u\,\frac{\mathrm{d}v}{\mathrm{d}x} \quad \text{since} \quad \lim_{\Delta x \to 0}\left(\frac{\Delta u}{\Delta x}\,\Delta v\right) = 0$$

$$\therefore\ y' = vu' + uv'.$$

8.3 QUOTIENTS

We now wish to consider functions of the type

$$y = \frac{e^x}{x^2} \quad \text{or generally} \quad y = \frac{u(x)}{v(x)} = \frac{u}{v}.$$

We could of course transform these to the same form as a product by writing

$$y = e^x(x^2)^{-1} = e^x x^{-2} \quad \text{or generally} \quad y = uv^{-1}$$

and using the results in paragraph 8.2. However, there is a general formula which can prove useful:

$$\text{If } y = \frac{u}{v} \quad \text{then} \quad y' = \frac{vu' - uv'}{v^2}$$

$$y(x) = \frac{u(x)}{v(x)} \quad \text{then} \quad \frac{dy}{dx} = \frac{v\dfrac{du}{dx} - u\dfrac{dv}{dx}}{v^2} = \frac{x^2e^x - e^x 2x}{x^4} = \frac{e^x(x-2)}{x^3}$$

for the example above.

8.4 THE CHAIN RULE

If we wish to differentiate a function such as $y = (x^2 + x)^4$ then our only available option with the techniques so far available would be to expand the expression, and differentiate each term; that is,

$$y = x^8 + 4x^7 + 6x^6 + 4x^5 + x^4$$

whence

$$\begin{aligned}\frac{dy}{dx} &= 8x^7 + 28x^6 + 36x^5 + 20x^4 + 4x^3 \\ &= 4(x^2 + x)^3(2x + 1).\end{aligned}$$

But we could not extend this method to such functions as $y = (x^2 + x)^{1/2}$ where expansion to a finite number of terms is not possible. Similarly, functions such as $y = \log_e(1 + 3x^2)$, or $y = e^{x^2}$ cannot be tackled with the rules derived so far. In order to deal with such functions we must make use of the so-called *chain rule*.

Suppose that $u = u(x)$ and $y = y(u)$. (In words, u is a function of x, and y is a function of u.) Then $y = y(u) = y(u(x))$, and y can also be treated as a function of x.

The *chain rule* states that

$$\frac{dy}{dx} = \frac{dy}{du}\frac{du}{dx}.$$

(Notice how, if we were to treat $\mathrm{d}y/\mathrm{d}x$ etc. as fractions, then cancellation would make sense, i.e.

$$\frac{\mathrm{d}y}{\mathrm{d}x} = \frac{\mathrm{d}y}{\cancel{\mathrm{d}u}}\frac{\cancel{\mathrm{d}u}}{\mathrm{d}x}.$$

This is intended only to help the memory, since such a procedure is, mathematically, illegal.)

Similarly, if $v = v(x)$, $u = u(v)$, and $y = y(u)$, then

$$y = y(u(v(x))) \quad \text{and} \quad \frac{\mathrm{d}y}{\mathrm{d}x} = \frac{\mathrm{d}y}{\mathrm{d}u}\frac{\mathrm{d}u}{\mathrm{d}v}\frac{\mathrm{d}v}{\mathrm{d}x}$$

and again cancellation still appears to work:

$$\frac{\mathrm{d}y}{\cancel{\mathrm{d}u}}\frac{\cancel{\mathrm{d}u}}{\cancel{\mathrm{d}v}}\frac{\cancel{\mathrm{d}v}}{\mathrm{d}x}.$$

This rule is helpful in the following way. If $y(x)$ is a complex function, it can in many cases (such as those introduced at the beginning of this paragraph) be treated as two simple functions $y = y(u)$ and $u = u(x)$ each of which can be differentiated more readily.

Example 1 $y = (x^2 + x)^4$.

Let $u = (x^2 + x)$, so $\mathrm{d}u/\mathrm{d}x = 2x + 1$ and $y = u^4$, so $\mathrm{d}y/\mathrm{d}u = 4u^3$

$$\therefore \frac{\mathrm{d}y}{\mathrm{d}x} = \frac{\mathrm{d}y}{\mathrm{d}u}\frac{\mathrm{d}u}{\mathrm{d}x} = 4u^3(2x + 1) = 4(x^2 + x)^3(2x + 1)$$

which agrees with the longer method used above.

Example 2 $y = \sqrt{x^2 + x} = (x^2 + x)^{1/2}$.

Let $u = (x^2 + x)$, so $\mathrm{d}u/\mathrm{d}x = 2x + 1$ and $y = u^{1/2}$, so $\mathrm{d}y/\mathrm{d}u = \frac{1}{2}u^{-1/2}$

$$\therefore \frac{\mathrm{d}y}{\mathrm{d}x} = \frac{\mathrm{d}y}{\mathrm{d}u}\frac{\mathrm{d}u}{\mathrm{d}x} = \tfrac{1}{2}(u^{-1/2})(2x + 1) = \frac{2x + 1}{2(x^2 + x)^{1/2}}.$$

Example 3 $y = \log_e(1 + 3x^2)$.

Let $u = (1 + 3x^2)$, so $\mathrm{d}u/\mathrm{d}x = 6x$ and $y = \log_e u$, so $\mathrm{d}y/\mathrm{d}u = 1/u$

$$\therefore \frac{\mathrm{d}y}{\mathrm{d}x} = \frac{\mathrm{d}y}{\mathrm{d}u}\frac{\mathrm{d}u}{\mathrm{d}x} = \frac{6x}{u} = \frac{6x}{(1 + 3x^2)}.$$

Example 4 $y = e^{x^2}$.

Let $u = x^2$, so $du/dx = 2x$ and $y = e^u$, so $dy/du = e^u$

$$\therefore \frac{dy}{dx} = \frac{dy}{du}\frac{du}{dx} = e^u\, 2x = 2xe^{x^2}.$$

Example 5 $y = \log_e \sqrt{x^3 + 4}$.

Let $u = x^3 + 4$, so $du/dx = 3x^2$. Let $v = \sqrt{u} = u^{1/2}$, so $dv/du = \frac{1}{2}u^{-1/2}$ and $y = \log_e v$, so $dy/dv = 1/v$

$$\therefore \frac{dy}{dx} = \frac{dy}{dv}\frac{dv}{du}\frac{du}{dx} = \frac{1}{v}\,\tfrac{1}{2}u^{-1/2}\,3x^2 = \frac{3x^2}{2(x^3 + 4)}.$$

(Note that we could have made this example a little simpler by using $\log_e \sqrt{x^3 + 4} = \log(x^3 + 4)^{1/2} = \frac{1}{2}\log_e(x^3 + 4)$, when only one *dummy variable*, u, need have been introduced.)

8.5 PARTIAL DIFFERENTIATION

Until now we have only considered functions of one variable, so that we have been seeking dy/dx for a given function $y(x)$.

Suppose now that we have a function of two variables:

$$f = f(x, y).$$

The question arises as to whether the concept of a *gradient* is at all meaningful. Again, a geometrical interpretation can help us here. We can imagine the x–y plane to be the horizontal plane, with the value of f plotted appropriately in the vertical direction. (In fact we are really generating ordered triples (x, y, f) in cartesian space, and as f is a *function*, then for each pair (x, y) there is at most one corresponding value of f.) This generates a surface in three-dimensional space, rather like the surface of a landscape with hills and valleys, etc.

The gradient of a *hill* clearly does have meaning, but now the direction of that gradient becomes important. A path which ascends a hill directly from bottom to top is very steep, while a spiralling or zigzag path rises more gently. Clearly there are many choices in the direction we could take but one simple question might be: what is the gradient if we continue to travel in a Northerly, or an Easterly direction? In algebraic terms we are asking: what is the gradient if we continue to travel in the y or in the x-direction?

The answer is quite simple for these two cases. To find the value of the gradient in the x-direction, we suppose that y remains constant, with value y_0 say, and we can then differentiate in the usual way with respect to x (only).

So, the gradient of $f = f(x, y)$ along $y = y_0$ is given by

$$\frac{\mathrm{d}f}{\mathrm{d}x}(y = y_0 \text{ always}) = \frac{\mathrm{d}f(x, y_0)}{\mathrm{d}x}.$$

We do, however, use a slightly different notation to indicate that f is strictly a function of x and y, and to emphasize that we are differentiating with respect to x only, thus:

$$\frac{\partial f(x, y)}{\partial x}$$

which is called the *partial derivative* of $f(x, y)$ with respect to x. There is a similar partial derivative of f, with respect to y, viz:

$$\frac{\partial f}{\partial y}.$$

Examples

(a) $f(x, y) = 3x + 2y, \dfrac{\partial f}{\partial x} = 3, \dfrac{\partial f}{\partial y} = 2.$

Notice that $2y$ is treated as a constant when differentiating (partially) with respect to x.

(b) $f(x, y) = xy^2, \dfrac{\partial f}{\partial x} = y^2, \dfrac{\partial f}{\partial y} = 2xy.$

(c) $f(x, y) = \dfrac{12}{3x^2 + y^2}, \dfrac{\partial f}{\partial x} = \dfrac{-12.6x}{(3x^2 + y^2)^2} = \dfrac{-72x}{(3x^2 + y^2)^2}$

and $\dfrac{\partial f}{\partial y} = \dfrac{-2.12y}{(3x^2 + y^2)^2} = \dfrac{-24y}{(3x^2 + y^2)^2}.$

The geometrical interpretation of this function

$$f(x, y) = 12/(3x^2 + y^2)$$

is shown in Figures 8.1(a) as a surface similar to a cone. The interpretation of the partial derivatives as gradients is shown in Figures 8.1(c) and (d).

8.6 HIGHER ORDER DERIVATIVES

The first derivative of a function $y(x)$ is $\mathrm{d}y/\mathrm{d}x$ and this gives us some information on the shape of the curve $y(x)$ in terms of its slope or gradient. We can differentiate further, however, and obtain the *higher order*

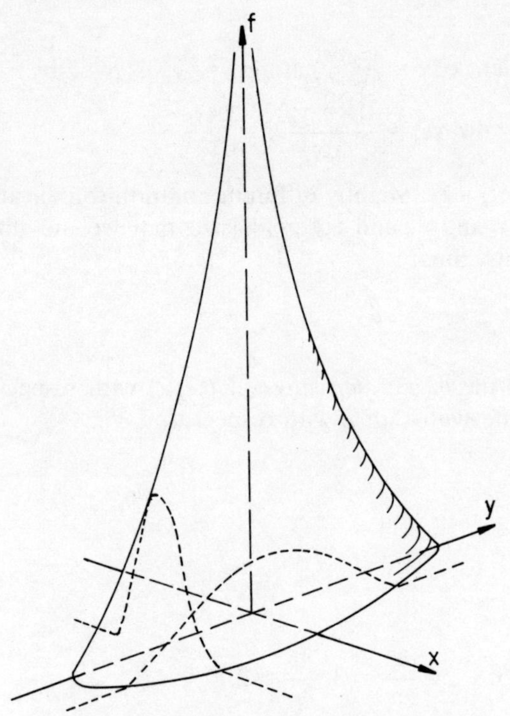

Figure 8.1(a) An illustration of the function $f = 12/(3x^2 + y^2)$ showing cuts through the planes x = constant and y = constant

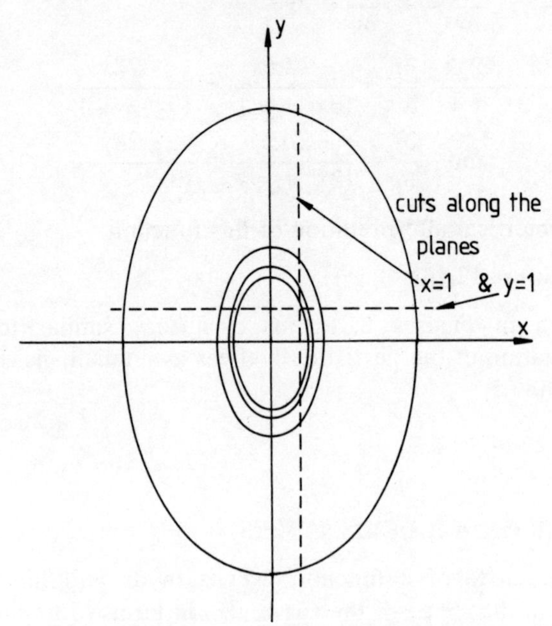

Figure 8.1(b) A contour map of the function f, looking down onto the x–y plane. (The central infinite peak has been omitted)

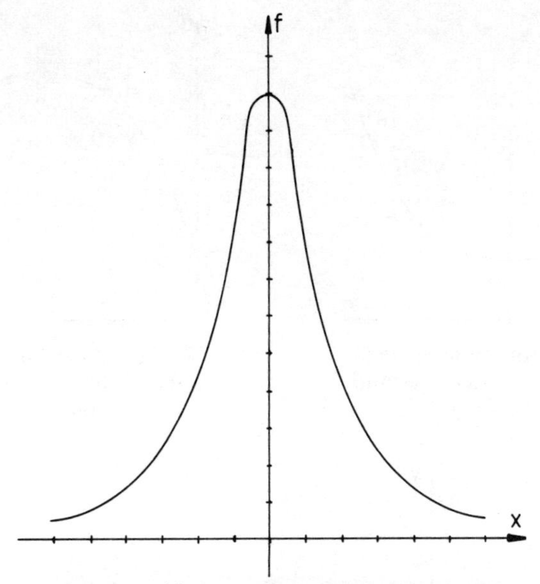

Figure 8.1(c) A cross-section along the plane y = constant through the function f. The gradient is given by $\partial f/\partial x$

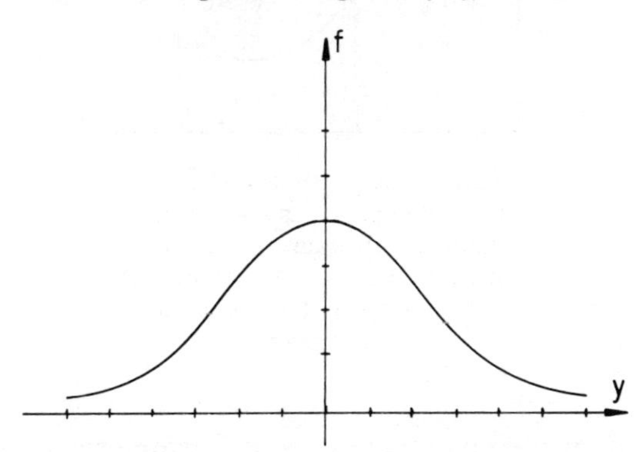

Figure 8.1(d) A cross-section along the plane x = constant through the function f. The gradient is given by $\partial f/\partial y$

derivatives of $y(x)$. The second derivative is therefore

$$y'' \quad \text{or} \quad \frac{\mathrm{d}}{\mathrm{d}x}\left(\frac{\mathrm{d}y}{\mathrm{d}x}\right) \quad \text{or} \quad \frac{\mathrm{d}^2y}{\mathrm{d}x^2} \quad (\text{d-two-}y \text{ by d-}x^2).$$

The third derivative is

$$y''' \quad \text{or} \quad \frac{\mathrm{d}}{\mathrm{d}x}\left(\frac{\mathrm{d}^2y}{\mathrm{d}x^2}\right) \quad \text{or} \quad \frac{\mathrm{d}^3y}{\mathrm{d}x^3} \quad (\text{d-three-}y \text{ by d-}x^3)$$

and so on.

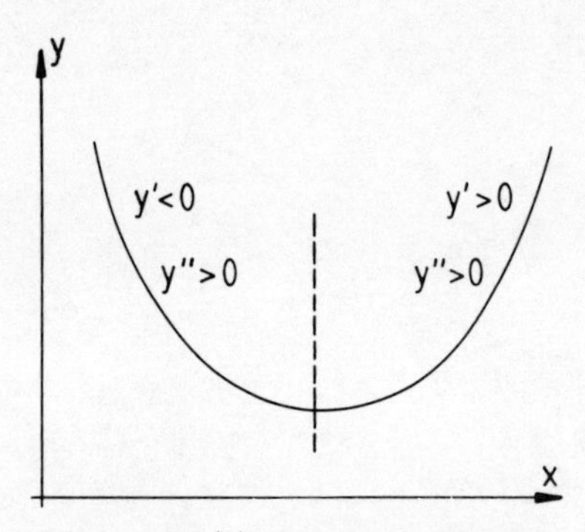

Figure 8.2(a) An example of a curve with a positive second derivative: $y'' > 0$

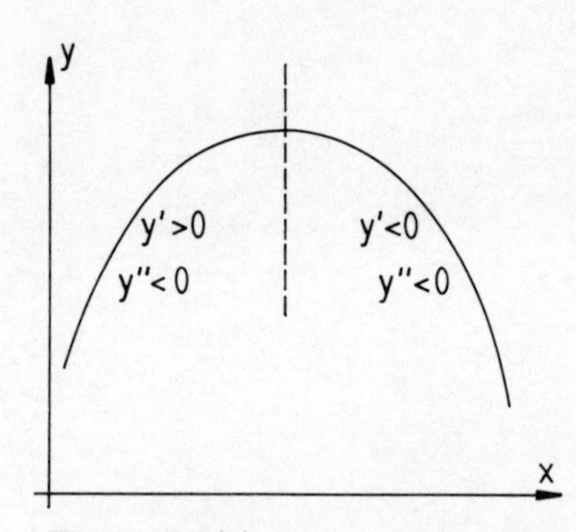

Figure 8.2(b) An example of a curve with a negative second derivative: $y'' < 0$

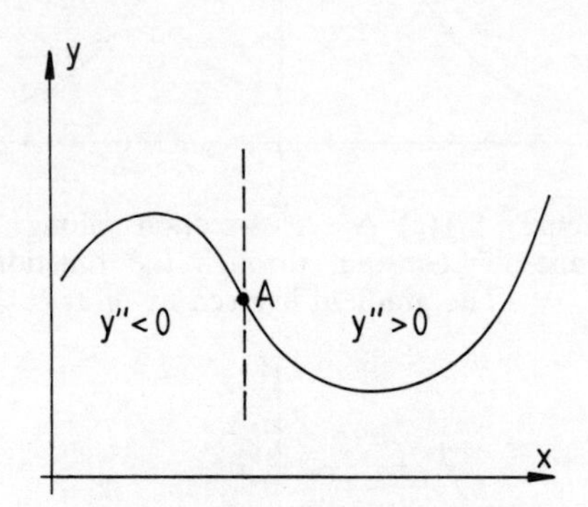

Figure 8.2(c) An example of a curve where y'' passes from a negative, through zero, to a positive value. At point A, $y'' = 0$ and we have a *point of inflexion*

The second derivative is related to the geometric curvature of the function $y(x)$. For example, if $y'(x)$ is positive and $y''(x)$ is positive then we have a curve with a positive gradient, and which gradient is changing positively as well. ($y''(x)$ can be described as the rate of change of the gradient, $y'(x)$.) This case is illustrated in Figure 8.2 along with other examples.

Notice that a local *minimum* in a continuous curve is associated with a *positive second derivative*, while a local *maximum* is associated with a *negative second derivative*, and this property can be used to distinguish between maxima and minima in curves.

A point of inflexion can coincide with a zero value of the gradient, and this occurs when

$$y'(x) = 0 \quad \text{and} \quad y''(x) = 0$$

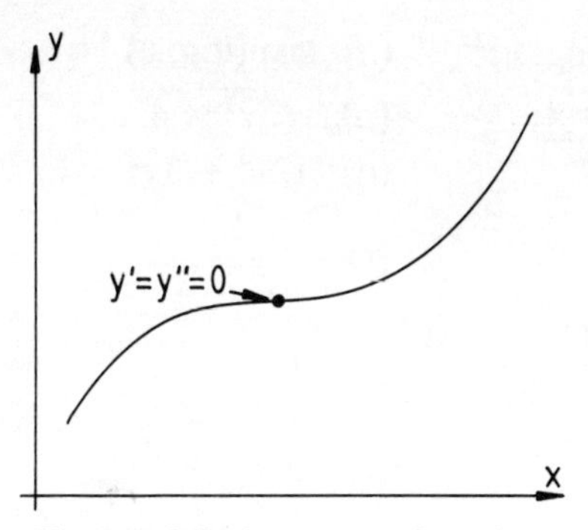

Figure 8.3 An example of a curve where $y' = y'' = 0$ at a point which is neither a local maximum nor a local minimum

for the same value of x. This is illustrated in Figure 8.3 and shows that $y'(x) = 0$ is not a good test for maxima and minima if $y''(x) = 0$ also.

The condition

$$y'(x) = 0$$

gives us, in general, a *stationary point* of the function $y(x)$. This term is useful when we do not have information concerning the second derivative, $y''(x)$, and cannot be as specific in our description as the terms maximum, minimum, or point of inflexion would imply. *Stationary point* covers them all.

8.7 HIGHER ORDER PARTIAL DERIVATIVES

Generally we will find higher order partial derivatives less useful, but their derivation follows from the rules for first derivatives (Section 8.5).

They are written

$$\frac{\partial^2 f}{\partial x^2} \quad \frac{\partial^2 f}{\partial y^2} \quad \frac{\partial^2 f}{\partial x\, \partial y} \quad \text{and} \quad \frac{\partial^2 f}{\partial y\, \partial x}.$$

The first of these, $\partial^2 f/\partial x^2$, will give information, for example, on the curvature of the section through $f(x, y)$ along the x-direction, as in Figure 8.1(c).

8.8 BASIC EXAMPLES FOR THE STUDENT

B8.1 Differentiate the following functions with respect to x.

(a) $x \log_e x$

(b) $x/\log_e x$

(c) $e^x \log_e x$

(d) $e^x/\log_e x$

(e) $(x + 3)^2 \log(x^2)$

(f) $(x \log_e x - x)$

(g) $(x + 3)^2(\log_e x)^2$

(h) $\dfrac{x^2 + 3x - 4}{x^2 - 1}$

(i) $\log_e(1/x)$

(j) $e^{2x}(e^{2x} - e^{-2x})$

(k) $\dfrac{x - 1}{(x + 2)(x + 3)}$

(l) $\log_e(x^3 + 4x)$

(m) $\log_e(e^x)$

(n) $\exp(x^2 - 2x + 1)$

(o) $\exp(\log_e x)$

(p) $\sqrt{3x^3 - 4}$

(q) $(2x^4 + 3x^2 - 1)^{-1/3}$

(r) $\dfrac{1}{\sqrt{x^2 + 4}}$

(s) $\left(x^2 + \dfrac{1}{x^2}\right)^{-1/5}$

(t) $\log_e 3 + \log_e y$

(u) $\log_e \sqrt{x^3 - x}$.

B8.2 Partially differentiate the following functions with respect to x.

(a) $x^3y^2 + 4y$

(b) $\log_e(xy)$

(c) $(x + y)^3$

(d) $\dfrac{x + y}{x - y}$

(e) $\sqrt{x^2 + y^2}$

(f) $\exp(x^2 + 2xy + y^2)$

(g) $y^6 + 2y + 4$

(h) $\log_e 3 + \log_e y + \log_e x$.

8.9 INTERMEDIATE EXAMPLES FOR THE STUDENT

I8.1 Differentiate the following

(a) $4\sqrt{x}$

(b) $(x + 3)^3/(x - 1)$

(c) $e^x(\log_e x)^4$

(d) $\log_e(x^4) + \log_e 4$

(e) $\log_e(1/x)$

(f) $e^{2x}(e^{2x} - e^{-2x})$

(g) $x + x \log_e x$

(h) $(x^3 - 3/x^3)^4$

(i) $\sqrt{2x^2 + 3x + 4}$

(j) $(x + 1)(x + 2)(x + 3)$.

I8.2 The total cost of manufacturing x units of product is

$$\text{TC} = 0{\cdot}01x^2 + 10x + 400.$$

(i) What is the average cost of each unit?
(ii) What is the marginal cost?
(iii) Find the value of x which minimizes the average unit cost.
(iv) Show that the minimum occurs when the average cost and the marginal cost are equal.

I8.3 100 metres of fencing are available to enclose a rectangular field. What is the size of the largest field which can be enclosed?

I8.4 Find the maxima, minima, and point of inflexion for the curve

$$y = x^3 - 9x + 6.$$

CHAPTER 9

Applications of Differential Calculus I: Economic Model

9.1 INTRODUCTION

Differential calculus is a powerful analytical tool, and we wish to demonstrate some of the ways in which it has been employed in some basic problems of economics. Again the equivalence of geometric (graphical) representation and algebraic representation should be noted in the examples that we introduce.

9.2 DEPENDENT AND INDEPENDENT VARIABLES

We firstly want to distinguish two types of variable, the *dependent* and the *independent*. When, for example, the equation of a straight line is written in the form

$$y = mx + c$$

then it is usually convenient to think that, for each and every value of x, we can calculate a value for y. (For *functions*, only one value of y will be possible.) Putting this a little more strongly, the value of x determines each value of y and so the y-value depends on the x-value. In this sense y is often considered to be the *dependent* variable, and x the *independent*. However a moment's thought will remind us that we could have written instead

$$x = \frac{1}{m}y - \frac{c}{m}$$

when it appears that y becomes the independent variable and x the dependent variable.

In fact, while we are talking in terms of purely *mathematical* variables, the distinction between dependent and independent is not very useful. When variables are taken from the real world, however, this may not always be the case. Some suggestion of cause and effect may creep in, and then it would be usual to treat the *cause* as the *independent* variable, and the *effect* as the *dependent* variable. In most sciences it is conventional to represent the independent variable along the horizontal axis (usually the x-axis), and the dependent variable along the vertical axis (usually the y-axis).

Except in economics.

When the curves for demand against price and supply against price are drawn economists prefer to use the vertical y-axis for price, even though it is usually the independent variable. The quantity which producers would like to supply, and the quantity which customers demand are both usually drawn along the horizontal x-axis even though they are, strictly, the dependent variables.

9.3 EXAMINATION OF AN ECONOMIC MODEL

In order to demonstrate some specific uses of graphical analysis, we shall use a simple economic model based on costs, prices, revenue, profits, etc. We are more interested here in developing the use of some mathematical techniques of analysis rather than a full economic theory, so that our economic assumptions will not be fully justified here except on grounds of reasonableness. We shall leave the full explanation to economic theory, but would just point out that the assumptions we introduce are applicable to the case of a *monopolist*, where we have only one producer in the market, and the market demand curve becomes the demand curve for that one producer.

Notation: We shall use the following notation throughout

p	the price charged for one unit of product
q	the quantity of the product sold per unit time
R	the revenue (general) collected by the producer on sales per unit time
TR	the total revenue per unit time
MR	the marginal revenue per unit product
AR	the average revenue per unit product
C	the costs (general) incurred in producing the product, per unit time
TC	the total costs per unit time
FC	the fixed costs per unit time
VC	the variable costs per unit time
MC	the marginal costs per unit product
AC	the average costs per unit product
Π	the gross profit generated per unit time
η	the elasticity of demand
a, b, f, h, g	are (positive) constants.

9.4 PRICES, QUANTITIES, AND REVENUE

Some elementary relationships between the quantities we wish to use can firstly be defined.

(i) At a given price level the total revenue collected must be given by (price per unit) × (number of units sold)

$$\therefore \text{TR} = pq.$$

(ii) The (instantaneous) marginal revenue is the rate of increase of total revenue with respect to changes in production level, at the particular production level considered and assuming that all production is sold (by price adjustment if necessary)

$$\therefore \text{MR} = \frac{\text{d(TR)}}{\text{d}q}.$$

(iii) The total revenue divided by the total production level gives the average revenue per unit product

$$\therefore \text{AR} = \frac{\text{TR}}{q} = p.$$

(So, the average revenue = price.)

The model

To develop the ideas further we need to introduce a suitable relationship between price and demand. We assume that as prices increase demand falls, and that demand is the same as market demand (this is where the *monopolist* case commences). Further, on grounds of simplicity we *assume* a *linear* relationship between price and demand level:

$$p = a - bq \quad \text{or} \quad q = \frac{a}{b} - \frac{1}{b}p. \tag{1}$$

Immediately, therefore, we can write down the total, marginal, and average revenues in terms of either price or quantity sold (Figure 9.1):

$$\begin{aligned} \text{TR} &= pq &&= aq - bq^2 = \frac{a}{b}p - \frac{p^2}{b} \\ \text{MR} &= \frac{\text{d(TR)}}{\text{d}q} &&= a - 2bq \ = 2p - a \\ \text{AR} &= \frac{\text{TR}}{q} &&= a - bq \quad = p. \end{aligned}$$

9.5 ELASTICITY OF DEMAND

From Figure 9.1(bii) it is immediately apparent that as the price is increased, the revenue of the producer also increases initially, but *subsequently reaches a maximum value*, and falls as the price is further increased.

Similarly, as the quantity sold is increased the revenue reaches a maximum value, and then declines (Figure 9.1(bi)). (As increased sales result on this model from decreasing prices, we are seeing the same maximum revenue point appear, but effectively as prices decrease monotonically rather than increase monotonically.)

Now TR = pq

$$\therefore \frac{d(TR)}{dq} = q\frac{dp}{dq} + p \qquad \frac{d(TR)}{dp} = p\frac{dq}{dp} + q.$$

At maximum (or minimum) values of the total revenue, therefore,

$$q\frac{dp}{dq} + p = 0 \quad \text{or} \quad p\frac{dq}{dp} + q = 0$$

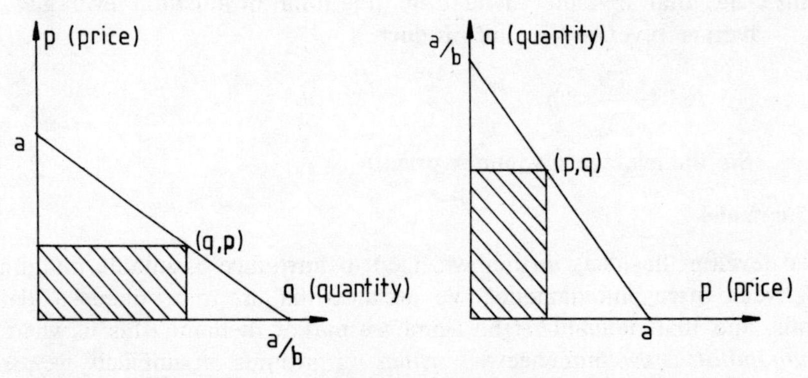

Figure 9.1(a) Price vs. quantity (economists' convention—LH figure). Price vs. quantity (mathematicians' convention—RH figure). The shaded area is pq and therefore represents the total revenue at any particular price level

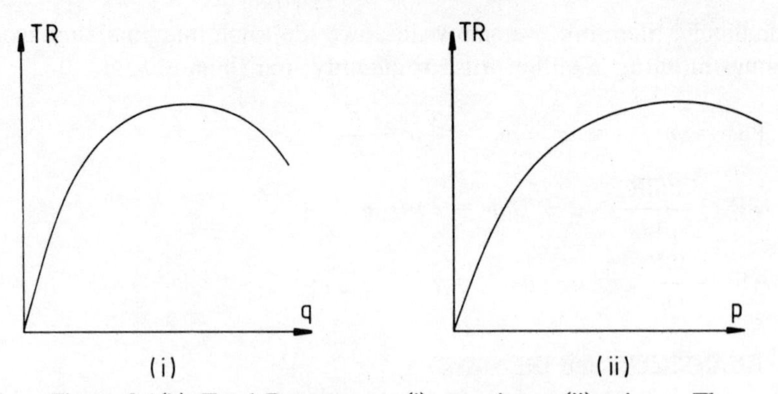

Figure 9.1(b) Total Revenue vs. (i) quantity q, (ii) price p. The gradient of curve (i) gives, at any point, the marginal revenue at that value of q

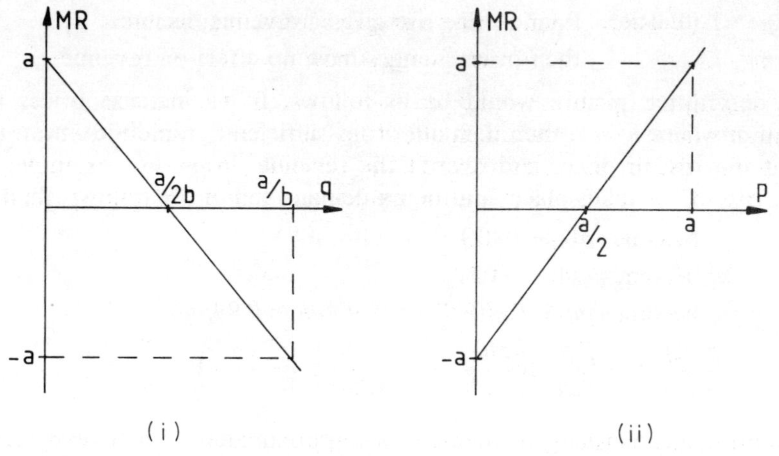

Figure 9.1(c) Marginal revenue vs. (i) quantity q, (ii) price p

which both reduce to the same criterion:

$$-\frac{p}{q}\frac{dq}{dp} = 1.$$

When an *increase* in prices results in an *increase* in revenue then

$$\frac{d(TR)}{dp} = \frac{d(pq)}{dp} = p\frac{dq}{dp} + q > 0$$

so that

$$\frac{p}{q}\frac{dq}{dp} > -1$$

$$\therefore -\frac{p}{q}\frac{dq}{dp} < 1.$$

Similarly, when an *increase* in prices results in a *decrease* in revenue then

$$\frac{d(TR)}{dp} < 0$$

and

$$-\frac{p}{q}\frac{dq}{dp} > 1.$$

This quantity, $\left(-\frac{p}{q}\frac{dq}{dp}\right)$, is called the *elasticity of demand*, is usually given the symbol η, and is clearly quite important in giving us some indication as to the response of revenue to price changes.

If $\eta < 1$ (inelastic) then a price rise gives a revenue rise.

If $\eta > 1$ (elastic) then a price rise gives a revenue decline.
If $\eta = 1$ then price changes have no effect on revenue.

A descriptive picture would be as follows. If we increase prices in a situation where $\eta > 1$, then demand drops sufficiently rapidly to more than offset the rise in price, and overall the revenue drops. For example, if a price rise of 2% takes place and then a demand fall of 3% follows, then

$$p \text{ becomes } p(1 + 0{\cdot}02)$$
$$q \text{ becomes } q(1 - 0{\cdot}03)$$
$$R = pq \text{ becomes } pq(1{\cdot}02)(0{\cdot}97) = 0{\cdot}989pq \cong 0{\cdot}99pq$$
$$\eta \cong -\frac{p}{q}\frac{\Delta q}{\Delta p} = -\left(\frac{p}{\Delta p}\right)\left(\frac{\Delta q}{q}\right) = -\left(\frac{1}{0{\cdot}02}\right)\left(-\frac{0{\cdot}03}{1}\right) = \frac{3}{2}.$$

Therefore, the elasticity of demand is (approximately) 1·5 (*elastic*) and a revenue fall of (approximately) (3 − 2)%, or 1%, results.

The need to use the word *approximately* arises because strictly speaking the elasticity is defined at a point and not over a range. However for *small* relative changes in p and q we can use the good approximation that

$$\frac{\Delta p}{\Delta q} \cong \frac{\mathrm{d}p}{\mathrm{d}q} \quad \text{etc.}$$

If, as in the above example, we use a finite interval and use the finite changes in price and quantity, Δp and Δq, then we are calculating the *arc elasticity*, rather than the instantaneous (point) elasticity.

So far, all our comment on elasticity has been very general since we have not introduced our particular (monopolist) model. If we now do so, then

$$\text{TR} = aq - bq^2$$
$$\frac{\mathrm{d(TR)}}{\mathrm{d}q} = a - 2bq = 0 \text{ at maximum}$$
$$\therefore q = a/2b \quad \text{when} \quad \text{TR} = a^2/4b.$$

Notice that this is also the value of q for which marginal revenue is zero.

Also notice that

(i) the gradient of the price vs. quantity line is $-b$ and the gradient of the marginal revenue vs. quantity line is given by $-2b$, since

$$\frac{\mathrm{d(MR)}}{\mathrm{d}q} = \frac{\mathrm{d}(a - 2bq)}{\mathrm{d}q} = -2b.$$

(ii) $\eta > 1$ when $-\dfrac{p}{q}\dfrac{\mathrm{d}q}{\mathrm{d}p} > 1$

$$\therefore -\frac{p}{q}\left(-\frac{1}{b}\right) > 1 \quad \therefore \frac{p}{b} > q \quad \therefore \frac{a - bq}{b} > q$$

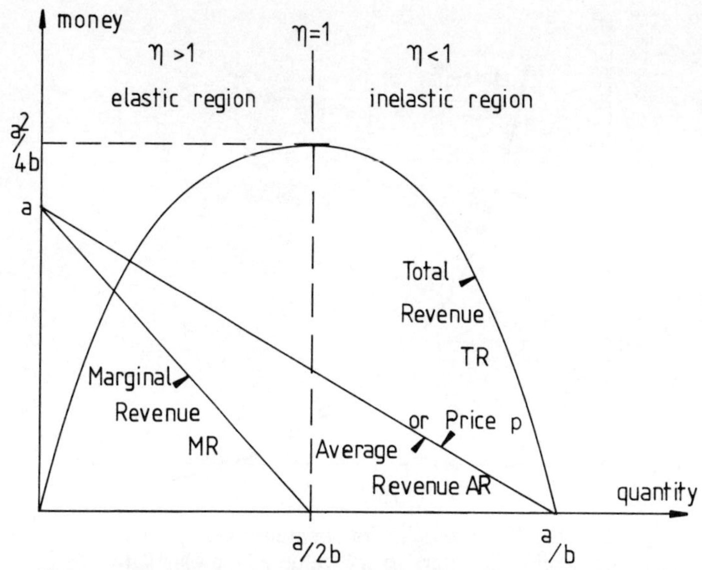

Figure 9.2 Total, marginal, and average revenues vs. quantity sold for the case where $p = a - bq$

$$\therefore a - bq > bq \quad \therefore a > 2bq$$

$$\therefore \eta > 1 \text{ when } q < a/2b \quad \text{and} \quad \eta < 1 \text{ when } q > a/2b.$$

All these results can be summarized on one diagram (Figure 9.2).

Although the elasticity includes the gradient of the p vs. q curve in its definition, it is not the same as the gradient. In this example the p vs. q curve is linear and has a constant gradient. However the elasticity varies, and takes the value $+\infty$ for $q = 0$ and falls to zero where $p = \text{AR} = \text{TR} = 0$.

For this model

$$\eta = -\frac{p}{q}\frac{\mathrm{d}q}{\mathrm{d}p} = \frac{a - bq}{bq} = \frac{a}{bq} - 1$$

so that

$$(\eta + 1)q = \frac{a}{b} \quad \text{(a rectangular hyperbola)}$$

which is shown in Figure 9.3.

9.6 PRODUCER'S COSTS

Again we begin this section by defining some quantities that we wish to use:

(i) The total costs, TC, at a given production level, q, can be treated as

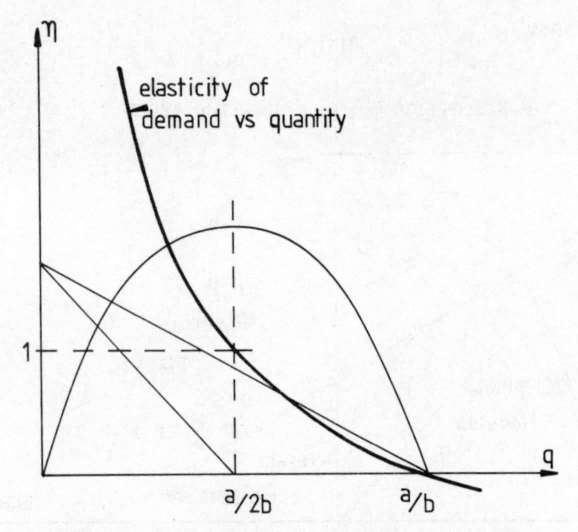

Figure 9.3 Elasticity of demand vs. q. The vertical scale refers to the value of the elasticity η only, to demonstrate the point at which it achieves the value 1. The curves for TR, MR, and AR are also indicated as in the previous figure

the sum of fixed costs FC (which are independent of the production level) and variable costs VC (which vary with production level).

$$TC = FC + VC.$$

(ii) The marginal cost is the rate of increase of total cost with respect to changes in the production level, at the particular production level considered

$$\therefore MC = \frac{d(TC)}{dq}.$$

(iii) The total costs divided by the total production level gives the average cost per unit product

$$\therefore AC = \frac{TC}{q}.$$

The model

Again, to develop the ideas further we wish to introduce a specific function for TC(q). A reasonable *assumption* to make is that variable costs are made up from

(a) a component proportional to the production level, gq
(b) a component which includes diseconomies of large scale production rates, hq^2.

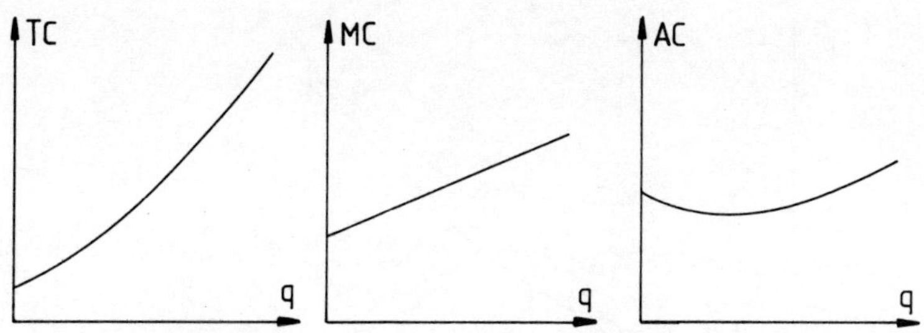

Figure 9.4 Total, marginal, and average costs on the assumption of parabolic total costs (equation 2)

If we set fixed costs at the constant value f, we have

$$TC = f + gq + hq^2 = FC + VC. \qquad (2)$$

Therefore

$$MC = \frac{d(TC)}{dq} = g + 2hq$$

$$AC = \frac{TC}{q} = \frac{f}{q} + g + hq.$$

(See Figure 9.4.)

Marginal and average costs can also be given useful geometric interpretations on the total cost curve, as in Figure 9.5.

By drawing in the equivalent of line OA for many points (q, t) it can be seen that the minimum average costs must occur when the line OA has its smallest gradient. This occurs when OA is in fact a *tangent* to the curve,

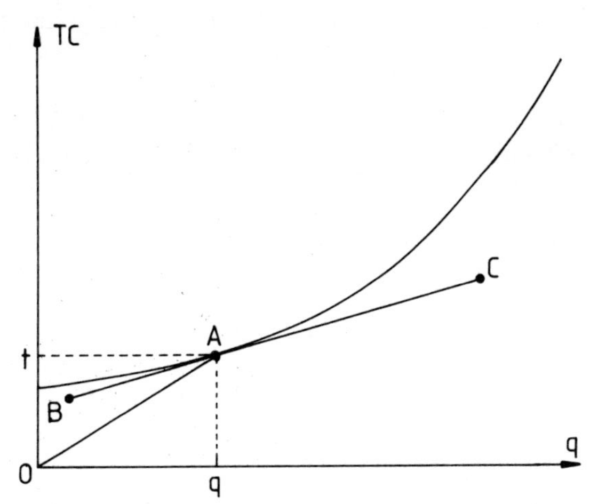

Figure 9.5 Marginal and average costs. For any production level q where total costs are t

$AC = t/q$ = gradient of OA
$MC = dt/dq$ = gradient of BC

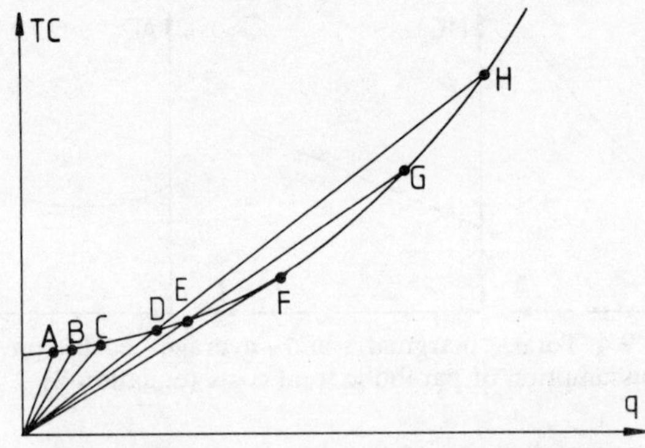

Figure 9.6 TC vs. q. At point H, for example, AC = gradient of OH. The minimum gradient is for line OF, which is also the gradient of TC, at F. At F, min.AC = MC

and then has the *same* gradient as the curve at that point. But the curve gradient is the marginal cost, so geometrically it appears that the minimum average cost occurs when the marginal cost equals the average cost (Figure 9.6).

Now AC = TC/q

$$\therefore \frac{\mathrm{d(AC)}}{\mathrm{d}q} = \left(q\,\frac{\mathrm{d(TC)}}{\mathrm{d}q} - \mathrm{TC}\right)\Big/ q^2 = 0 \text{ at max/min.}$$

$$\therefore \frac{\mathrm{d(TC)}}{\mathrm{d}q} = \frac{\mathrm{TC}}{q}$$

$\therefore$ MC = AC when AC is minimized with respect to q.

So this result is general, and is not restricted to the assumptions of this model.

For this model

$$\mathrm{AC} = \frac{f}{q} + g + hq$$

$$\therefore \frac{\mathrm{d(AC)}}{\mathrm{d}q} = \frac{-f}{q^2} + h = 0 \quad \text{when } q^2 = f/h$$

$$\left.\begin{aligned} \therefore \mathrm{MC} &= g + 2hq = g + 2h\sqrt{f/h} = g + 2\sqrt{fh} \\ \mathrm{AC} &= \frac{f}{q} + g + hq = f\sqrt{h/f} + g + h\sqrt{f/h} = g + 2\sqrt{fh} \end{aligned}\right\} \text{at point of minimum average cost}$$

These results are summarized in Figure 9.7.

9.7 GROSS PROFIT

We are now in a position to be able to study the relationship between *gross profit* and level of production. This is of particular interest because

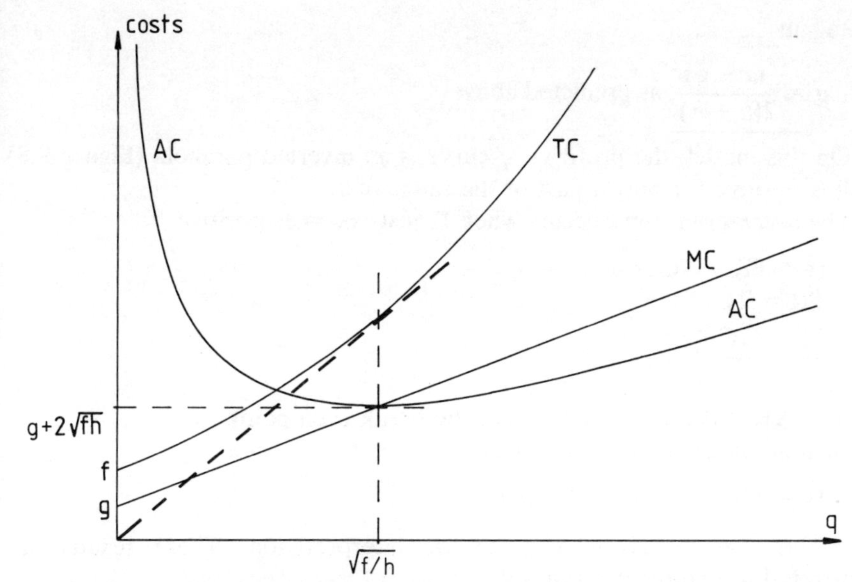

Figure 9.7 The relationship between TC, MC, and AC showing AC = MC when AC is at its minimum value

the economist would claim that the level of production which should be adopted is that value of q which maximizes the profit.

We shall define the profit, Π, to be the difference between the total revenue and the total costs.

$$\therefore \Pi = \text{TR} - \text{TC}$$

$$\therefore \frac{d\Pi}{dq} = \frac{d(\text{TR})}{dq} - \frac{d(\text{TC})}{dq} = 0 \text{ at max/min.}$$

$$\therefore \frac{d(\text{TR})}{dq} = \frac{d(\text{TC})}{dq}$$

$\therefore$ MR = MC, so marginal revenue = marginal costs at the point of profit maximization.

Combining our models for revenue and costs we have

$$\begin{aligned}\Pi = \text{TR} - \text{TC} &= (aq - bq^2) - (f + gq + hq^2)\\ &= -(h + b)q^2 + (a - g)q - f\end{aligned}$$

$$\therefore \frac{d\Pi}{dq} = -2(h + b)q + (a - g) = 0 \text{ at max/min.}$$

$$\therefore q = \frac{(a - g)}{2(b + h)} \text{ for profit maximization.}$$

MR = MC when $a - 2bq = g + 2hq$

$$\therefore q(2h + 2b) = a - g$$

so, again

$$q = \frac{(a - g)}{2(b + h)}, \text{ as predicted above.}$$

On this model, the profit vs. q curve is an inverted parabola (Figure 9.8) and is positive for only a part of the range of q.

The *break-even point* occurs when Π just becomes positive

$$\therefore \Pi = \text{TR} - \text{TC} = 0$$
$$\therefore \text{TR} = \text{TC}$$
$$\therefore \frac{\text{TR}}{q} = \frac{\text{TC}}{q}.$$

Hence AR = AC and TR = TC at the break-even point.

In general, the profit is given by

$$\Pi = \text{TR} - \text{TC} = q\text{AR} - q\text{AC}$$

and this can be given a geometric interpretation. These results are illustrated in Figures 9.8 and 9.9.

9.8 SECOND ORDER DERIVATIVES

In this analysis we have not shown rigorously in each case whether a zero first derivative signifies a maximum or a minimum value of the appropriate variable, but have simply asserted one or other to be the case. In principle this is easily rectified using second order derivatives. For example, to

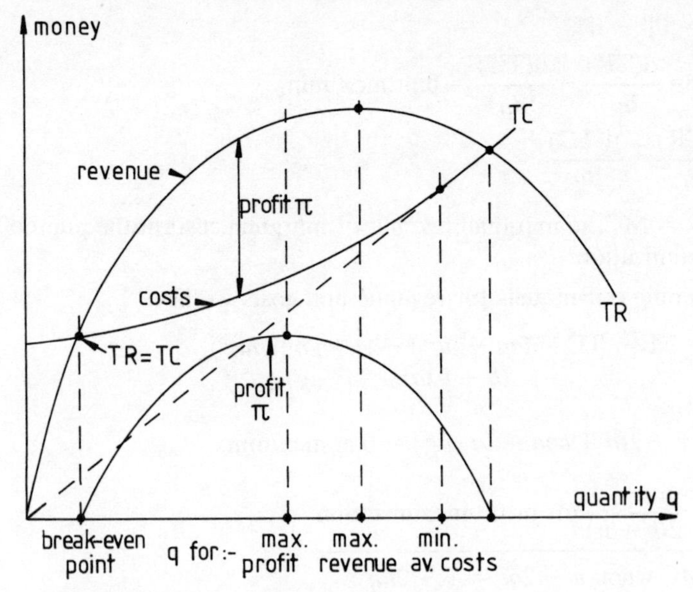

Figure 9.8 The relationship between TC, TR, Π as functions of q

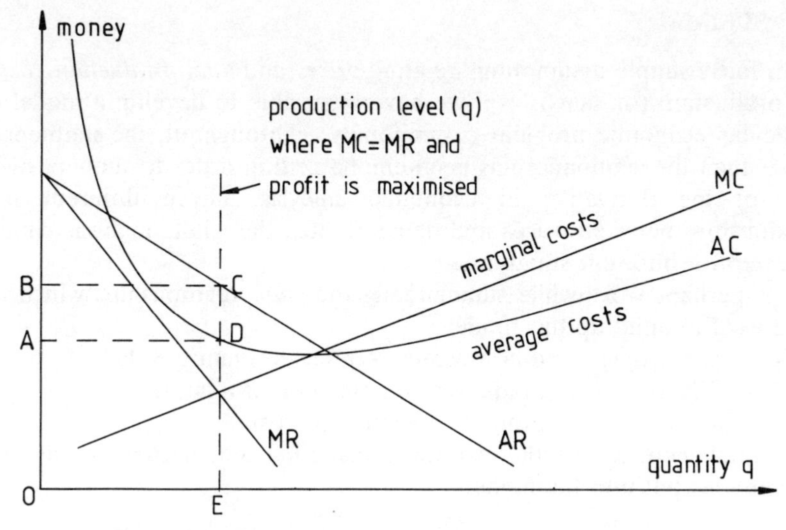

Figure 9.9 Geometric interpretation of gross profit, Π.

Area of OBCE $= q$ AR = TR
Area of OADE $= q$AC = TC
Area of ABCD = TR − TC = Π

define the point of maximum profit it is sufficient to have

$$\frac{\mathrm{d}\Pi}{\mathrm{d}q} = 0 \quad \text{and} \quad \frac{\mathrm{d}^2\Pi}{\mathrm{d}q^2} < 0.$$

Other cases follow similarly.

This condition for maximum profit can be interpreted in the following way. Since

$$\Pi = \mathrm{TR} - \mathrm{TC}$$

and

$$\frac{\mathrm{d}\Pi}{\mathrm{d}q} = \mathrm{MR} - \mathrm{MC}$$

then the condition for maximum profit becomes

$$\mathrm{MR} - \mathrm{MC} = 0 \quad \text{and} \quad \frac{\mathrm{dMR}}{\mathrm{d}q} - \frac{\mathrm{dMC}}{\mathrm{d}q} < 0$$

$$\therefore \mathrm{MR} = \mathrm{MC} \qquad \text{and} \quad \frac{\mathrm{dMR}}{\mathrm{d}q} < \frac{\mathrm{dMC}}{\mathrm{d}q}.$$

That is,

marginal revenue = marginal costs

and marginal revenue is increasing less rapidly than marginal costs.

9.9 SUMMARY

From fairly simple assumptions relating *prices* and *total production costs* to the production (or sales) level we have been able to develop a model of a particular economic problem (a monopoly). Throughout, the mathematics rather than the economics has been emphasized in order to show particular uses of the derivative in economic analysis. Given different initial relationships between prices and demand, etc., the whole analysis could be repeated for different situations.

It is perhaps worthwhile summarizing the *main* relationships which have been used to build up this model:

$\text{TR} = pq = p(q)q$	total revenue = price × quantity sold
$\text{TC} = \text{TC}(q)$	total costs as a function of quantity
$\Pi = \text{TR} - \text{TC}$	profit = revenue less costs.

The initial choices for the two functions $p(q)$ and $\text{TC}(q)$ are the main *assumptions* put into the model.

9.10 BASIC EXAMPLES FOR THE STUDENT

B9.1 Assume that the price vs. quantity sold curve for a particular product is given by $p = 2500 - 0{\cdot}8q$. Fixed costs are £22 500. Variable costs are given by $\text{VC} = 2000q + 0{\cdot}2q^2$. Find

(a) An expression for the total revenue.
(b) An expression for the total costs.
(c) An expression for the gross profit.
(d) The level of output for which the profit is maximized, and find the profit level achieved.
(e) The value of the marginal costs and revenues at this output.

B9.2 Imagine that for a particular product, the price-demand relationship is of the form

$$p = 1000/q.$$

Find expressions for the elasticity of demand, and the total revenue. Also imagine that the fixed costs are 100, and the variable costs are given by $\text{VC} = 0{\cdot}1q$. Find an expression for the gross profit and determine at what level of output (or sales) the maximum profit is achieved. At what output does the profit become zero?

9.11 INTERMEDIATE EXAMPLES FOR THE STUDENT

I9.1 Write down expressions for the marginal and average revenues in terms of the total revenue TR, and for the marginal and average costs in terms of the total costs TC, if q is produced and sold. Show

(i) that the maximum total revenue occurs when the marginal revenue is zero,
(ii) that the average revenue is the price per unit,

(iii) that the average cost per unit is minimized when average costs equal marginal costs (if this occurs),

(iv) that the average cost per unit equals the average revenue per unit at the break-even point,

(v) that the profit is maximized when marginal revenues equal marginal costs,

if these conditions exist, for any functions TR(q) and TC(q). Do *not* assume any relationship between p and q for this question.

I9.2 Assume that the price–demand curve is given by

$$p = 10 - 0{\cdot}005q$$

where p = price per unit, and q is demand per unit time. Also assume that fixed costs are 100, and variable costs are $4q + 0{\cdot}01q^2$ in appropriate units.

(a) Give expressions for the following, as functions of q,

(i) total revenue,

(ii) total costs,

(iii) marginal and average revenues per unit of product,

(iv) marginal and average costs per unit of product,

(v) the elasticity of demand.

(b) Find the values of

(i) the maximum total revenue, and the corresponding quantity q,

(ii) the marginal revenue for the level of production found in (i),

(iii) the minimum average cost, and the marginal cost both at the same level of q,

(iv) the level of production for the break-even point, and the average costs and revenues at this production level,

(v) the level of production for maximum profit, and the marginal revenues and marginal costs at this production level.

(Note that each part of this question requires a different production level q to be calculated unless otherwise stated.)

CHAPTER 10

Applications of Differential Calculus II: Curve Sketching

10.1 INTRODUCTION

It is the aim of this chapter to draw on some of the analytical techniques already covered and to use them to help us interpret algebraic relations in terms of geometry on a graph. In other words, we wish to be able to sketch a graph of a particular relation $y = y(x)$ with the least effort, but including all the interesting or principal features of the shape of the curve.

The ability to sketch a curve quickly can be useful in two main ways. Firstly, we may have some *empirical* (i.e. experimental) data which we have plotted on a graph and on the basis of which we wish to develop a theory or a (mathematical) model of the real situation. For example, we might have some real figures on the demand for a product at various price levels and we would like to examine the price–demand curve more carefully, perhaps to determine the elasticity of demand. In this case we need to be able to select a suitable mathematical function which, by suitable choice of coefficients, etc., will most closely match the data. Having found this curve, we can analyse the mathematical model rather than the original data, and perhaps draw some predictive conclusions from it. Now one set of data will often permit us to select from several types of curve, all of which may be made to fit the data quite well but not all of which will make sense in the context of the real world. By having some knowledge of the basic characteristics of the commonest types of curve the most appropriate selection might be made.

Suppose, for example, that we have the following two pieces of data relating price and demand, and wish to build a simple model. When

$$p = £1.10, \quad q = 1600 \text{ units}$$

and when

$$p = £1.20, \quad q = 1500 \text{ units.}$$

For different assumptions on the type of curve we obtain the following relations, all of which pass through the two points for (p, q), viz. (1·10, 1600) and (1·20, 1500).

(a) Linear $p = -0{\cdot}001q + 2{\cdot}7$ Figure 10.1(a)

(b) Hyperbolic (1) $p = \dfrac{2400}{q} - 0{\cdot}4$ Figure 10.1(b)

(c) Hyperbolic (2) $p = \dfrac{1320}{q - 400}$ Figure 10.1(c)

(d) Cubic $p = 10^{-9} q^3 - 2{\cdot}648 \times 10^{-6} q^2 + 3{\cdot}783$ Figure 10.1(d).

Although we have over-simplified the situation by using only two points (and thus giving ourselves a wide choice of curves) it can still be seen that the choice drastically affects the overall character of the eventual model *particularly well away from the original data*. In particular model (d) (the cubic) is unsatisfactory since it shows

(i) increasing demand with increasing price for high values

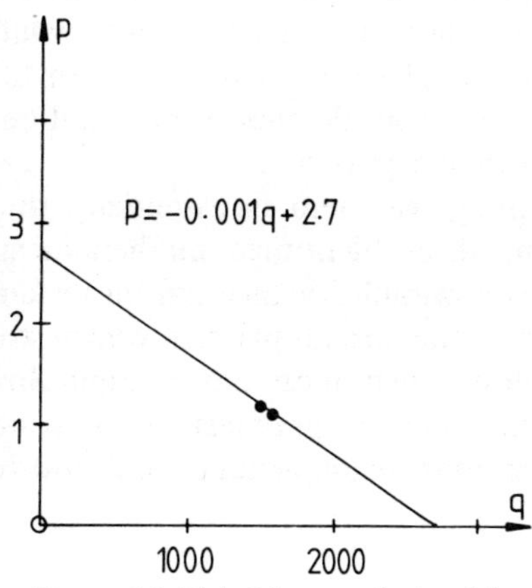

Figure 10.1(a) Linear relationship

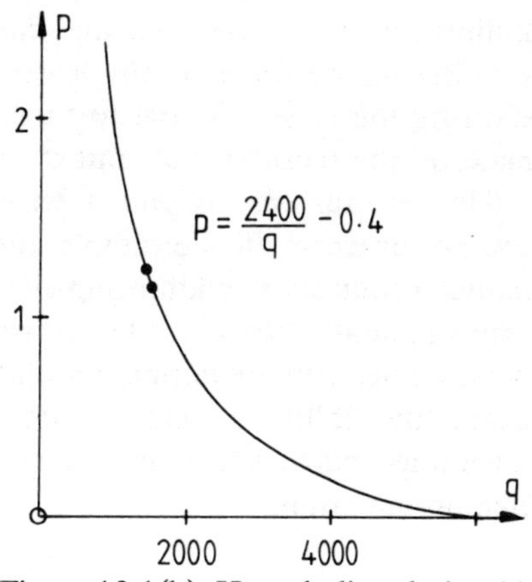

Figure 10.1(b) Hyperbolic relationship

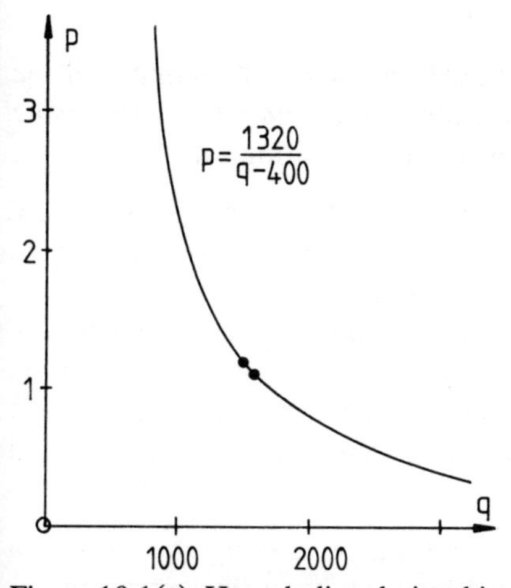

Figure 10.1(c) Hyperbolic relationship

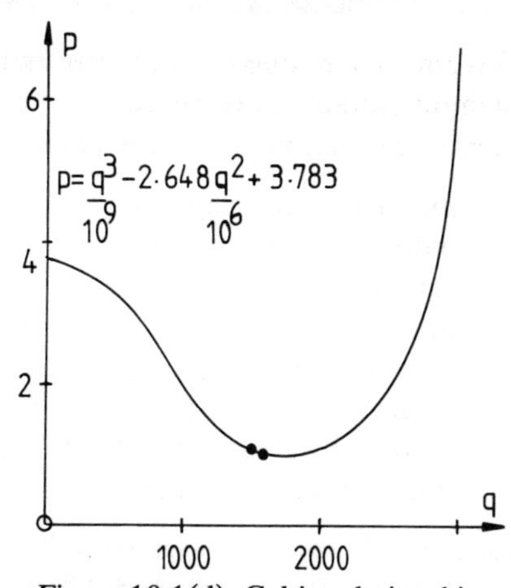

Figure 10.1(d) Cubic relationship

and

(ii) two possible demand levels for prices in the range

$$1{\cdot}03 \leqslant p \leqslant 3{\cdot}78$$

(i.e. this curve is not a *function* in the sense that $q(p)$ is not always single-valued.)

To summarize, therefore, the original choice of curve should be sensible, and the ability to sketch the broad characteristics of any curve quickly and economically is an advantage in making this choice.

It should be mentioned that in situations where there is an adequate amount of information (not just two points, as here) the aim is rarely to find a curve which will fit *all* points *exactly*. Instead, it is generally the situation that statistical variation (arising probably from several sources) will produce scatter or *dispersion*. In this case, what is required is a curve which follows the broad shape generated by the data. The problem of finding the best curve (i.e. the closest fit of data and curve) involves what is called *regression analysis*. Even so, one is still left with the problem of selecting the type of curve to fit the data, so a sensible choice must still be made on the basis of the context of the particular problem.

The second way in which curve sketching can help is essentially the reverse process of the above discussion. If a theoretical mathematical model produces a relationship between two variables which can be tested experimentally, then a sketch will indicate the broad pattern which we would expect the experimental data to follow, when plotted as a graph. So, again, the ability to read an algebraic expression in graphical terms is an advantage, particularly since a graph is simpler to understand and absorb than an equation.

10.2 APPROACHING CURVE SKETCHING

Several techniques which can be employed in curve sketching will be demonstrated. Nevertheless it is important to stress that the exact approaches adopted for different curves will vary according to:

(i) the basic characteristics of the algebraic equation itself. It is usually better to begin with the more obvious features and to increase subsequently the sophistication of the analysis;

(ii) the purpose of the exercise. Always bear in mind the question *what are we seeking to find out*, because a full analysis may not always be necessary. A lot of time and effort can be wasted by pursuing detailed features of curves which may be quite irrelevant to the overall question being asked.

One might also ask why straightforward plotting of any equation on to a graph should be inadequate. There are two main faults in this approach:

(i) it can be extremely tedious and time-consuming, particularly when an exact graph is not required; and
(ii) the main features can often be missed if one examines the less interesting regions of the curve. If the curve

$$y = 2x^3 - 27\,000x^2 + 20\,000\,000x$$

is plotted by calculating the y-values for $x = 1, 2, 3$, etc. then not much of interest occurs until $x = 4000$ or so; though it can be done—given 3 years and a roll of wallpaper!

10.3 CURVES OF THE TYPE $y = ax^n$ AND SYMMETRY

Curves of the type $y = ax^n$ fall into two main groups: the even-powered *symmetric* functions; and the odd-powered *anti-symmetric* functions (Figure 10.2).

In the case of *symmetric* functions, the y-axis is effectively a mirror and the curve in the $+x$ region is *reflected* in the $-x$ region. For *anti-symmetric* functions, the curve in the $+x$ region is *rotated* through 180° about the origin into the $-x$ region. For symmetric functions, therefore, we should usually specify *the line* in which reflection occurs, such as $x = 0$ (the y-axis). For anti-symmetric functions, however, we should specify the point about which rotation occurs, such as $x = 0$, $y = 0$ (the origin). Formally, we have

for symmetric functions (about $x = 0$) $\qquad f(-x) = +f(x)$
for anti-symmetric functions (about (0, 0)) $\qquad f(-x) = -f(x)$.

These properties can sometimes be of considerable use in sketching. The sum of symmetric functions is itself a symmetric function (and likewise for

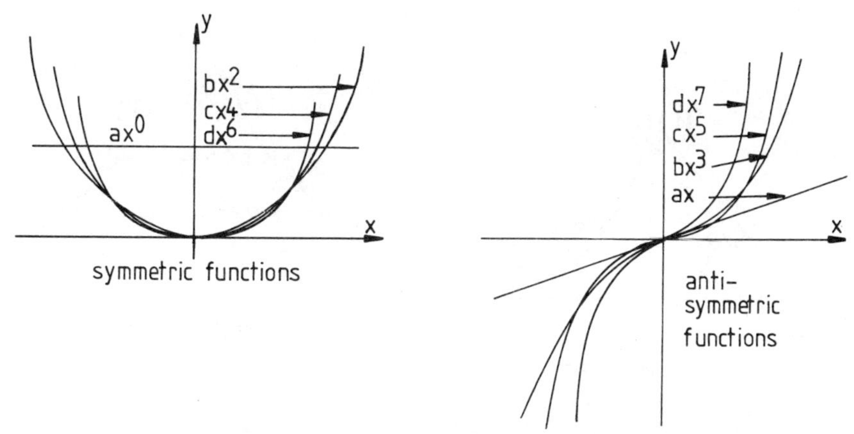

Figure 10.2 Symmetric functions with positive powers. Constants a, b, and c, etc., have all been taken as positive. Negative coefficients would have the effect of reflecting the curves in the x-axis. Properties of symmetry are then unchanged

anti-symmetric functions). So, for example,

$y = ax^5 + bx^3 + cx$ is anti-symmetric about (0, 0)
$y = dx^4 - ex^2 + f$ is symmetric about $x = 0$.

These properties can all be extended to negative powers. So, for example,

$$y = ax^{-1} = \frac{a}{x} \qquad y = \frac{b}{x^3} \qquad y = \frac{c}{x^5}$$

are anti-symmetric about (0, 0) and

$$y = \frac{a}{x^2} \qquad y = \frac{b}{x^4} \quad \text{and} \quad y = \frac{c}{x^6}$$

are symmetric about $x = 0$ (Figure 10.3).

Hence the function

$$y = ax^3 + b/x$$

is anti-symmetric about (0, 0) and the function

$$y = ax^4 - b/x^2 + c$$

is symmetric about $x = 0$.

Some functions are neither symmetric nor anti-symmetric about any axis, and are termed *asymmetric*. Functions of mixed powers (odd and even) are often of this type. For example,

$$y = x^4 + x^3$$

is an asymmetric function.

Some functions which appear to be asymmetric can be transformed into symmetric or anti-symmetric functions by shifting the axes. So, for

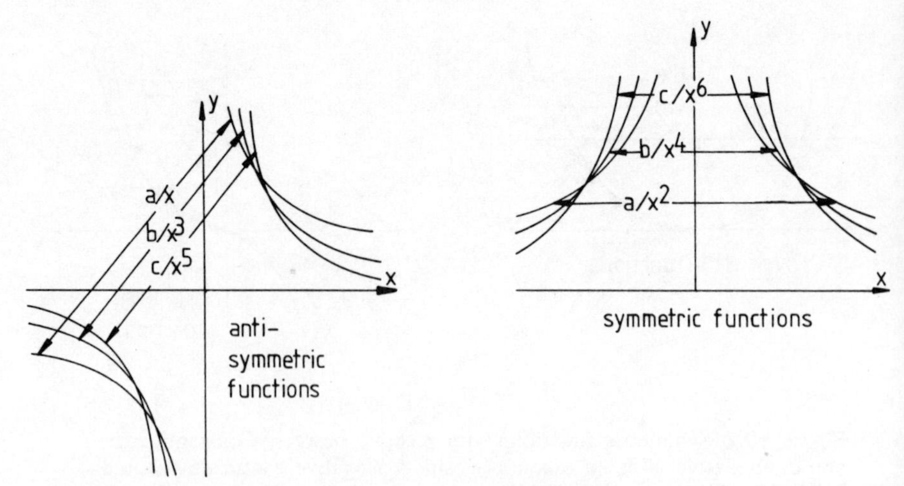

Figure 10.3 Symmetric functions with negative powers. Again, negative coefficients would reflect the functions in the x-axis

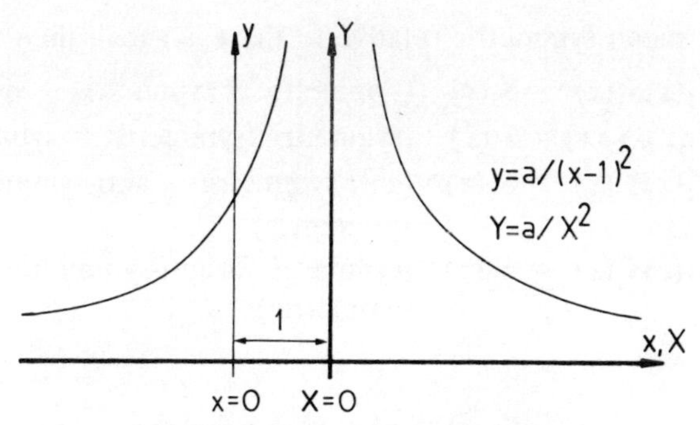

Figure 10.4 The curve is symmetric about the Y-axis though not about the y-axis

example,

$$y = \frac{a}{(x-1)^2}$$

can be transformed by putting $Y = y$, $X = x - 1$, when we have

$$Y = \frac{a}{x^2},$$

a symmetric function when referred to the X and Y coordinates (Figure 10.4).

It is possible to show very easily by an appropriate shift of axes that any cubic equation is anti-symmetric about its point of inflexion. The general cubic equation takes the form

$$y = ax^3 + bx^2 + cx + d$$

If the new axes (X, Y) are chosen so that the point of inflexion lies at the origin of the new axes, the form of the equation becomes

$$Y = eX^3 + fX$$

which is anti-symmetric about $X = 0$, $Y = 0$ as only odd powers of X are involved. (The reader might like to show this, once he has understood the following sections of this chapter.)

10.4 PRODUCTS AND SYMMETRY

If

$y = S_i(x)$ is any symmetric function of x,
$y = S_i'(x)$ is any anti-symmetric function of x, and
$y = A_i(x)$ is any asymmetric function of x

(by which we mean symmetric relative to the x, y axes), then

$$S_1(x)S_2(x) = S_3(x) \quad \text{(symmetric × symmetric = symmetric)}$$
$$\text{(and} \therefore S_3(x)/S_2(x) = S_1(x) \quad \text{(symmetric/symmetric = symmetric))}$$
$$S_1'(x)S_2'(x) = S_1(x) \quad \text{(anti-symmetric × anti-symmetric = symmetric)}$$
$$S_1(x)S_1'(x) = S_2'(x) \quad \text{(symmetric × anti-symmetric = anti-symmetric).}$$

For example,

$$\underset{\text{anti-symmetric}}{(x^3 + x)} \quad \underset{\text{symmetric}}{(x^4 - 1/x^2)} = \underset{\text{anti-symmetric}}{x^7 - x^5 - x - 1/x.}$$

Again, appropriate shifts of axes may reveal properties of symmetry not obvious in the original form; for example

$$\underset{\substack{\text{anti-symmetric}\\ \text{about } x = 1}}{(x - 1)} \quad \times \quad \underset{\substack{\text{anti-symmetric}\\ \text{about } x = 3}}{(x - 3)} \quad = \quad \underset{\substack{\text{asymmetric}\\ \text{about } x = 0, 1, \text{ or } 3}}{x^2 - 4x + 3.}$$

but

$$\begin{aligned} x^2 - 4x + 3 &= (x - 1)(x - 3) \\ &= (X + 1)(X - 1) \quad \text{if} \quad X = x - 2 \\ &= X^2 - 1. \end{aligned}$$

This final form includes only even powers of X and is thus symmetric about $X = 0$, when $x = 2$. Hence,

$$y = x^2 - 4x + 3$$

is symmetric about the line $x = 2$.

10.5 ORDERS OF MAGNITUDE

Consider the terms in the series below

$$ax^3 \quad bx^2 \quad cx \quad d \quad e/x \quad f/x^2 \quad g/x^3.$$

For very large values of x we find that

$$ax^3 \gg bx^2 \gg cx \gg d \gg e/x \gg f/x^2 \gg g/x^3$$

and for very small values of x we find that

$$ax^3 \ll bx^2 \ll cx \ll d \ll e/x \ll f/x^2 \ll g/x^3.$$

The meaning of *large* and *small* depends on the exact values we give to the constants a, b, c, etc. So for example,

$$ax^3 \gg bx^2$$

implies that $x \gg b/a$, and b/a is the standard by which we judge large and

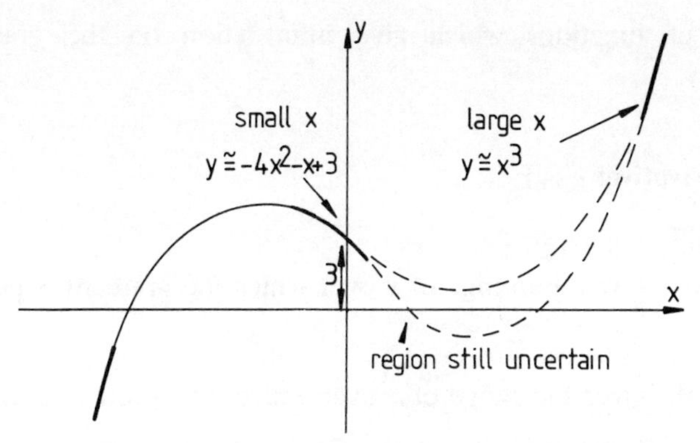

Figure 10.5 A quick sketch of the curve $y = x^2 - 4x^2 - x + 3$ using inspection of the coefficients only

small for this pair of terms. When we have

$$ax^3 \gg bx^2 \quad \text{or} \quad ax^3 \ll bx^2$$

then we can use the expression ax^3 *is an order of magnitude larger (or smaller) than* bx^2, since the relative sizes depend, largely, on the relative powers of x.

This property is extremely valuable in sketching curves, particularly in regions of high or low x-values.

For example, consider the cubic curve

$$y = 1x^3 - 4x^2 - 1x + 3$$

For very large values of x	$y \cong x^3$	anti-symmetric cubic.
For very small values of x	$y \cong -x + 3$	linear.
For smallish values of x	$y \cong -4x^2 - x + 3$	inverted parabola.

So, just by inspection of the coefficients, a very quick picture of the curve can be drawn (Figure 10.5).

The coefficient of x^3: $+1$ $\therefore$ curve moves from top RHS to lower LHS.
The coefficient of x^2: -4 $\therefore$ inverted parabolic shape near y-axis.
The coefficient of x: -1 $\therefore$ negative slope at the y-axis
Constant of $+3$ $\therefore$ y-intercept of $+3$.

Further analysis is needed before we can tell whether or not the curve cuts the x-axis once only (negative x) or three times (one negative and two positive values of x).

A more complete analysis for the cubic is given in the first example of Section 10.9.

10.6 THE USE OF DERIVATIVES

So far, we have discussed techniques which examine the equation of a curve directly. However considerable insight can be gained from the derivatives of functions which give information on the gradients and curvature.

The first derivative: $y'(x)$

The condition

$y'(x) > 0$ gives the range of x over which the gradient is positive

and

$y'(x) < 0$ gives the range of x over which the gradient is negative.

Over a range where the *sign* of the first derivative does not change, the function is said to be increasing/decreasing monotonically, i.e. always increasing or always decreasing, even though the rate of change may vary.

The condition

$y'(x) = 0$ gives the values of x for which the gradient is zero, which indicate local maxima, minima, or points of inflexion (see below).

The second derivative: $y''(x)$

The condition

$y''(x) > 0$ gives the range of x over which the *gradient* is always increasing (such as in the upright ∪ parabola).

and

$y''(x) < 0$ gives the range of x over which the *gradient* is always decreasing (such as in the inverted ∩ parabola).

The sign of the second derivative therefore distinguishes between minimum values ($y'' > 0$) and maximum values ($y'' < 0$).

The double condition

$y''(x) = 0$

and

$y'(x) \neq 0$ gives the values of x for which points of inflexion occur.

The double condition

$y''(x) = 0$

and

$y'(x) = 0$ may give a maximum, minimum, or point of inflexion.

In this last case, we need to inspect higher derivatives to distinguish the three cases. Suppose the first derivative is zero when $x = x_0$, so that

$$y'(x_0) = 0.$$

The higher derivatives of y are then evaluated at $x = x_0$, until one is found (the nth derivative say) which is not zero: i.e.

$$\begin{aligned}
y''(x_0) &= 0 \\
y'''(x_0) &= 0 \text{ (say)} \\
&\vdots \\
y^{(n-1)}(x_0) &= 0 \\
y^{(n)}(x_0) &\neq 0 \\
y^{(n+1)}(x_0) &= 0.
\end{aligned}$$

If n is *odd* then a point of inflexion occurs at $x = x_0$.
If n is *even* then a *maximum* occurs

$$\text{if } y^{(n)}(x_0) < 0$$

and a *minimum* occurs

$$\text{if } y^{(n)}(x_0) > 0.$$

This may seem a long procedure, but it is very simple to apply in practice. A simple example will illustrate this point.

Let

$$\begin{aligned}
y &= x^6 \\
y' &= 6x^5 = 0 \text{ when } x = x_0 = 0.
\end{aligned}$$

Hence

$$\begin{aligned}
y''(x_0) &= 30x^4 = 0 \text{ when } x = x_0 = 0 \\
y'''(x_0) &= 120x^3 = 0 \text{ when } x = x_0 = 0 \\
y^{(4)}(x_0) &= 360x^2 = 0 \text{ when } x = x_0 = 0 \\
y^{(5)}(x_0) &= 720x = 0 \text{ when } x = x_0 = 0 \\
y^{(6)}(x) &= 720 \neq 0 \text{ for all } x \\
y^{(7)}(x) &= 0 \qquad \text{for all } x
\end{aligned}$$

So, $n = 6$ in this case, which is *even*.

Also $y^{(6)}(x) = 720 > 0$, and this implies a *minimum* value of $y(x)$. This is entirely consistent with the graphical representation in Figure 10.2(a).

10.7 THE USE OF LIMITS

In Section 10.5 we introduced the idea of taking large and small values of x to understand the behaviour of a curve in certain regions. This can be

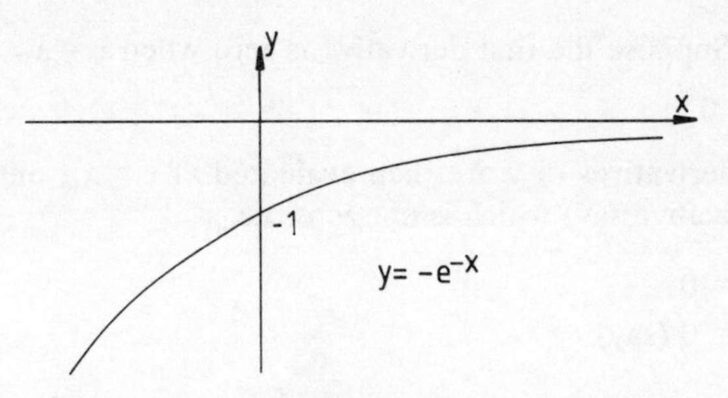

Figure 10.6

done slightly more formally by using limits. For example, we can let the x value tend towards some value (usually 0 or $\pm\infty$) and find what behaviour y, y', or y'' follow. This procedure is generally useful, but particularly so for functions involving e^x.

For example, consider the function $y = -e^{-x}$ (Figure 10.6), where $y' = e^{-x}$.

As $x \to +\infty$ $y \to 0$ $y' \to 0$ (As x tends towards $+\infty$, then y and y' tend towards zero.)

As $x \to -\infty$ $y \to -\infty$ $y' \to +\infty$.

As $x \to 0$ $y \to -1$ $y' \to 1$.

Notice that $y < 0$ always but $y' > 0$ always

10.8 INTERCEPTS

For many curves it is simple and worthwhile to find the values of the x and y intercepts. For functions in the form $y = f(x)$ it is usually easier to find the y intercepts ($f(0)$) than it is to find the x intercepts. These are the *roots* of the equation

$$f(x) = 0.$$

(i) For straight lines this is straightforward.
(ii) For parabolas, we are faced with the problem of solving a *quadratic* equation of the general type.

$$ax^2 + bx + c = 0.$$

This can be solved by factorization, or by using the general formula

$$x = \frac{-b \pm \sqrt{b^2 - 4ac}}{2a}.$$

When $b^2 > 4ac$, there are two roots (x intercepts) Figure 10.7(a).
When $b^2 = 4ac$, there is one root given by $x = -b/2a$ Figure 10.7(b).
When $b^2 < 4ac$, there are no real roots (x intercepts) Figure 10.7(c).

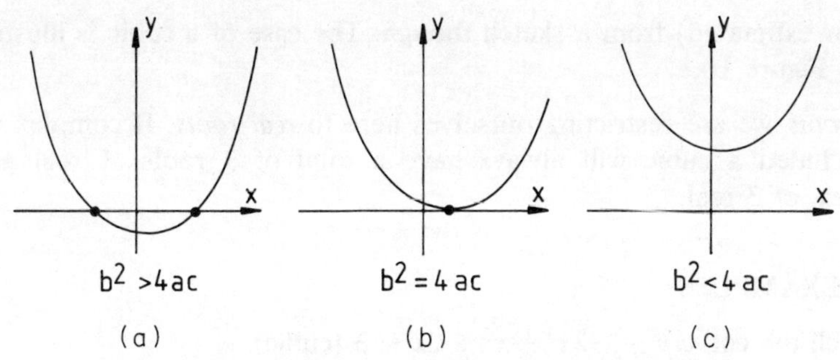

Figure 10.7 Solutions to the equation $ax^2 + bx + c = 0$ (a) Two real roots. (b) One real root. (c) No real roots

(iii) For cubics, finding the roots becomes a more tricky problem in general, and will not be demonstrated here (particularly since it involves the use of complex numbers). Generally, in fact, we are more likely to use the sketch to estimate the roots, rather than use the roots to help sketch the curve. The number of real roots can be determined

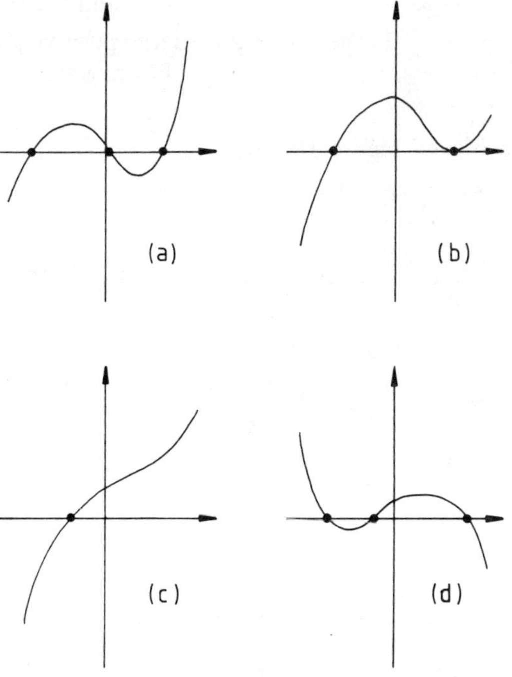

Figure 10.8 Sketches of the typical cubic $y = ax^3 + bx^2 + cx + d$ (order n). (a) and (d) three real roots; (b) two real roots; (c) one real root. Maximum number of turning points is two $(n - 1)$; maximum number of real roots is three (n)

(or estimated) from a sketch though. The case of a cubic is illustrated in Figure 10.8.

By *roots* we are restricting ourselves here to *real roots*. If complex roots are included a cubic will always have a total of 3 roots: 1 real and 2 complex, or 3 real.

10.9 EXAMPLES

I Sketch the curve $y = -2x^3 + x^2 + 2x - 3$ (cubic)

1. $y = -2x^3 + x^2 + 2x - 3$
   $y' = -6x^2 + 2x + 2$
   $y'' = -12x + 2.$
2. $x = 0$ $\quad y = -3$ $\quad$ y intercept of -3
   $y' = +2$ $\quad$ +ve slope at y intercept
   $y'' = +2$ $\quad$ +ve curvature at y intercept.
3. $x \to +\infty$ $\quad y \to -\infty$ $\quad$ Large +ve x gives large negative y
   $y' \to -\infty$ $\quad$ and y'
   $y'' \to -\infty$ $\quad$ and y''.
4. $x \to -\infty$ $\quad y \to +\infty$ $\quad$ Large −ve x gives large positive y and y''
   $y' \to -\infty$ $\quad$ but negative y'.
   $y'' \to +\infty$

So far we have achieved Figure 10.9(a).

5. $y' = 0$ when $-6x^2 + 2x + 2 = 0$,

$$x = \frac{-2 \pm \sqrt{2^2 + 48}}{-12} = \frac{-2 \pm \sqrt{52}}{-12} = -0{\cdot}43 \text{ or } 0{\cdot}77.$$

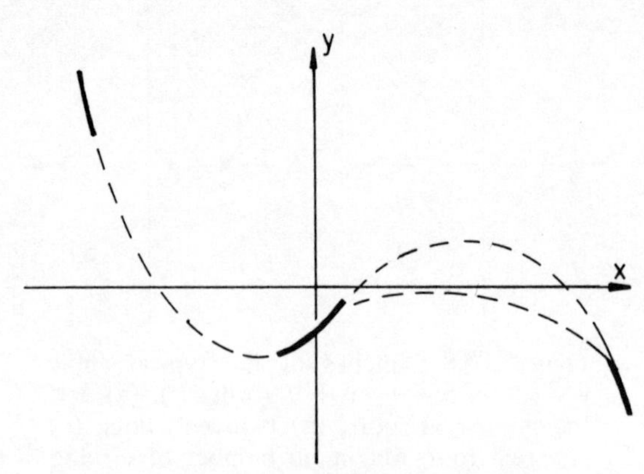

Figure 10.9(a) Results from the analysis 1 to 4

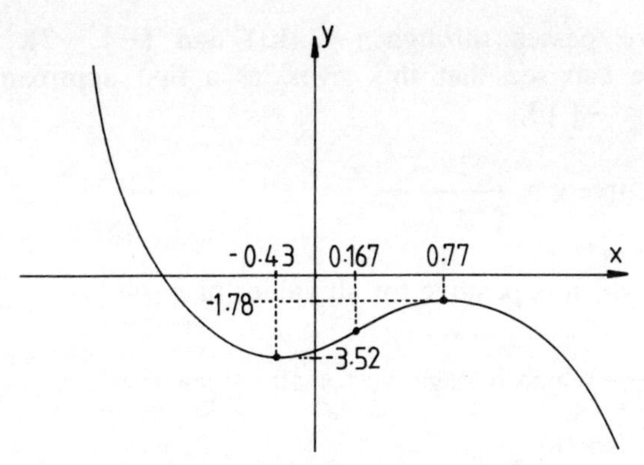

Figure 10.9(b) Results from the analysis 1 to 7

It can already be seen that a maximum must occur when $x = +0{\cdot}77$, and that a minimum must occur when $x = -0{\cdot}43$.

6. $y(0{\cdot}77) = -2(0{\cdot}77)^3 + (0{\cdot}77)^2 + 2(0{\cdot}77) - 3 = -1{\cdot}78$.
   $y(-0{\cdot}43) = -2(-0{\cdot}43)^3 + (-0{\cdot}43)^2 + 2(-0{\cdot}43) - 3 = -3{\cdot}52$.
   So it is now clear that the curve does not cut the x-axis more than once, since $y(0{\cdot}77) < 0$.
7. $y'' = 0$ when $-12x + 2 = 0$ $\therefore$ $x = 2/12 = 1/6 = 0{\cdot}167$. We have now arrived at Figure 10.9(b).
8. Finally, we might wish to estimate the value of the x-intercept. We shall do this by approximating to a straight line between $x = -1$ and $x = -2$.

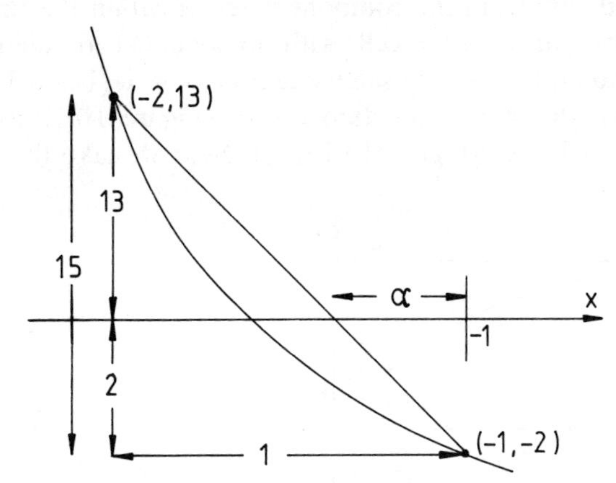

Figure 10.10 The x-intercept for $y = -2x^3 + x^2 + 2x + 3$. By similar triangles

$$\frac{15}{1} = \frac{2}{\alpha} \quad \therefore \alpha = 0{\cdot}1\dot{3}.$$

The x-intercept of the straight line is therefore $x = -1{\cdot}1\dot{3}$

The curve passes through $(-2, 13)$ and $(-1, -2)$. From Figure 10.10(a) we can see that this gives, as a first approximation, $-2 <$ x-intercept $< -1{\cdot}1\dot{3}$.

II Sketch the curve $y = \dfrac{1}{e^x + 1}$

1. $y = \dfrac{1}{e^x + 1}$ which is positive for all values of x

$y' = \dfrac{-e^x}{(e^x + 1)^2}$ which is negative for all values of x.

$$\left.\begin{array}{lll} \text{2. } x \to +\infty & y \to 0 \\ & y' \to 0 \\ \text{3. } x \to -\infty & y \to 1 \\ & y' \to 0 \end{array}\right\} \therefore \text{Gradient tends towards zero for both positive and negative large values of } x.$$

4. $x \to 0 \quad y \to \frac{1}{2}$
$\qquad\qquad y' \to -\frac{1}{4}$

5. $y - \dfrac{1}{2} = \dfrac{1}{e^x + 1} - \dfrac{1}{2} = \dfrac{2 - e^x - 1}{2(e^x + 1)} = \dfrac{1 - e^x}{2(1 + e^x)} = \dfrac{e^{-x/2} - e^{x/2}}{2(e^{-x/2} + e^{x/2})} = Y$ say

$Y(x) = -Y(-x)$ so the curve is also anti-symmetric about the axes $y = \frac{1}{2}$, $x = 0$ (Figure 10.11).

An alternative approach to sketching several types of function is to build up a picture gradually from the component terms within the function. Since this often works particularly well with exponential functions, we shall illustrate this technique with the same example, $y = 1/(1 + e^x)$.

We first sketch the simple function $y = e^x$ (Figure 10.12(a)). From this we proceed to $y = 1 + e^x$ (Figure 10.12(b)). Next we take the reciprocal of

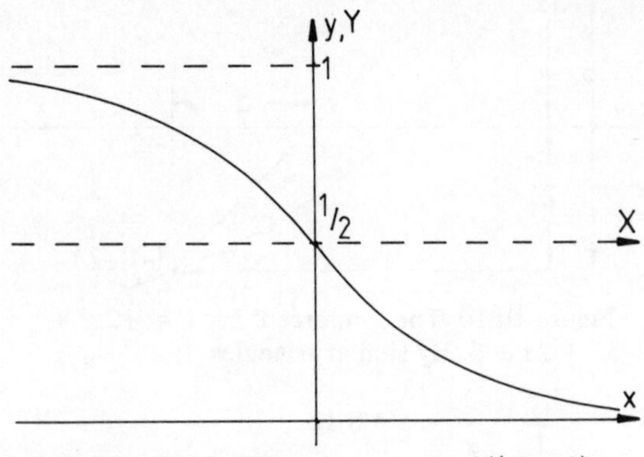

Figure 10.11 The curve $y = 1/(e^x + 1)$, which is anti-symmetric about the axes $x = 0$, $y = \frac{1}{2}$

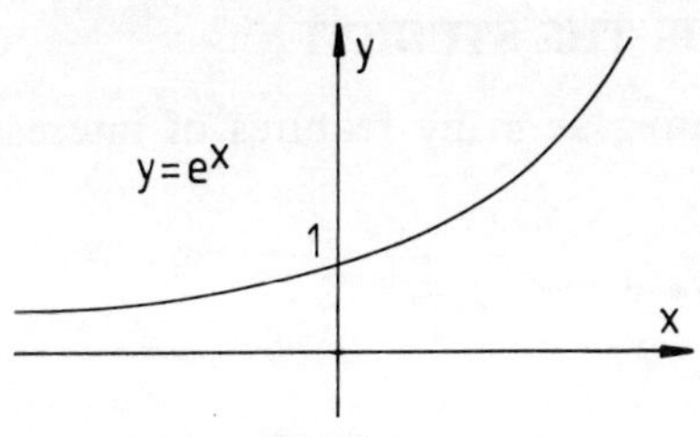

Figure 10.12(a) $y = e^x$

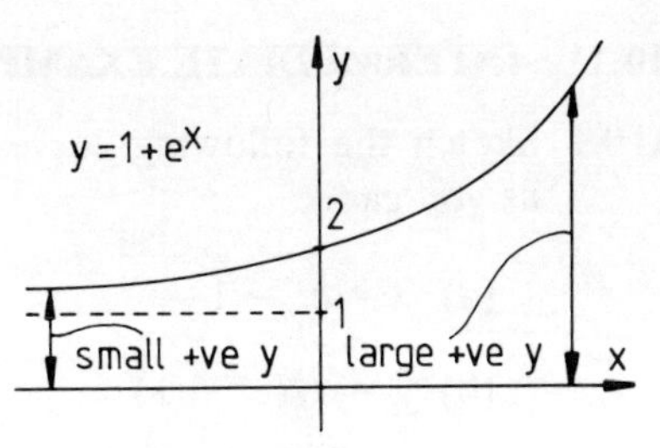

Figure 10.12(b) $y = 1 + e^x$

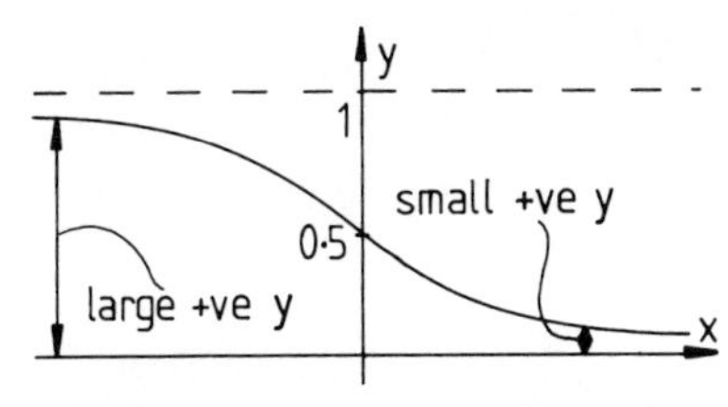

Figure 10.12(c) $y = 1/(1 + e^x)$

the y-value of this function, for all values of x, to obtain instead the required function $y = 1/(1 + e^x)$. When finding the reciprocal $(1/z)$ of a term z the following basic principles apply:

(i) small values of z give large values of $1/z$;
(ii) large values of z give small values of $1/z$;
(iii) the smallest value of z gives the largest $1/z$;
(iv) the largest value of z gives the smallest $1/z$;
(v) the sign is conserved so that

$$z > 0, \qquad 1/z > 0,$$
$$z < 0, \qquad 1/z < 0.$$

Applying these principles to $y = 1 + e^x$, we obtain $y = 1/(1 + e^x)$, as in Figure 10.12(c). Notice that the descriptions as *large* or *small* are *relative*. Since the *smallest* value of $1 + e^x$ is 1, the *largest* value reached by $1/(1 + e^x)$ is also 1. This has been described as *large* in this *relative* sense.

10.10 BASIC EXAMPLES FOR THE STUDENT

B10.1 Sketch the following curves

(a) $2y = x^2 + 7x - 4$
(b) $3y = 3x^3 - 14x^2 + 32$
(c) $y = ax + b + c/x$
(d) $y = e^x$
(e) $y = 1/(e^x - 1)$
(f) $y = x^4 - x^2$.

10.11 INTERMEDIATE EXAMPLES FOR THE STUDENT

I10.1 Sketch the following curves, indicating as many features of interest as you can.

(a) $y = e^x - 1$

(b) $y = 1/(e^x - 1)$

(c) $y = x^2 + 2x - 4$

(d) $y = x^3 + 3x^2 - 2x - 1$

(e) $y = x^4 - \dfrac{1}{x^2} + 2$

(f) $y = x^4 - x^3$

(g) $y = \dfrac{1}{(x-1)(x-3)}$.

CHAPTER 11

Further Curves: Exponential, Logarithmic, Hyperbolic

11.1 INTRODUCTION

In this chapter it is proposed to analyse three particular types of curve, using the curve-sketching techniques developed in the previous chapter. The three types, exponential, logarithmic, and hyperbolic, have been chosen since they are often useful in developing mathematical models, and some simple applications will be presented.

11.2 THE EXPONENTIAL CURVE $y = e^x$

Most of the properties of e^x become more apparent when it is written as the sum of an infinite series of terms, thus:

$$e^x = 1 + \frac{x}{1!} + \frac{x^2}{2!} + \frac{x^3}{3!} + \frac{x^4}{4!} + \ldots$$

(A proof of this is given in Section 11.5.) Hence

(i) $$e = 1 + \frac{1}{1!} + \frac{1}{2!} + \frac{1}{3!} + \frac{1}{4!} + \ldots$$
$$= 1 + 1 + 0{\cdot}5 + 0{\cdot}1\dot{6} + 0{\cdot}041\dot{6} + \ldots$$
$$= 2{\cdot}71828\ldots, \text{ a positive constant.}$$

(ii) $e^x > 0$ for all values of x, since any positive constant raised to any power is always positive; for example,

$$e^3 = (2{\cdot}71828)^3 = 20{\cdot}0855 > 0$$
$$e^{-3} = 1/(2{\cdot}71828)^3 = 0{\cdot}04979 > 0.$$

(iii) $e^0 = 1$ since any number raised to the power zero equals one. This also follows from the infinite series, putting $x = 0$.

(iv) For the range $x > 0$, $e^x > 1$ since all the terms in the infinite series are positive.

For $x < 0$, we can put $z = -x$ so that $z > 0$. Then

$$e^x = e^{-z} = \frac{1}{e^z}$$

but $e^z > 1$ as $z > 0$, so that

$$\frac{1}{e^z} < 1 \quad \text{if} \quad e^z > 1$$

$$\therefore e^x < 1 \quad \text{for} \quad x < 0.$$

Hence, for $x > 0$, $e^x > 1$; and for $x < 0$, $0 < e^x < 1$.

(v) $e^x = y = 1 + \dfrac{x}{1!} + \dfrac{x^2}{2!} + \dfrac{x^3}{3!} + \dfrac{x^4}{4!} + \ldots$

$$\therefore \frac{dy}{dx} = 0 + \frac{1}{1!} + \frac{2x}{2!} + \frac{3x^2}{3!} + \frac{4x^3}{4!} + \ldots$$

$$= 1 + \frac{x}{1!} + \frac{x^2}{2!} + \frac{x^3}{3!} + \ldots$$

$$= e^x = y.$$

Now if $y'(x) = y(x)$, then by continued differentiation it follows immediately that

$$y = \frac{dy}{dx} = \frac{d^2y}{dx^2} = \frac{d^3y}{dx^3} = \frac{d^4y}{dy^4}, \text{ etc.}$$

Therefore all derivatives of e^x also equal e^x.

(vi) It follows from (v) above that the gradient of e^x has the same properties as e^x.

$$\therefore \frac{dy}{dx} > 0 \quad \frac{dy}{dx} = 1 \text{ at } x = 0 \quad \frac{dy}{dx} > 1 \text{ for } x > 0 \quad 0 < \frac{dy}{dx} < 1 \text{ for } x < 0.$$

From these results (i) to (vi) a sketch of $y = e^x$ and $y = e^{-x}$ can be drawn as in Figure 11.1.

11.3 FUNCTIONS OF THE FORM $ae^{f(x)}$

An important class of functions takes the general form

$$y = ae^{f(x)} = a \exp(f(x)).$$

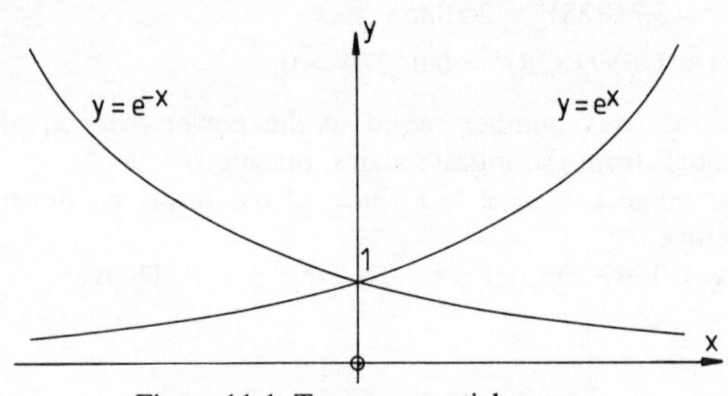

Figure 11.1 Two exponential curves

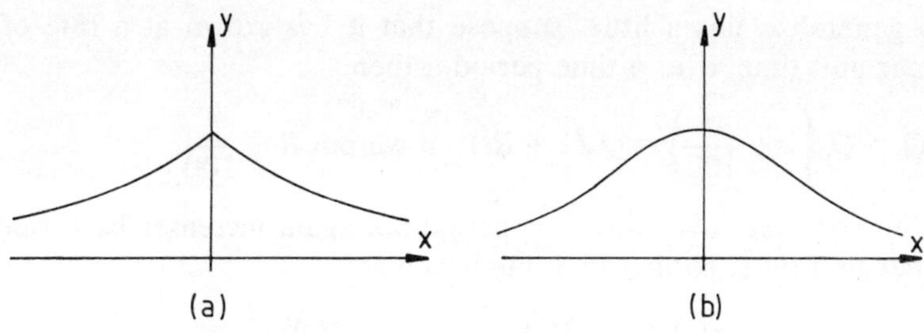

Figure 11.2(a) $y = a\,e^{-b|x|}$. (b) $y = a\,e^{-bx^2}$

For example, the Normal distribution in statistical theory takes the form

$$y = \frac{1}{\sigma\sqrt{2\pi}}\,e^{-(x-\mu)^2/2\sigma^2}$$

where μ, σ are constants, $\pi = 3{\cdot}14159\ldots$

The shape of these curves clearly depends on the form of the function $f(x)$. In the case of the standard Normal distribution for example, the only power of x involved is x^2, and so the function is symmetric, and is not so far different from the shape we might expect from the function

$$y = ae^{-b|x|}.$$

(See Figure 11.2.)

The first derivative of $y = e^{f(x)}$ can be found using the *chain rule*:

$$\frac{dy}{dx} = \frac{dy}{df}\frac{df}{dx} = e^{f(x)}\frac{df}{dx} = f'(x)e^{f(x)}$$

i.e. $e^{f(x)}$ multiplied by the first derivative of f. For example,

if $y(x) = e^x$ then $y'(x) = 1e^x$;
if $y(x) = e^{3x}$ then $y'(x) = 3e^{3x}$;
if $y(x) = e^{-x^2}$ then $y'(x) = -2xe^{-x^2}$;
if $y(x) = \exp(1/(x + x^2))$ then $y'(x) = \dfrac{-(1+2x)}{(x+x^2)^2}\exp(1/(x + x^2))$.

(N.B. the derivative of $(x + x^2)^{-1}$ is $-(1 + 2x)/(x + x^2)^2$ using the chain rule.)

11.4 AN EXAMPLE OF THE EXPONENTIAL FUNCTION

Continuous Rates of Growth

Suppose that we have a quantity, original value Q_0 at time $t = 0$, and that it grows at a uniform rate so that at the end of one year ($t = 1$) it has increased in magnitude by 10 per cent. At the end of this year its value is therefore given by

$$Q = Q_0(1 + (0{\cdot}1)(1)) = Q_0(1 + 0{\cdot}1) = 1{\cdot}1Q_0.$$

If we generalize this a little, suppose that it has grown at a rate of r per cent per unit time, over a time period t, then

$$Q = Q_0\left(1 + \frac{rt}{100}\right) = Q_0(1 + Rt) \quad \text{if we put } R \equiv \frac{r}{100}.$$

Now suppose that over the next period t it again increases by a rate r per cent per unit time, so that its value becomes

$$Q = Q_0\left(1 + \frac{rt}{100}\right)\left(1 + \frac{rt}{100}\right) = Q_0\left(1 + \frac{rt}{100}\right)^2 = Q_0(1 + Rt)^2.$$

Over n such periods, the value would therefore become

$$Q = Q_0\left(1 + \frac{rt}{100}\right)^n = Q_0(1 + Rt)^n.$$

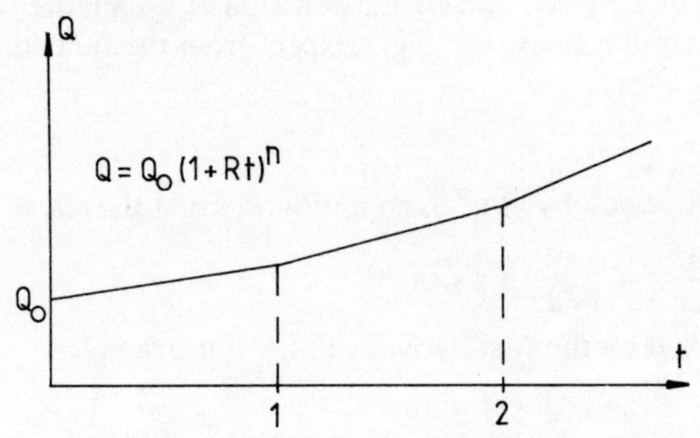

Figure 11.3(a) Interest compounded at t = 1, 2, 3, etc.

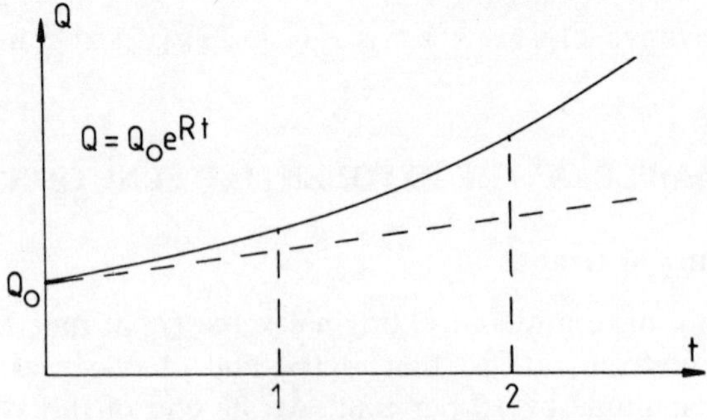

Figure 11.3(b) Interest compounded continually

On a graph (Figure 11.3(a)) we see that, because we have *compounded* the *interest* or extra growth, there are several kinks in the growth curve. This might be realistic where a sum of money is invested and the interest is compounded at regular intervals of 6 months or a year. However, when talking about economic growth, or rates of inflation, this picture is unrealistic and we should expect to see a smooth curve as in Figure 11.3(b). This smooth curve, again expressing a growth rate of r per cent per unit time but with *continuous compounding*, is represented by the equation

$$Q = Q_0 e^{rt/100} = Q_0 e^{Rt}.$$

In other words, continuous growth at a fixed rate of interest, but with all interest generated itself accruing interest immediately, gives an exponential rate of growth.

This follows from the result that e^x can be expressed in the form

$$e^x = \lim_{n \to \infty} \left(1 + \frac{x}{n}\right)^n \tag{11.1}$$

If we take an interval of one unit, for example, then we can write, with a single compounding of interest,

$$Q = Q_0(1 + Rt).$$

If we divide this interval into two parts, and compound twice, then the growth factor in each half-interval will be

$$\left(1 + \frac{Rt}{2}\right)$$

but this will be applied twice, so that the new final sum Q would be given instead by

$$Q = Q_0\left(1 + \frac{Rt}{2}\right)^2.$$

If we had divided the original interval into 3 equal parts, then a similar argument would have given

$$Q = Q_0\left(1 + \frac{Rt}{3}\right)^3$$

Generalizing to n divisions of the original interval we obtain

$$Q = Q_0\left(1 + \frac{Rt}{n}\right)^n$$

If now we take the limit as n approaches infinity, we are compounding the interest an infinite number of times (in effect, continuously). So, with continuous compounding

$$Q = \lim_{n\to\infty} Q_0\left(1 + \frac{Rt}{n}\right)^n$$

$$= Q_0 \quad \lim_{n\to\infty} \left(1 + \frac{Rt}{n}\right)^n$$

$$= Q_0 e^{Rt}$$

using equation (11.1).

So, for our original figures of $t = 1$, $r = 10$, $R = 0{\cdot}10$,

$$Q = Q_0 e^{0\cdot10} = 1{\cdot}1052 Q_0$$

which gives an effective rate of (uncompounded) interest of 10·52%, over one year.

11.5 THE EXPANSION OF e^x AS A SERIES (Optional)

Define the function of x, $f(x)$, as the sum of the infinite series

$$f(x) = 1 + \frac{x}{1!} + \frac{x^2}{2!} + \frac{x^3}{3!} + \ldots$$

$$\therefore f(y) = 1 + \frac{y}{1!} + \frac{y^2}{2!} + \frac{y^3}{3!} + \ldots \text{ (the same series with a new variable)}$$

$$\therefore f(x)f(y) = \left(1 + \frac{x}{1!} + \frac{x^2}{2!} + \frac{x^3}{3!} + \ldots\right)\left(1 + \frac{y}{1!} + \frac{y^2}{2!} + \frac{y^3}{3!} + \ldots\right)$$

$$= 1 + \frac{x}{1!} + \frac{y}{1!} + \frac{x^2}{2!} + \frac{xy}{1!1!} + \frac{y^2}{2!} + \frac{x^3}{3!} + \frac{x^2y}{2!1!} + \frac{xy^2}{1!2!} + \frac{y^3}{3!} + \ldots$$

$$= 1 + \frac{(x + y)}{1!} + \frac{(x^2 + 2xy + y^2)}{2!} + \frac{(x^3 + 3x^2y + 3xy^2 + y^3)}{3!} + \ldots$$

$$= 1 + \frac{(x + y)}{1!} + \frac{(x + y)^2}{2!} + \frac{(x + y)^3}{3!} + \ldots$$

$\therefore f(x)f(y) = f(x + y)$ for any x and y.

Now

$$f(1) = 1 + \frac{1}{1!} + \frac{1}{2!} + \frac{1}{3!} + \ldots = 2{\cdot}71828 \ldots$$

an irrational number which is given the symbol 'e'. But $f(1) \times f(1) = f(2)$ from above and $f(1) = e$

$$\therefore e \times e = e^2 = f(2)$$

$$\therefore e \times e^2 = f(1) \times f(2) = f(3) = e^3$$

$$\therefore f(1) = e \quad f(2) = e^2 \quad f(3) = e^3 \text{ etc.} \quad \text{so that } f(x) = e^x$$

$$\therefore e^x = f(x) = 1 + \frac{x}{1!} + \frac{x^2}{2!} + \frac{x^3}{3!} + \ldots$$

Strictly, we have only demonstrated that $f(x) = e^x$ for integral values of x. We can extend the proof to those cases where x is a rational number along the lines of the following example:

$$f(\tfrac{1}{2}) \times f(\tfrac{1}{2}) = f(1) = e$$

but

$$e^{1/2} \times e^{1/2} = e$$

$$\therefore f(\tfrac{1}{2}) = e^{1/2}.$$

(The reader might like to attempt the more general case of $f(m/n)$.)

Another useful starting point for the function e^x is to use instead the formula

$$e^x = \lim_{n\to\infty}\left(1 + \frac{x}{n}\right)^n.$$

Expansion of the power using the binomial theorem (Chapter 20) and taking the limit $n \to \infty$ gives the same series as $f(x)$, employed in this section.

11.6 THE LOGARITHMIC CURVE $y = \log_e x$

The properties of the logarithmic curve can be found immediately from the properties of the exponential curve in the following way.

If $x = e^y$ then $y = \log_e x$; these are equivalent statements. The curve $x = e^y$ is shown in Figure 11.4(a), and thus the curve for $y = \log_e x$ can be found immediately (Figure 11.4(b)).

In fact the same curve has been drawn for both figures. Figure 11.4(b) is the same as Figure 11.4(a) except it has been rotated through a right angle, and then reflected (as in a mirror). This is what we mean when we say $\log_e x$ is the *inverse function* of e^x (see Chapter 3).

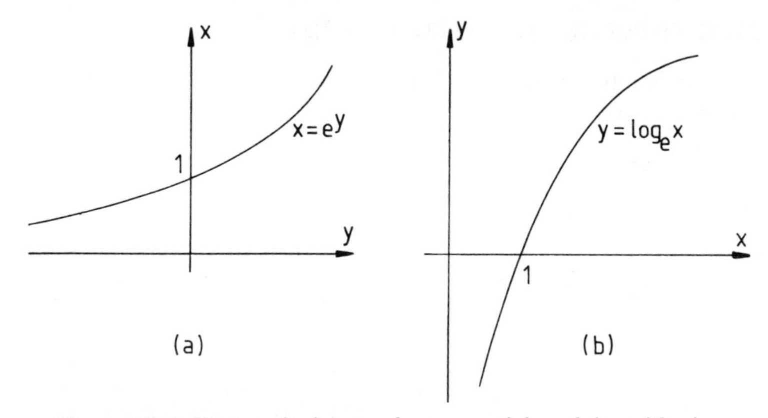

Figure 11.4 The equivalence of exponential and logarithmic functions: (a) exponential form; (b) logarithmic form

The derivative of $y = \log_e x$ can also be found fairly simply using the result already given for $y = e^x$.

$$y = \log_e x \quad \text{so that} \quad x = e^y$$

$$\therefore \frac{dx}{dy} = e^y$$

(Note the positions of x and y carefully, since we normally use $y = e^x$, whence $dy/dx = e^x$.) Using the result shown at the end of Section 7.3,

$$\frac{dy}{dx} = 1 \Big/ \frac{dx}{dy} = \frac{1}{e^y} = \frac{1}{x}.$$

Hence if $y(x) = \log_e(x)$ then $y'(x) = 1/x$.

Other properties of the logarithmic curve also follow from the exponential (comparison between Figures 11.4(a) and 11.4(b) make these points clearer).

(a) $\log_e x$ is real only if $x > 0$
(b) $\log_e x > 0$ for $x > 1$
$\log_e x < 0$ for $0 < x < 1$
} true for *any* base

(c) $\log_e 1 = 0$ (this is true for *any* base since $b^0 = 1$ so that $\log_b 1 = 0$ for any b)

(d) $\dfrac{dy}{dx} < 1$ for $x > 1$

$\dfrac{dy}{dx} > 1$ for $0 < x < 1$

These results follow immediately using $y'(x) = 1/x$, $x > 0$.

11.7 FUNCTIONS OF THE FORM $a \log_e f(x)$

We are principally interested here in the method of differentiating functions of the type

$$y = a \log_e[f(x)]$$

using the *chain rule*.

If

$$y = a \log_e f(x) = a \log_e f,$$

then

$$\frac{dy}{df} = a \times \frac{1}{f} = \frac{a}{f}$$

and

$$\frac{dy}{dx} = \frac{dy}{df} \cdot \frac{df}{dx} = \frac{a}{f} f'(x).$$

Hence if $y = a \log_e f(x)$, then

$$\frac{dy}{dx} = \frac{af'(x)}{f(x)}.$$

For example,

$$\text{if } y(x) = 3 \log_e x \quad \text{then} \quad y'(x) = \frac{3}{x} \cdot 1 = \frac{3}{x},$$

$$\text{if } y(x) = 2 \log_e 3x \quad \text{then} \quad y'(x) = \frac{2}{3x} \cdot 3 = \frac{2}{x}.$$

(N.B. $\log_e 3x = \log_e 3 + \log_e x$ so this example is the same as $y = 2 \log_e 3 + 2 \log_e x$.)

$$\text{If } y(x) = \log_e(x^2 + 3x + 2) \quad \text{then} \quad y'(x) = \frac{2x + 3}{(x^2 + 3x + 2)},$$

$$\text{if } y(x) = \log_e(1/(x + x^2)) \quad \text{then} \quad y'(x) = -\ (x + x^2)\frac{(1 + 2x)}{(x + x^2)^2}$$

$$= -\ \frac{(1 + 2x)}{(x + x^2)}.$$

(N.B. $\log_e(1/(x + x^2)) = \log_e(x + x^2)^{-1} = -\log_e(x + x^2)$ which can then be differentiated more easily to give $-(1 + 2x)/(x + x^2)$.)

11.8 USE OF LOGARITHMIC FUNCTIONS IN DIFFERENTIAL CALCULUS

If we put $z = \log_e y$, then

$$\frac{dz}{dx} = \frac{dz}{dy}\frac{dy}{dx} = \frac{1}{y}\frac{dy}{dx}$$

as long as y is itself a function of x (otherwise, $dy/dx = 0$); that is,

$$\frac{d}{dx}(\log_e y) = \frac{1}{y}\frac{dy}{dx}$$

and this is useful when differentiating functions such as

$$y = \sqrt[3]{\frac{(a + x)(b + x)}{(c + x)(d + x)}}.$$

This could be differentiated using the usual rules, treating it as a quotient of the form u/v. However, if we take logarithms on both sides and then differentiate we obtain the following result:

$$y = \sqrt[3]{\frac{(a+x)(b+x)}{(c+x)(d+x)}} = \frac{(a+x)^{1/3}(b+x)^{1/3}}{(c+x)^{1/3}(d+x)^{1/3}}$$

$$\therefore \log_e y = \log_e(a+x)^{1/3} + \log_e(b+x)^{1/3} - \log_e(c+x)^{1/3} - \log_e(d+x)^{1/3}$$

$$= \tfrac{1}{3}[\log_e(a+x) + \log_e(b+x) - \log_e(c+x) - \log_e(d+x)]$$

$$\therefore \frac{1}{y}\frac{dy}{dx} = \frac{1}{3}\left[\frac{1}{(a+x)} + \frac{1}{(b+x)} - \frac{1}{(c+x)} - \frac{1}{(d+x)}\right]$$

$$\therefore \frac{dy}{dx} = \frac{y}{3}\left[\frac{1}{(a+x)} + \frac{1}{(b+x)} - \frac{1}{(c+x)} - \frac{1}{(d+x)}\right]$$

$$\therefore \frac{dy}{dx} = \frac{1}{3}\sqrt[3]{\frac{(a+x)(b+x)}{(c+x)(d+x)}}\left[\frac{1}{a+x} + \frac{1}{b+x} - \frac{1}{c+x} - \frac{1}{d+x}\right].$$

This is a far neater technique than the quotient method.

11.9 THE RECTANGULAR HYPERBOLA $y = c/x$

The function xy = constant or $y = c/x$ is called a rectangular hyperbola and can be a useful curve for employment in mathematical models. Its

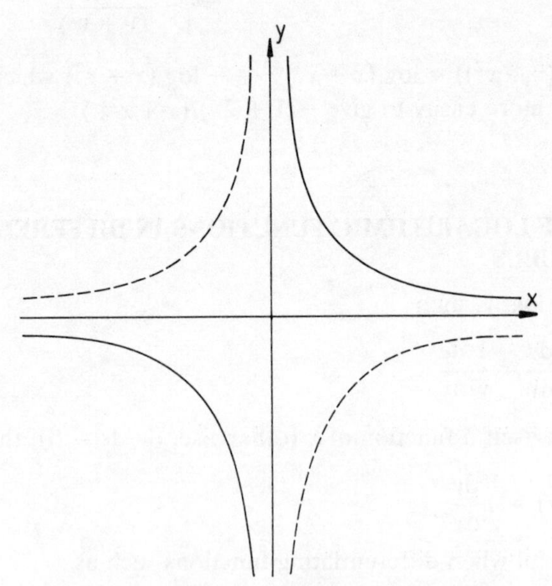

Figure 11.5 The rectangular hyperbola $xy = c$. The axes are asymptotes to the curve (tangents at infinity); the solid line is the curve for $c > 0$; the broken line is the curve for $c < 0$

properties can be fairly readily derived, and we shall assume that c is a *positive* constant for the moment. (See Figure 11.5.)

(i) $y > 0$ for $x > 0$
$y < 0$ for $x < 0$

(ii) $y \to +\infty$ as $x \to 0+$
$y \to -\infty$ as $x \to 0-$

(iii) $y \to 0+$ as $x \to +\infty$
$y \to 0-$ as $x \to -\infty$

(iv) $\dfrac{dy}{dx} = -\dfrac{c}{x^2}$ which is always negative

(v) $\dfrac{dy}{dx} \to -\infty$ as $x \to 0$ $\dfrac{dy}{dx} \to 0$ as $x \to \infty$.

11.10 VARIATIONS ON THE BASIC CURVE

The curve can be shifted up and down, or sideways, by altering the basic equation, although it does not then assume such a simple form (rotation will not be considered here).

Suppose that the whole curve is moved upwards a distance a relative to the axes. Whatever value y took for any x-value previously, then it must

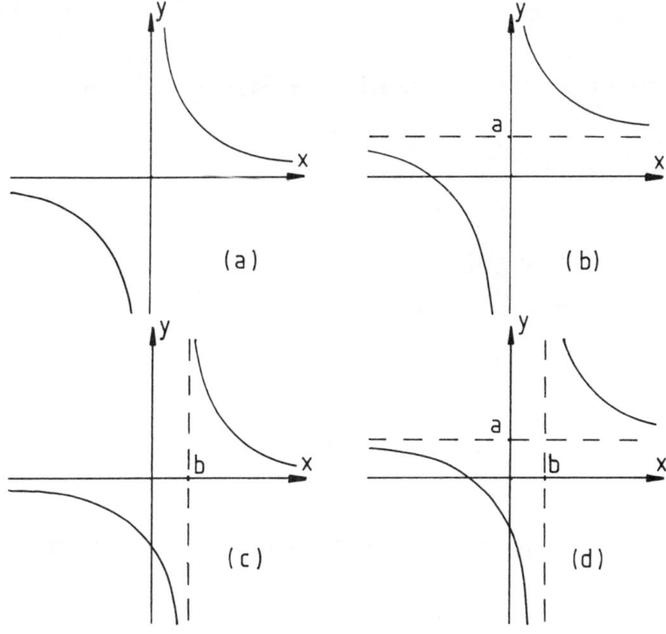

Figure 11.6 Variations on the basic hyperbolic curve. (a) $xy = c$ (standard form), (b) $x(y - a) = c$, (c) $(x - b)y = c$, (d) $(x - b)(y - a) = c$

now take a value larger by a (Figure 11.6(b)). Hence $y = c/x$ becomes $y = c/x + a$

$$\therefore (y - a)x = c$$

is the new form.

Now suppose that the original curve ($xy = c$) is moved to the right by a distance b. In this case every x-value must increase by b (Figure 11.6(c)). Hence $x = c/y$ becomes $x = c/y + b$

$$\therefore (x - b)y = c$$

is the new form.

If the curve is moved to the right by b, and then upwards by a (Figure 11.6(d)) then

$$y = \frac{c}{x - b} \quad \text{becomes} \quad y = \frac{c}{x - b} + a$$

$$\therefore (y - a)(x - b) = c$$

is the new form.

Therefore the general form of the rectangular hyperbola is

$$(x - b)(y - a) = c \quad \text{or} \quad xy - ax - by = d$$

where the asymptotes are given by $x = b$, $y = a$.

Example

Find the asymptotes of the hyperbola $xy - 3x - 4y + 6 = 0$.

If

$$xy - ax - by = d$$

and

$$xy - 3x - 4y + 6 = 0$$

then $a = 3$ and $b = 4$. Now

$$(x - 4)(y - 3) = xy - 4x - 3y + 12.$$

Hence

$$xy - 4x - 3y + 6 = (x - 4)(y - 3) - 6 = 0.$$

$$\therefore (x - 4)(y - 3) = 6$$

is the equivalent equation of the hyperbola, which has asymptotes $x = 4$, $y = 3$ and constant 6.

11.11 USE OF THE HYPERBOLA

When it is desired to employ a hyperbolic curve in a model, the exact choice will clearly depend on the type of problem being considered.

Nevertheless, the choice will be partly affected by the following considerations.

The form $xy = c$ has only one constant to determine (c) and so only one choice is possible for any point (x, y).

The form $(y - a)x = c$ and $y(x - b) = c$ both allow the choice of two parameters (a and c, or b and c), while the form $(x - b)(y - a) = c$ allows three parameters: a, b, and c.

In each case the positions of the asymptotes will have some bearing on the choice. For example, as $x \to 0$ do we wish $y \to \infty$ or $y \to$ finite constant?

If $y \to \infty$ as $x \to 0$ then one asymptote is the y-axis and $b = 0$.

If $y \to d$ as $x \to 0$ then the y-axis is no longer an asymptote and $b \neq 0$.

11.12 BASIC EXAMPLES FOR THE STUDENT

B11.1 Sketch the curve $y = \mathrm{e}^x$. Now, on the same graph sketch the curves $y = a^x$ and $y = b^x$ where $a > \mathrm{e}$ and $1 < b < \mathrm{e}$. Where do the curves intersect?

B11.2 From your answers for question B11.1, sketch the curves $y = \log_a x$ and $y = \log_b x$ where, again, $a > \mathrm{e}$ and $1 < b < \mathrm{e}$. Where do the curves intersect?

If $\log_p x > \log_q x$ when $x > 1, p > 1, q > 1$, what can be deduced about p and q?

B11.3 Sketch the curves

(a) $y = A(1 - \mathrm{e}^{-x})$ (c) $y = (\mathrm{e}^x + \mathrm{e}^{-x})(\mathrm{e}^x - \mathrm{e}^{-x})$

(b) $y = \frac{1}{2}(\mathrm{e}^x + \mathrm{e}^{-x})$ (d) $y = \log(x^3)$.

B 11.4 Differentiate the functions

(a) $y = \mathrm{e}^{3x^4}$ (c) $y = \log_\mathrm{e}[(1 + x)/(1 - x)]$

(b) $y = \exp\sqrt{x^2 - 1}$ (d) $y = \log_\mathrm{e}(x + \sqrt{1 + x^2})$.

B 11.5 Sketch the functions

(a) $y = \dfrac{2x + 4}{x + 1}$ (c) $xy + 2x + 3y = 0$

(b) $y = -\dfrac{2x + 6}{x}$ (d) $xy = 0$

11.13 INTERMEDIATE EXAMPLES FOR THE STUDENT

I11.1 Sketch, on the same diagram, the functions

$$y = \log_\mathrm{e} x, \qquad y = \log_\mathrm{e} x^2, \quad \text{and} \quad y = \log_\mathrm{e}(1/x).$$

I11.2 Show that

$$\frac{(\log_\mathrm{e} x)}{(\log_{10} x)} = c \text{ (a constant)}$$

and find the value of this constant in terms of e and 10.

Why do we use logarithms to base 10 for calculations but logarithms to base e in calculus?

I11.3 Show that a^x can be expressed as the sum of an infinite series in powers of x, where a is a positive constant. (You may *assume* a series for e^x.)

I11.4 Differentiate the functions below

(a) $y = \log_e(x - \sqrt{1 + x^2})$

(b) $y = \log_e \sqrt[43]{1 + x^2}$ (i.e. forty-third root)

(c) $y = \exp(x^3 - 2x^2 + x + 6)$

(d) $y = \sqrt{e^x}$

(e) $y = \dfrac{1}{(x - a)(x - b)(x - c)(x - d)}$

(f) $y = a^x$.

I11.5 Sketch the functions below

(a) $y = \dfrac{x + 3}{x + 6}$

(b) $y = \log_e(-x)$

(c) $y = A(1 - e^{-Bx})$ (indicate A and B on your sketch).

CHAPTER 12

Calculus: Integration I

12.1 INTRODUCTION

Calculus was introduced, in chapters 7 and 8, through the technique of *differentiation*. In this and the following chapter, calculus technique will be extended to include *integration*. In fact the extension is only a limited one, since integration is essentially *the process of differentiation in reverse*. Since this is the case, the topic will be introduced in just that way—as anti-differentiation—and the applications introduced subsequently.

12.2 ANTI-DIFFERENTIATION

Suppose $y = x^3 - 3x + 2$, so that $dy/dx = 3x^2 - 3$. It follows that if $dy/dx = 3x^2 - 3$, then $y = x^3 - 3x + \text{constant}$, which is essentially a process of anti-differentiation, or differentiation in reverse. From a study of this example, the following points can be made:

(a) As for differentiation, the anti-differentiation process can be applied term by term. $(3x^2)$ gives the anti-derivative (x^3), and (-3) gives the anti-derivative $(-3x)$; so $(3x^2 - 3)$ gives the anti-derivative $(x^3 - 3x)$.

(b) We cannot determine the value of the constant in the anti-derivative. Both the functions $y = x^3 - 3x + 2$ and $y = x^3 - 3x - 16$ give the same derivative: $dy/dx = 3x^2 - 3$. Hence when the process is reversed we cannot specify the final constant (+2, −16, or anything else) unless further information is provided on $y(x)$.

(c) When $dy/dx = 3x^2$, $y = x^3(+c)$. In general, if

$$\frac{dy}{dx} = ax^n \quad \text{then} \quad y = \frac{ax^{n+1}}{n+1} \quad (+c).$$

(d) Similarly, if

$$\frac{dy}{dx} = \frac{a}{x} \quad \text{then} \quad y = a\log_e x(+c)$$

and if

$$\frac{dy}{dx} = a\,e^x \quad \text{then} \quad y = a\,e^x(+c)$$

which both follow from the rules for differentiating $y = \log_e x$ and $y = e^x$.

Example

If

$$\frac{dy}{dx} = 3x^4 + 4x^3 + 6x^2 - 2x + 10 \tag{12.1}$$

then

$$y = 3\frac{x^5}{5} + 4\frac{x^4}{4} + 6\frac{x^3}{3} - 2\frac{x^2}{2} + 10x + c$$

$$\therefore y = \frac{3}{5}x^5 + x^4 + 2x^3 - x^2 + 10x + c.$$

Check by differentiation,

$$\frac{dy}{dx} = \frac{3}{5}5x^4 + 4x^3 + 2.3.x^2 - 2.x^1 + 10$$

$$\therefore \frac{dy}{dx} = 3x^4 + 4x^3 + 6x^2 - 2x + 10 \tag{12.2}$$

Equations (12.1) and (12.2) are seen to be identical.

Examples for the Student

Given the following functions for dy/dx, find the corresponding functions $y(x)$.

(a) $\dfrac{dy}{dx} = x^3 + \dfrac{1}{x}$

(b) $\dfrac{dy}{dx} = e^x + \dfrac{4}{x^2} + 2$

(c) $\dfrac{dy}{dx} = \sqrt{x} + \dfrac{3}{x^4}$

(d) $\dfrac{dy}{dx} = \dfrac{6}{x} + \dfrac{4}{\sqrt{x}} + 6x^2$

12.3 INTEGRATION AS A PROCESS OF SUMMATION

At point A, the value of the y-coordinate is given by

$$y = \Delta y_1 + \Delta y_2 + \Delta y_3 + \ldots + \Delta y_5 + c = \sum_{i=1}^{5} \Delta y_i + c$$

(Figure 12.1).

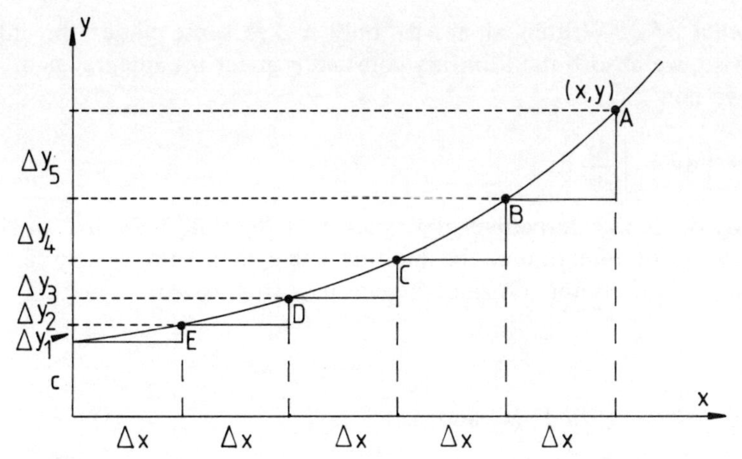

Figure 12.1 A curve cut into sections of equal width (in the x-direction)

Suppose that our information on the curve is in terms of *gradients*, rather than absolute values of y. We may therefore prefer to write

$$\Delta y_1 = \frac{\Delta y_1}{\Delta x}\,\Delta x,\ \ \Delta y_2 = \frac{\Delta y_2}{\Delta x}\,\Delta x,\ \ \Delta y_3 = \frac{\Delta y_3}{\Delta x}\,\Delta x \text{ etc.}$$

since we have introduced the gradients of the chords AB, BC, CD, etc. ($\Delta y_1/\Delta x$, $\Delta y_2/\Delta x$, $\Delta y_3/\Delta x$, etc.). Hence

$$y = \sum_{i=1}^{5} \frac{\Delta y_i}{\Delta x} \cdot \Delta x + c. \tag{12.3}$$

However, gradients of the *chords* are less convenient than the gradient of the *curve* at each point (i.e. $\mathrm{d}y/\mathrm{d}x$). Also such a formula as (12.3) can only apply *at* points A, B, C, D, etc., and not at intermediate points. The solution to these objections is to reduce the size of the intervals, and let $\Delta x \to 0$, or the number of intervals $n \to \infty$.

$$\therefore y = \lim_{\substack{\Delta x \to 0 \\ \text{or } n \to \infty}} \sum_{i=1}^{n} \frac{\Delta y_i}{\Delta x} \cdot \Delta x + c.$$

We write

$$\lim_{\Delta x \to 0} \left(\frac{\Delta y_i}{\Delta x}\right) = \frac{\mathrm{d}y}{\mathrm{d}x}$$

and

$$\lim_{\Delta x \to 0} (\Sigma \Delta x) = \int \mathrm{d}x.$$

The integral sign $\int$, a long S, is the limiting value of a sum, and $\mathrm{d}x$ is the *differential of x*. Written alone, dx only makes sense under the integral sign. Also, we absorb the arbitrary constant c under the integral sign.

So, we now have

$$y = \int dy = \int \frac{dy}{dx}\,dx.$$

Since dy/dx is the derivative of y, then y is the anti-derivative of dy/dx. The process of integration, the limiting value of a sum, is therefore the same as anti-differentiation since it gives us y from dy/dx.

Example

If $dy/dx = 2x - 1$, find $y(x)$ and sketch it.

Solution: $y = \int \frac{dy}{dx}\,dx = \int (2x - 1)dx = 2\,\frac{x^2}{2} - 1\cdot\frac{x^1}{1} + c$

$\therefore \underline{y = x^2 - x + c}$ (a parabola).

It can be seen from Figure 12.2 that the arbitrary value of c produces uncertainty in the *position*, but not in the *shape*, of the curve $y(x)$. For different values of c, the curve shifts in the y-direction (and for this particular example, c is the y-intercept; this is not always so). This sort of uncertainty is not surprising since our original information was $dy/dx = 2x - 1$, that is, the gradient as a function of x. Each of the three curves illustrated in Figure 12.2 has the same gradient at the same value of x (e.g. $dy/dx = 0$ for $x = 0{\cdot}5$ for all three).

Notice that the constant c emerges when the actual integration process (anti-differentiation) is carried out.

In this example, the function $(2x - 1)$ is the *integrand*.

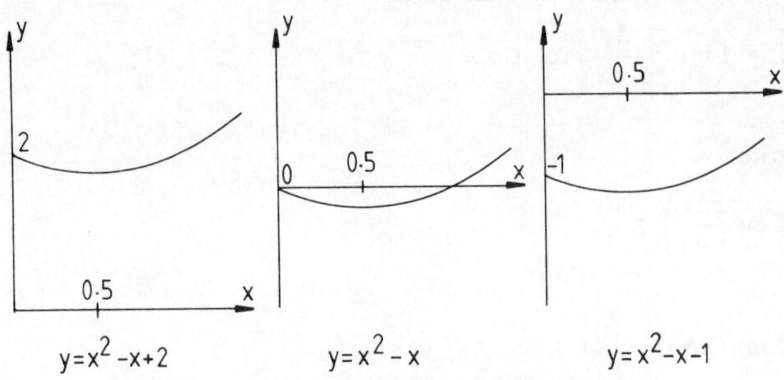

Figure 12.2 3 parabolas with the same form for $y'(x)$

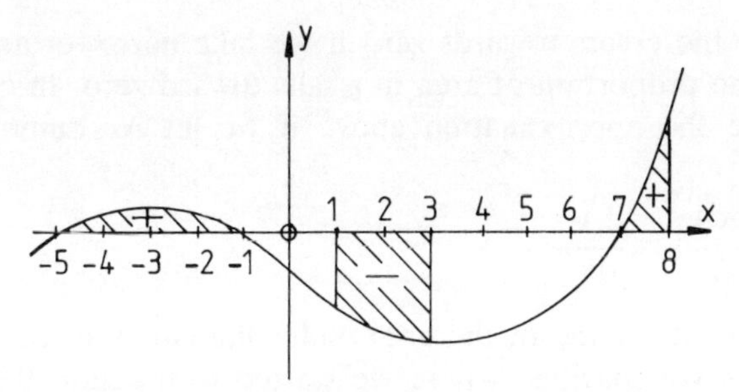

Figure 12.3 The area under a curve for the 3 ranges: $-5 \leqslant x \leqslant -1$, $1 \leqslant x \leqslant 3$, and $7 \leqslant x \leqslant 8$

12.4 THE AREA UNDER A CURVE

One of the most important uses of integration is in the determination of the geometric area underneath a curve. By *underneath* we mean between the curve and the x-axis as in Figure 12.3, where the convention of positive and negative areas is illustrated.

To find the area underneath a curve, we first make the approximation illustrated in Figure 12.4 of dividing the area into a series of vertical rectangular strips, one of which has been shaded in Figure 12.4(a) and enlarged in Figure 12.4(b). The area underneath the curve is then approximately equal to the sum of the small elements (or increments) of area within each rectangle. This equality is only approximate because of the small errors near the curve itself, marked as ϵ in Figure 12.4(b).

So, as a first approximation, the shaded element of area ΔA is given by

$$\Delta A \cong y\Delta x$$

$$\therefore \quad y \cong \frac{\Delta A}{\Delta x}.$$

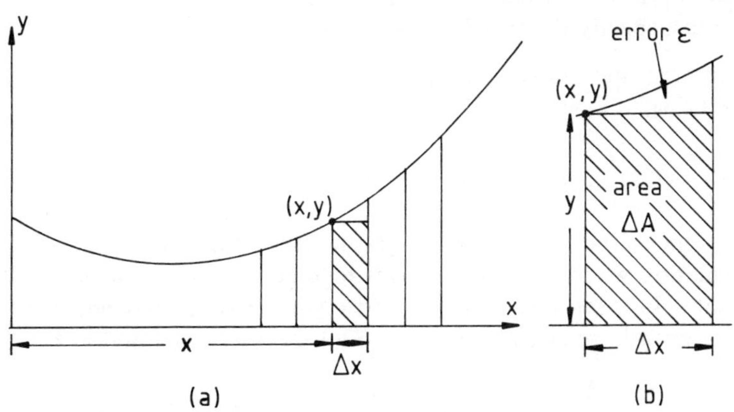

Figure 12.4 First approximation to the area under a curve

We can reduce the errors towards zero if we take narrower and narrower strips so that the proportion of area in ε falls toward zero. In other words, we can remove the approximation above if we let Δx diminish towards zero.

$$\therefore y = \lim_{\Delta x \to 0} \left(\frac{\Delta A}{\Delta x} \right) = \frac{\mathrm{d}A}{\mathrm{d}x}$$

where $A(x)$ gives the value of the area under the curve as a function of x. We have not as yet specified where we started to measure the area from (e.g. $x = 0$, or $x = 3$, or . . .) and this gives rise to the ubiquitous arbitrary constant c. We will sort out this point below.

However, from Section 12.3 we can write

$$A = \int \mathrm{d}A = \int \frac{\mathrm{d}A}{\mathrm{d}x} \, \mathrm{d}x$$

and therefore

$$A = \int \frac{\mathrm{d}A}{\mathrm{d}x} \, \mathrm{d}x = \int y \, \mathrm{d}x.$$

So, the area underneath the curve $y = y(x)$ is given simply by the integral of y (or anti-derivative of y) with respect to x.

Example

Find the area underneath the curve $y = x^2 - x + 2$

Solution: $A = \int y \, \mathrm{d}x = \int (x^2 - x + 2) \, \mathrm{d}x = \dfrac{x^3}{3} - \dfrac{x^2}{2} + 2x + c$

But this gives us
(i) a function of x—what is x?
(ii) an arbitrary constant c—what is c?
So, we must specify the problem more carefully, thus:

Find the area underneath the curve $y = x^2 - x + 2$ in the range $1 \leqslant x \leqslant 3$ (Figure 12.5).

Solution: $A = \displaystyle\int_1^3 y \, \mathrm{d}x = \int_1^3 (x^2 - x + 2) \, \mathrm{d}x = \left[\frac{x^3}{3} - \frac{x^2}{2} + 2x + c \right]_1^3.$

Area is measured from left to right (this follows from our earlier approximation where we added strips together, letting x *increase*). We are not absolutely clear where we started counting from, but when we reach $x = 3$ we must have $A(3) = (x^3/3 - x^2/2 + 2x + c)$ evaluated at $x = 3$.

$$\therefore A(3) = \frac{3^3}{3} - \frac{3^2}{2} + 2.3 + c = 9 - \frac{9}{2} + 6 + c = \frac{21}{2} + c.$$

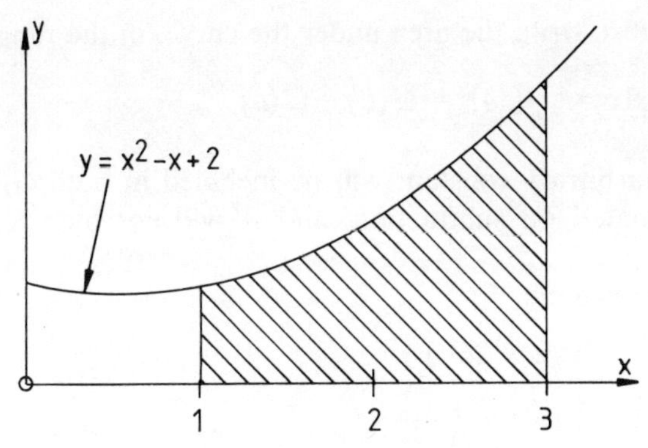

Figure 12.5 The shaded area falls in the range $1 \leqslant x \leqslant 3$

Similarly, when we reached $x = 1$, we must have

$$A(1) = \frac{1^3}{3} - \frac{1^2}{2} + 2.1 + c = \frac{1}{3} - \frac{1}{2} + 2 + c = \frac{11}{6} + c.$$

Hence the area between $x = 1$ and $x = 3$ is given by

$$A(3) - A(1) = \left(\frac{21}{2} + c\right) - \left(\frac{11}{6} + c\right) = \frac{52}{6} = \frac{26}{3}.$$

Hence

$$\int_1^3 y \, \mathrm{d}x = \frac{26}{3}.$$

The form $\int_1^3 y \, \mathrm{d}x$ is called the *definite integral* and is a number, whose value depends on the form of the integrand y; while the form $\int y \, \mathrm{d}x$ is called the *indefinite integral*, and is a function of x, including an arbitrary constant c.

12.5 THE DEFINITE INTEGRAL

In this section we shall write down the results of the previous section in a more general form.

Suppose we have a function $y(x)$ and wish to determine the area A under the curve between $x = a$ and $x = b$; that is,

$$A = \int_a^b y(x) \, \mathrm{d}x \quad (y = x^2 - x + 2 \text{ above}).$$

Also suppose that we know how to integrate (anti-differentiate) the function $y(x)$, and that the indefinite integral of $y(x)$ is $G(x)$; that is,

$$G(x) = \int y(x) \, \mathrm{d}x \quad \text{or} \quad G'(x) = y(x).$$

(In the above example $y = x^2 - x + 2$, and $G(x) = x^3/3 - x^2/2 + 2x + c$.)

We can therefore write the area under the curve, in the range $a \leqslant x \leqslant b$,

$$A = \int_a^b y(x)\,\mathrm{d}x = [G(x)]_a^b = G(b) - G(a).$$

Since the *same* arbitrary constant will be included in both $G(b)$ and $G(a)$, it will be eliminated on subtraction, and A will *not* include an arbitrary constant.

(In the above example,

$$G(3) = 3^3/3 - 3^2/2 + 2.3 + c$$

$$G(1) = 1^3/3 - 1^2/2 + 2.1 + c$$

and

$$A = G(3) - G(1) = 21/2 - 11/6.)$$

It is also easy to show that

$$\int_b^a y\,\mathrm{d}x = -\int_a^b y\,\mathrm{d}x$$

as follows.

$$\int_b^a y\,\mathrm{d}x = [G(x)]_b^a = G(a) - G(b) = -(G(b) - G(a))$$

$$= -[G(x)]_a^b = -\int_a^b y\,\mathrm{d}x.$$

12.6 EXAMPLES

(i) Find the area underneath the curve $y = x^3 - 3x^2 - x + 3$ over the ranges:
  (a) $-1 \leqslant x \leqslant 1$
  (b) $1 \leqslant x \leqslant 3$
  (c) $-1 \leqslant x \leqslant 3$.

The curve is illustrated in Figure 12.6. The cubic function $y = x^3 - 3x^2 - x + 3$ is the integrand.

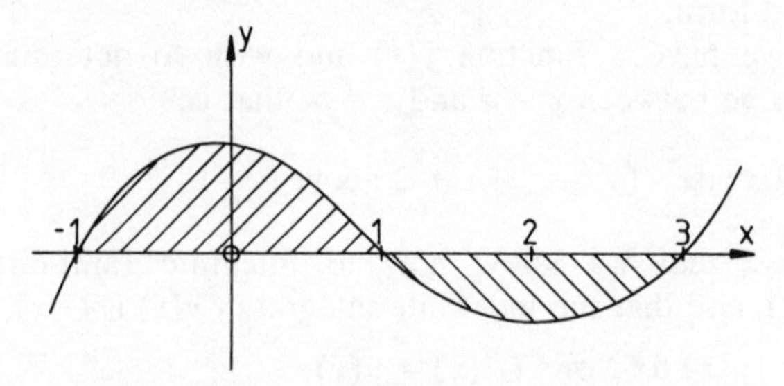

Figure 12.6 $y = x^3 - 3x^2 - x + 3$ (not to scale)

Solution $G(x) = \int y\,dx = \int (x^3 - 3x^2 - x + 3)\,dx$

$$= \frac{x^4}{4} - 3\frac{x^3}{3} - \frac{x^2}{2} + 3x + c \quad \text{(the indefinite integral).}$$

(a) $$\int_{-1}^{1} y\,dx = \left[\frac{x^4}{4} - x^3 - \frac{x^2}{2} + 3x + c\right]_{-1}^{1} = G(1) - G(-1)$$

$$= \left(\frac{1}{4} - 1 - \frac{1}{2} + 3 + c\right) - \left(\frac{1}{4} + 1 - \frac{1}{2} - 3 + c\right)$$

$$= \frac{7}{4} - \left(-\frac{9}{4}\right) = \frac{16}{4} = 4$$

$\therefore$ Area shaded thus [hatched box] is $\int_{-1}^{1} y\,.\,dx = 4$ (square units).

(b) $$\int_{1}^{3} y\,dx = \left[\frac{x^4}{4} - x^3 - \frac{x^2}{2} + 3x + c\right]_{1}^{3} = G(3) - G(1)$$

$$= \left(\frac{81}{4} - 27 - \frac{9}{2} + 9 + c\right) - \left(\frac{1}{4} - 1 - \frac{1}{2} + 3 + c\right)$$

$$= \left(-\frac{9}{4}\right) - \left(\frac{7}{4}\right) = -\frac{16}{4} = -4$$

$\therefore$ Area shaded thus [hatched box] is $\int_{1}^{3} y\,dx = -4$ (square units).

(c) It should now be intuitively obvious that if we calculate the area over the range $-1 \leqslant x \leqslant 3$, we should obtain the result zero, since an area $+4$ occurs in the first section, $-1 \leqslant x \leqslant 1$, and an area -4 occurs over the range $1 \leqslant x \leqslant 3$. In other words, since areas add together in the same way as numbers (*scalar* quantities), we can write

$$\int_{-1}^{3} y\,dx = \int_{-1}^{1} y\,dx + \int_{1}^{3} y\,dx = 4 + (-4) = 0.$$

It is easily shown that any definite integral can be split into two (or more) sections by dividing the range into two parts.

$$\int_{a}^{b} y\,dx = [G(x)]_a^b = G(b) - G(a) = G(b) - G(a) + G(c) - G(c)$$
$$= G(c) - G(a) + G(b) - G(c) = [G(x)]_a^c + [G(x)]_c^b$$

$$\therefore \int_{a}^{b} y\,dx = \int_{a}^{c} y\,dx + \int_{c}^{b} y\,dx.$$

(Range a to b) equals (range a to c) + (range c to b).

This result also enables us to calculate the *arithmetic* area over the range $-1 < x < 3$. By arithmetic area we mean the total area, treating sections beneath the x-axis as positive.

So, the total shaded (arithmetic) area, [hatched box] and [hatched box], is given by

$$\left|\int_{-1}^{1} y\,\mathrm{d}x\right| + \left|\int_{1}^{3} y\,\mathrm{d}x\right| = |\,4\,| + |\,-4\,| = 4 + 4 = 8 \text{ square unit}$$

The algebraic area is

$$\int_{-1}^{3} y\,\mathrm{d}x = \int_{-1}^{1} y\,\mathrm{d}x + \int_{1}^{3} y\,\mathrm{d}x = 4 + (-4) = 0.$$

This last result can be confirmed directly

$$\int_{-1}^{3} y\,\mathrm{d}x = \left[\frac{x^4}{4} - x^3 - \frac{x^2}{2} + 3x + c\right]_{-1}^{3} = G(3) - G(-1)$$

$$= \left(\frac{81}{4} - 27 - \frac{9}{2} + 9 + c\right) - \left(\frac{1}{4} + 1 - \frac{1}{2} - 3 + c\right)$$

$$= \left(-\frac{9}{4}\right) - \left(-\frac{9}{4}\right) = -\frac{9}{4} + \frac{9}{4} = 0.$$

(ii) Find the area enclosed between the two curves

$$y = x^2 - 4x + 5 \quad \text{and} \quad y = -x^2 + 4x - 1.$$

Solution The principle of the solution is illustrated in Figure 12.7. The two curves intersect at A and B and we want the area shaded [III].

This is given by the difference of the two other shaded areas: [I] less [II].

Firstly we must find A and B. At the intersection, $y = x^2 - 4x + 5 = -x^2 + 4x - 1$ (simultaneous equations)

$$\therefore 2x^2 - 8x + 6 = 0$$

$$\therefore x^2 - 4x + 3 = 0$$

$$\therefore x = \frac{4 \pm \sqrt{16 - 12}}{2} \qquad \left(\frac{-b \pm \sqrt{b^2 - 4ac}}{2a}\right)$$

$$\therefore x = \frac{4 \pm \sqrt{4}}{2} = \frac{4 \pm 2}{2} = 3 \text{ or } 1.$$

Hence A and B have x coordinates of 1 and 3 respectively.

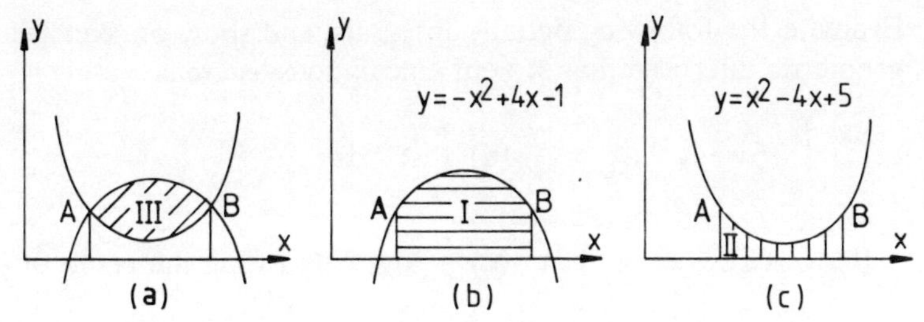

Figure 12.7 Area between two curves, III = I − II

∴ The area I

$$= \int_1^3 (-x^2 + 4x - 1)\,dx = \left[-\frac{x^3}{3} + \frac{4x^2}{2} - x + c\right]_1^3$$

$$= (-9 + 18 - 3 + c) - (-\tfrac{1}{3} + 2 - 1 + c)$$

$$= (6 + c) - \left(\frac{2}{3} + c\right) = \frac{16}{3}$$

and the area II

$$= \int_1^3 (x^2 - 4x + 5) = \left[\frac{x^3}{3} - \frac{4x^2}{2} + 5x + c\right]_1^3$$

$$= (9 - 18 + 15 + c) - (\tfrac{1}{3} - 2 + 5 + c)$$

$$= (6 + c) - (10/3 + c) = \tfrac{8}{3}$$

Hence, the area III = area I − area II

$$= \frac{16}{3} - \frac{8}{3} = \frac{8}{3}.$$

If any part of area III had fallen below the x-axis, we would have had to consider the question more carefully. The above method would have treated this section as *negative*. If we had wished all the area to be treated as positive (the arithmetic area) then the integral for area II (or perhaps I) would have had to be split into several portions, in much the same way as in the example in Section 12.5.

12.7 BASIC EXAMPLES FOR THE STUDENT

B12.1 Find the anti-derivatives of the following functions

(a) $2x^3 + \dfrac{3}{x^2} + e^x$ (c) $\dfrac{x^2 - 1}{x + 1}$

(b) $\dfrac{1}{x} + \sqrt{x} + e^{-x}$ (d) $\dfrac{1}{x^3} + \dfrac{1}{x^2} + \dfrac{1}{x} + \dfrac{1}{\sqrt{x}}$.

B12.2 Evaluate the following definite integrals, and show on sketches the geometric interpretation of your calculations as areas

(a) $\int_0^6 (6x - x^2)\,dx$

(b) $\int_0^6 (6x - x^2 - 5)\,dx$

(c) $\int_0^6 (6x - x^2 - 9)\,dx$

(d) $\int_0^\infty e^{-x}\,dx$

(e) $\int_0^1 \log_e x\,dx$ (using the result of part (d)).

B12.3 The marginal cost of production for a firm's product is given by

$$MC = 0{\cdot}02x + 10$$

where x is the production level. Fixed costs are 75 (in suitable units). Find an expression for the total costs as a function of x. Sketch the curves representing total costs and marginal costs. How can

(i) marginal costs be interpreted geometrically on the total cost curve, and

(ii) total costs be interpreted geometrically on the marginal cost curve?

12.8 INTERMEDIATE EXAMPLES FOR THE STUDENT

I12.1 Evaluate the following integrals

(a) $\int (x^2 + x + 2)\,dx$

(b) $\int \left(\frac{1}{x^2} + \frac{1}{x} + \frac{1}{2}\right) dx$

(c) $\int \left(\sqrt{x} + \frac{1}{\sqrt{x}}\right) dx$

(d) $\int dx$

(e) $\int (1 + e^{-x})^2\,dx$

(f) $\int_1^\infty \frac{dx}{x^3}$

(g) $\int_{-1}^1 \frac{dx}{x^3}$

(h) $\int_{-1}^1 \left(e^x + \frac{1}{e^x}\right) dx$

(i) $\int_1^e \frac{dx}{x}$

(j) $\int_0^4 (x^3 - 6x^2 + 8x)\,dx$

I12.2 Find the equation of the curve satisfying the conditions in each of the following

(a) The gradient at (x, y) is $3x^2$. The curve passes through $(1, 5)$.

(b) The rate of change of the value of the function y with respect to the variable x is $(2x - 5)$. Also, when $x = -3$, $y = -3$.

(c) The slope is zero at every point. Also $y = 5$ when $x = 2$.

I12.3 A business organisation found from an analysis of its production methods that with the present equipment and no additional workers the capacity production level is 10 000 units per day. They also estimated that the rate of change of production level P could be represented by

$$\frac{dP}{dx} = 200 - 3\sqrt{x}$$

where x is the number of additional workers over and above their present workforce. What would their capacity production level be with an additional 25 workers? Sketch a graph of production level P versus x.

CHAPTER 13

Calculus: Integration II

13.1 INTRODUCTION

The previous chapter introduced the topic of integration and its graphical interpretation. This chapter is concerned with the principal methods of integration which are available, and with a few more applications of the technique.

When integration was first discussed it was treated as being essentially a process of anti-differentiation. Two observations can then be made from this viewpoint.

(i) If we wish to find the integral of a function $y(x)$, for example

$$\int y(x)\,\mathrm{d}x = G(x)$$

then we are searching for a function $G(x)$, which, on differentiation, gives us $y(x)$

$$\frac{\mathrm{d}G(x)}{\mathrm{d}x} = y(x).$$

So the key problem in integration is really that of rearranging $y(x)$ into such a form that we can spot, or guess, the appropriate function $G(x)$.

(ii) While rules exist to enable us to differentiate *any* well-behaved function, this is not the case with integration. The function $y(x)$ can only be integrated if it is the result of a process of differentiation; that is, only if

$$y(x) = \frac{\mathrm{d}G(x)}{\mathrm{d}x} \quad \text{(a first derivative)}$$

can we find

$$\int y(x)\,\mathrm{d}x = \int \frac{\mathrm{d}G(x)}{\mathrm{d}x}\,\mathrm{d}x = G(x).$$

It should be noted that we are talking here about *indefinite* integrals. Definite integrals can be interpreted as areas, and these can be calculated using numerical rather than analytical methods.

The first step, therefore, is to identify and tabulate the types of functions which are actually first derivatives of the form $\mathrm{d}G/\mathrm{d}x$ and which can therefore be integrated (to give $G(x)$).

13.2 SIMPLE FUNCTIONS

(i) Powers of x

If we have a function of the form

$$G(x) = \frac{ax^{n+1}}{(n+1)}$$

where n is a constant (+ve, −ve, including fractions), then

$$\frac{\mathrm{d}G}{\mathrm{d}x} = \frac{a(n+1)}{(n+1)} x^n = ax^n = y(x).$$

Hence all powers of x are of the form $\mathrm{d}G/\mathrm{d}x$, and we can write

$$\int \frac{\mathrm{d}G}{\mathrm{d}x}\,\mathrm{d}x = \int ax^n\,\mathrm{d}x = \frac{ax^{n+1}}{(n+1)} + c. \tag{13.1}$$

(ii) The function $y(x) = 1/x$

The exception to the above section is the function $y = x^{-1}$, since this cannot be generated by differentiating another power of x. However, it can be generated by differentiating $\log_e x$; that is, if

$$G(x) = a \log_e x$$

then

$$\frac{\mathrm{d}G}{\mathrm{d}x} = \frac{a}{x} = y(x)$$

$$\therefore \int \frac{\mathrm{d}G}{\mathrm{d}x}\,\mathrm{d}x = \int \frac{a}{x}\,\mathrm{d}x = a \log_e x + c. \tag{13.2}$$

(iii) The function $y(x) = a\mathrm{e}^x$

If

$$G(x) = a\mathrm{e}^x$$

then

$$\frac{\mathrm{d}G}{\mathrm{d}x} = a\mathrm{e}^x = y(x)$$

$$\therefore \int \frac{\mathrm{d}G}{\mathrm{d}x}\,\mathrm{d}x = \int a\mathrm{e}^x\,\mathrm{d}x = a\mathrm{e}^x + c. \tag{13.3}$$

Examples

(i) $\int 8\sqrt{x}\,\mathrm{d}x = \int 8x^{1/2}\,\mathrm{d}x = \dfrac{8x^{3/2}}{(3/2)} + c = \dfrac{8.2}{3}x^{3/2} + c = \dfrac{16}{3}x^{3/2} + c.$

(ii) $\int \dfrac{5}{\sqrt{x}}\,\mathrm{d}x = \int 5x^{-1/2}\,\mathrm{d}x = \dfrac{5x^{1/2}}{(1/2)} + c = 5.2x^{1/2} + c = 10\sqrt{x} + c.$

(iii) $\int \left(x^3 + \dfrac{1}{x} + \dfrac{1}{x^2} + \mathrm{e}^x\right)\mathrm{d}x = \int (x^3 + x^{-1} + x^{-2} + \mathrm{e}^x)\,\mathrm{d}x$

$$= \frac{x^4}{4} + \log_e x + \frac{x^{-1}}{(-1)} + \mathrm{e}^x + c = \frac{1}{4}x^4 + \log_e x - \frac{1}{x} + \mathrm{e}^x + c.$$

13.3 CHAIN RULE IN REVERSE

There are two ways of approaching the chain rule reversed. In this section we shall focus on the identification of those types of function which, as *outputs of differentiation* by the chain rule, can be integrated directly by reversing the rule. In the next section we show how substitution methods give the same results. Suppose that B is any function of x which we can differentiate straightforwardly; for example,

$$B = x^3 - 2x + \log_e x + \mathrm{e}^x + \frac{1}{x}, \quad \frac{\mathrm{d}B}{\mathrm{d}x} = 3x^2 - 2 + \frac{1}{x} + \mathrm{e}^x - \frac{1}{x^2}.$$

We use B here, as it reminds one that usually such an expression will be enclosed in a bracket; for example,

$$\log_e(x^3 - 2x + \log_e x + \mathrm{e}^x + 1/x) \quad \text{or} \quad (x^3 - 2x + \log_e x + \mathrm{e}^x + 1/x)^3$$

or

$$\exp(x^3 - 2x + \log_e x + \mathrm{e}^x + 1/x) \quad \text{or} \quad \sqrt{(x^3 - 2x + \log_e x + \mathrm{e}^x + 1/x)}.$$

The chain rule for differentiation gives us the following results:

If $y = B^n$ then $\dfrac{\mathrm{d}y}{\mathrm{d}x} = nB^{n-1}\dfrac{\mathrm{d}B}{\mathrm{d}x}$.

If $y = \log_e B$ then $\dfrac{\mathrm{d}y}{\mathrm{d}x} = \dfrac{1}{B}\dfrac{\mathrm{d}B}{\mathrm{d}x}$.

If $y = \mathrm{e}^B = \exp(B)$ then $\dfrac{\mathrm{d}y}{\mathrm{d}x} = \mathrm{e}^B\dfrac{\mathrm{d}B}{\mathrm{d}x}$.

By studying the forms of the expressions on the right we can therefore integrate functions of the type

$$B^n\frac{\mathrm{d}B}{\mathrm{d}x}, \quad \frac{1}{B}\frac{\mathrm{d}B}{\mathrm{d}x} \quad \text{and} \quad \mathrm{e}^B\frac{\mathrm{d}B}{\mathrm{d}x}.$$

In fact, we have the rules:

$$\int B^n \frac{dB}{dx}\,dx = \frac{B^{n+1}}{(n+1)} + c \tag{13.4}$$

$$\int \frac{1}{B}\frac{dB}{dx}\,dx = \log_e B + c \tag{13.5}$$

and

$$\int e^B \frac{dB}{dx}\,dx = e^B + c. \tag{13.6}$$

Examples

(i) $\displaystyle\int \sqrt{(x^3 - 4x^2 + 8)(3x^2 - 8x)}\,dx = \int (x^3 - 4x^2 + 8)^{1/2}(3x^2 - 8x)\,dx.$

Here,

$$B = (x^3 - 4x^2 + 8), \quad \frac{dB}{dx} = (3x^2 - 8x) \quad \text{and} \quad n = \tfrac{1}{2}$$

$$\therefore \int B^{1/2}\frac{dB}{dx}\,dx = \frac{B^{3/2}}{(3/2)} + c = \underline{\tfrac{2}{3}(x^3 - 4x^2 + 8)^{3/2} + c}.$$

(ii) $\displaystyle\int \frac{3x^5 + 2x^3 + x}{(x^6 + x^4 + x^2)}\,dx = \frac{1}{2}\int \frac{6x^5 + 4x^3 + 2x}{x^6 + x^4 + x^2}\,dx.$

Here

$$B = (x^6 + x^4 + x^2), \quad \frac{dB}{dx} = 6x^5 + 4x^3 + 2x$$

$$\therefore \frac{1}{2}\int \frac{1}{B}\frac{dB}{dx}\,dx = \tfrac{1}{2}\log_e B + c = \underline{\tfrac{1}{2}\log_e(x^6 + x^4 + x^2) + c.}$$

(iii) $\displaystyle\int \left(\frac{1}{\sqrt{x}} + \frac{2}{x}\right)\exp(\sqrt{x} + \log_e x)\,dx = 2\int \left(\frac{1}{2}x^{-1/2} + \frac{1}{x}\right)\exp(x^{1/2} + \log_e x)\,dx.$

Here

$$B = (x^{1/2} + \log_e x), \quad \frac{dB}{dx} = \frac{1}{2}x^{-1/2} + \frac{1}{x}$$

$$\therefore 2\int \frac{dB}{dx} e^B\,dx = 2e^B + c = \underline{2\exp(x^{1/2} + \log_e x) + c}$$

The key difficulty, therefore, is in spotting that the integral can be pushed into one of the three basic forms (13.4), (13.5), or (13.6).

13.4 SUBSTITUTION METHODS

It will be remembered that when we introduced integration we used the result

$$y = \int dy = \int \frac{dy}{dx}\, dx.$$

It can be seen that under the integral sign, the dy/dx behaves rather like a vulgar fraction. This is generally true, and so the three basic rules (13.4), (13.5), and (13.6) can be slightly changed to give results analogous to the simple functions (13.1), (13.2), and (13.3).

$$(13.4) \quad \int B^n \frac{dB}{dx}\, dx = \int B^n\, dB = \frac{B^{n+1}}{n+1} + c \qquad \text{c.f.} \int x^n dx \quad (13.1)$$

$$(13.5) \quad \int \frac{1}{B}\frac{dB}{dx}\, dx = \int \frac{1}{B}\, dB = \log_e B + c \qquad \text{c.f.} \int \frac{1}{x}\, dx \quad (13.2)$$

$$(13.6) \quad \int e^B \frac{dB}{dx}\, dx = \int e^B\, dB = e^B + c \qquad \text{c.f.} \int e^x dx \quad (13.3)$$

To illustrate the method of substitution, therefore, example (ii) of the previous section will be repeated. Note that substitution is *not*, strictly speaking, another method of integration but is the chain rule in disguise, and applies to the same sorts of functions. However it is often clearer to use, as it allows a 'try it and see' approach. If the correct substitution is made (and this is not usually too difficult) the original function should reduce to a simple form. Judgement as to whether integration may or may not be done can then be based on the simpler functions; for example,

$$\int \frac{3x^5 + 2x^3 + x}{(x^6 + x^4 + x^2)}\, dx \qquad \text{try } B = x^6 + x^4 + x^2$$

so that

$$\frac{dB}{dx} = 6x^5 + 4x^3 + 2x$$

$$\therefore \quad \int \frac{3x^5 + 2x^3 + x}{x^6 + x^4 + x^2}\, dx = \int \frac{\frac{1}{2}\, dB/dx}{B}\, dx = \frac{1}{2}\int \frac{1}{B}\frac{dB}{dx}\, dx = \frac{1}{2}\int \frac{dB}{B}$$

which can be integrated directly to give $\frac{1}{2}\log_e B + c$

$$\therefore \quad \int \frac{3x^5 + 2x^3 + x}{x^6 + x^4 + x^2}\, dx = \tfrac{1}{2}\log_e(x^6 + x^4 + x^2) + c.$$

13.5 INTEGRATION BY PARTS

We have stated earlier that the key problem in integration is to shuffle the original integrand into an *integrable* form. The method of integration by parts is not, strictly, a method of integration at all, but is a way of replacing one integration by another; if the correct choice is made the second integration can be carried out while the original form could not.

From the rule for differentiating a product (Section 8.2)

$$\frac{d}{dx}(uv) = v\frac{du}{dx} + u\frac{dv}{dx}$$

$$\therefore uv = \int\left(v\frac{du}{dx} + u\frac{dv}{dx}\right)dx \text{ (as integration is differentiation in reverse)}$$

$$= \int v\frac{du}{dx}dx + \int u\frac{dv}{dx}dx$$

$$\therefore \int v\frac{du}{dx}dx = uv - \int u\frac{dv}{dx}dx \qquad (13.7)$$

Therefore

(i) the integral $\int v\frac{du}{dx}dx$ has been replaced by the integral $\int u\frac{dv}{dx}dx$

(ii) the original form of the integrand, $v(du/dx)$, suggests that the method is useful for products, though example (ii) below shows that the product is not always obvious.

Examples

(i) $\int xe^x dx$

Let $v = x$ so that $\frac{dv}{dx} = 1.$

Let $\frac{du}{dx} = e^x$ so that $u = e^x$

$$\therefore \int xe^x dx = uv - \int u\frac{dv}{dx}dx$$

$$= xe^x - \int e^x dx = xe^x - e^x + c$$

$$= \underline{e^x(x-1) + c.}$$

(ii) $\int \log_e x\,dx$

Let $v = \log_e x$ so that $\frac{dv}{dx} = \frac{1}{x}.$

Let $\frac{du}{dx} = 1$ so that $u = x$

$$\therefore \int \log_e x \, dx = uv - \int u \frac{dv}{dx} dx$$

$$= x \log_e x - \int x \frac{1}{x} dx$$

$$= x \log_e x - \int 1 \, dx$$

$$= x \log_e x - x + c.$$

(iii) $\int x^3 \sqrt{x^2 + 3} \, dx$

Let $v = x^2$, so that $\dfrac{dv}{dx} = 2x$.

Let $\dfrac{du}{dx} = x(x^2 + 3)^{1/2}$, so that by rule (13.4) $u = \frac{1}{3}(x^2 + 3)^{3/2}$

$$\therefore I = \int x^3 \sqrt{x^2 + 3} \, dx = \int x^2 x (x^2 + 3)^{1/2} \, dx$$

$$= uv - \int u \frac{dv}{dx} dx$$

$$= \frac{1}{3} x^2 (x^2 + 3)^{3/2} - \int \frac{2}{3} x (x^2 + 3)^{3/2} \, dx$$

and this second integration can be carried out using rule (13.4) again

$$\therefore I = \frac{1}{3} x^2 (x^2 + 3)^{3/2} - \frac{2}{15} (x^2 + 3)^{5/2} + c$$

$$= (x^2 + 3)^{3/2} \left(\frac{1}{3} x^2 - \frac{2}{15} (x^2 + 3) \right) + c$$

$$= \frac{1}{5} (x^2 - 2)(x^2 + 3)^{3/2} + c.$$

13.6 THE METHOD OF PARTIAL FRACTIONS

This technique is really little more than a neat trick of algebraic rearrangement to permit integration of functions of the general type

$$\int \frac{ax^2 + bx + c}{(x + d)(x + e)(x + f)} dx$$

(note that the highest power in the numerator must be at least one less than the highest power in the denominator of the integrand). The method will be demonstrated by means of an example.

Find $\int \frac{dx}{6 - x - x^2}$

Solution: $\frac{1}{6 - x + x^2} = \frac{1}{(x + 3)(2 - x)} \equiv \frac{p}{(x + 3)} + \frac{q}{(2 - x)}$

p and q are chosen so that the equality holds for *all* x

$$\therefore \frac{1}{6 - x - x^2} \equiv \frac{p}{(x + 3)} + \frac{q}{(2 - x)} = \frac{p(2 - x) + q(x + 3)}{(x + 3)(2 - x)}$$

$$= \frac{x(q - p) + (3q + 2p)}{6 - x - x^2}.$$

So, for all possible values of x, $(3q + 2p) + x(q - p) \equiv 1$

$$\therefore 3q + 2p = 1 \quad \text{and} \quad q - p = 0$$

$$\therefore p = q = \frac{1}{5}$$

$$\therefore \frac{1}{6 - x - x^2} = \frac{1}{5(x + 3)} + \frac{1}{5(2 - x)}$$

$$\therefore \int \frac{dx}{6 - x - x^2} = \frac{1}{5}\int \frac{dx}{(x + 3)} + \frac{1}{5}\int \frac{dx}{(2 - x)}$$

$$= \frac{1}{5}\log_e(x + 3) - \frac{1}{5}\log_e(2 - x) + c$$

$$= \underline{\frac{1}{5}\log_e\left(\frac{x + 3}{2 - x}\right) + c} \quad \text{(N.B. } \log\frac{A}{B} = \log A - \log B\text{)}.$$

13.7 THE USE OF INTEGRATION IN STATISTICAL CALCULATIONS

In statistics we are often concerned with a *random variable*, X say, which can take any value x over some range (which may be infinite in extent).

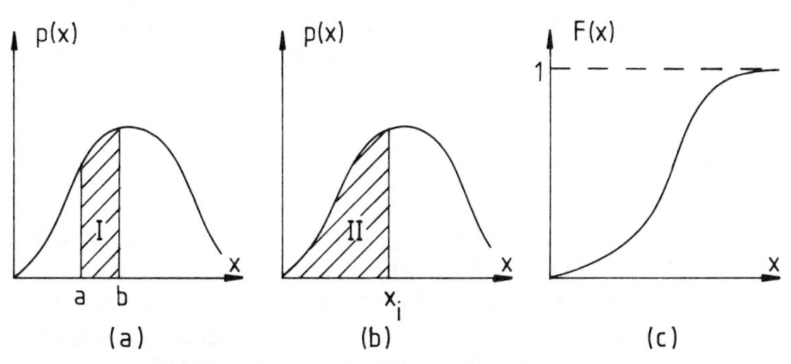

Figure 13.1 Probability density functions

The probability of its taking any value x varies with x however, and this fact is most easily illustrated on a graph showing the probability distribution. We define the probability that the variable will take a value between any two specified limits, a and b say, to be the area underneath the probability distribution function within the range $a \leqslant x \leqslant b$ (Figure 13.1(a)).

$$\therefore P(a \leqslant X \leqslant b) = \int_a^b p(x)\,\mathrm{d}x = \text{area I in Figure 13.1(a).}$$

Here $p(x)$ is the probability distribution function which we have taken to be *normalized*, i.e. the *total* area under the curve is arranged to be *one* by appropriate adjustment of the y-scale; that is,

$$\int_{-\infty}^{\infty} p(x)\,\mathrm{d}x = 1.$$

It follows that the probability that the random variable X will take any value up to and including x is given by $F(x)$ where

$$F(x_i) = P(X \leqslant x_i) = \int_{-\infty}^{x_i} p(x)\,\mathrm{d}x = \text{area II in Figure 13.1(b).}$$

$F(x)$ can be represented either by the shaded area II in Figure 13.1(b), or directly as a function as in Figure 13.1(c). Since the total area under the curve $p(x)$ is unity, then $F(x_i) \rightarrow 1$ as $x_i \rightarrow \infty$; that is,

$$\lim_{x_i \to \infty} F(x_i) = \lim_{x_i \to \infty} \int_{-\infty}^{x_i} p(x)\,\mathrm{d}x = \int_{-\infty}^{\infty} p(x)\,\mathrm{d}x = 1.$$

$F(x)$ is called the *cumulative probability function* (or cumulative distribution function, cumulative density function, or cumulative probability density function).

13.8 EXPECTED VALUES

If we have observed N values $(x_1, x_2, x_3, \ldots, x_N)$ for a measurable quantity X (perhaps a random variable, such as a height of a person), then the *mean* value, $\bar{x}$ (or arithmetic average), of these observations is found by taking the sum of all these values, and dividing this sum by the number of observations. So, we define $\bar{x}$ by

$$\bar{x} = \frac{x_1 + x_2 + x_3 + \ldots + x_N}{N} = \frac{1}{N}\sum_{i=1}^{N} x_i.$$

Suppose that some measurements appear several times in our series of observations. Rather than adding them in individually, we could add them in as a group. For example, if the height 1·8 m appeared seven times in the list, it would make sense to add in 7 × 1·8 m once, rather than add in 1·8 m seven times. More generally, if reading x_i is observed f_i times in a series of N measurements in which the values $x_1, x_2, x_3, \ldots, x_n$ appear, then

$$\bar{x} = \frac{f_1 x_1 + f_2 x_2 + \ldots, f_n x_n}{f_1 + f_2 + \ldots + f_n}$$

$$= \frac{\sum_{i=1}^{n} f_i x_i}{\sum_{i=1}^{n} f_i} = \frac{\sum_{i=1}^{n} f_i x_i}{N}$$

$$= \sum_{i=1}^{n} \left(\frac{f_i}{N}\right) x_i.$$

The last form has been introduced since f_i/N is the relative frequency of the measurement x_i, which proves useful below. Before proceeding, however, it is worth remarking that if we were seeking the mean value of x^2, say (or indeed any other function of x), then the above analysis would only be altered by the inclusion of the terms in x^2 rather than x. So, x_3 would be replaced by x_3^2, etc.

Thus, as an example

$$\overline{x^2} = \frac{x_1^2 + x_2^2 + x_3^2 + \ldots + x_N^2}{N}$$

$$\sum_{i=1}^{n} \left(\frac{f_i}{N}\right) x_i^2$$

where we have again grouped the N individual measurements into n classes, as above.

Now, as the number of measurements becomes very large, we can often replace a discrete distribution (as in a histogram) by a continuous distribution. In addition the sum becomes an integral. This last point will not be proved here, but it will be remembered that the integral sign $\int$ was introduced as the limiting value of a sum Σ, when the number of elements in the sum approached infinity. Thus

$$\bar{x} = \int_{-\infty}^{\infty} p(x) x \, dx$$

since the relative frequency f_i/N becomes the probability if N becomes large. Similar results apply to functions of the random variable.

For example,

$$\overline{x^2} = \int_{-\infty}^{\infty} x^2 p(x) \, dx, \quad \overline{x^3} = \int_{-\infty}^{\infty} x^3 p(x) \, dx \quad \text{etc.}$$

The term *expected value* is often used in this context, since $\overline{x^2}$ for example is effectively the long-run average value of x^2, and is therefore the *expected value of* x^2. Likewise for any other function of x.

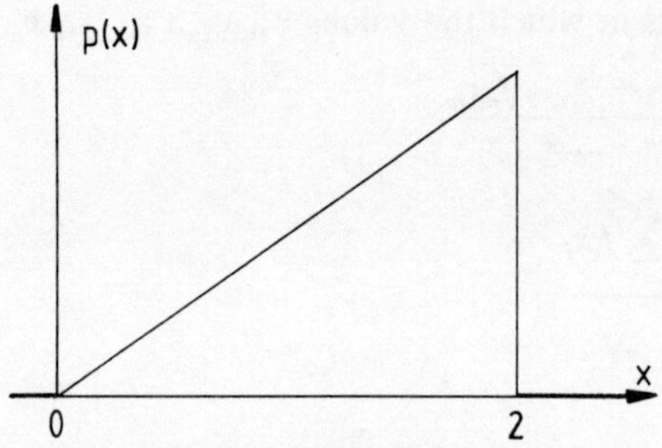

Figure 13.2 The probability distribution function

Example

A random variable has a triangular probability distribution given by the observed probability density function

$$p(x) = \begin{cases} Kx & \text{when } 0 \leqslant x \leqslant 2 \\ 0 & \text{otherwise (Figure 13.2).} \end{cases}$$

Find

(i) the value of K
(ii) the mean value of x
(iii) the standard deviation s
(iv) the probability that $x \leqslant 1$.

Solution

(i) $\int_{-\infty}^{\infty} p(x)\,\mathrm{d}x = 1$ for a normalized function

$$\therefore \int_{-\infty}^{0} p(x)\,\mathrm{d}x + \int_{0}^{2} p(x)\,\mathrm{d}x + \int_{2}^{\infty} p(x)\,\mathrm{d}x = 1$$

$$\therefore 0 + \int_{0}^{2} Kx\,\mathrm{d}x + 0 = 1$$

$$\therefore \left[K\frac{x^2}{2}\right]_0^2 = 2K - 0 = 1 \qquad \therefore K = \frac{1}{2}.$$

(ii) $$\bar{x} = \int_{-\infty}^{\infty} xp(x)\,\mathrm{d}x = \int_{0}^{2} x\,Kx\,\mathrm{d}x$$

$$= \frac{1}{2}\int_{0}^{2} x^2\,\mathrm{d}x = \frac{1}{2}\left[\frac{x^3}{3}\right]_0^2 = \frac{1}{2}\left(\frac{8}{3} - 0\right) = \frac{4}{3}.$$

$\therefore$ The mean value of the variable is 4/3.

(iii) The standard deviation, s, of a set of measurements $(x_1, x_2, \ldots, x_i, \ldots, x_N)$ is defined as the *root-mean-squared deviation* of these measurements; that is,

$$s = \sqrt{\overline{(x_i - \bar{x})^2}}$$

or

$$s^2 = \overline{(x_i - \bar{x})^2}.$$

It can be shown that this is equivalent to the form

$$s^2 = \overline{x^2} - \bar{x}^2. \tag{13.8}$$

The standard deviation is a useful measure of the *dispersion*, or amount of *spread* of a set of observations of a quantity. We do not explore the full significance of this quantity here, since it more properly belongs to any introductory text in statistics. For our purposes, we can proceed from the definition given by (13.8) above.

$$\overline{x^2} = \int_{-\infty}^{\infty} x^2 p(x)\, dx = \int_0^2 x^2 Kx\, dx = \frac{1}{2}\int_0^2 x^3\, dx = \frac{1}{2}\left[\frac{x^4}{4}\right]_0^2 = 2$$

$$\therefore s^2 = \overline{x^2} - \bar{x}^2 = 2 - \left(\frac{4}{3}\right)^2 = \frac{2}{9}$$

$\therefore$ the standard deviation s of the variable is $\sqrt{2/9} = 0{\cdot}47$.

(iv) The probability that the variable will give results in the range $x \leqslant 1$ is given by

$$P(X \leqslant 1) = \int_{-\infty}^{1} p(x)\, dx = \int_0^1 Kx\, dx = \frac{1}{2}\int_0^1 x\, dx = \frac{1}{2}\left[\frac{x^2}{2}\right]_0^1 = \frac{1}{4}$$

i.e. 25% of all results in the long run will fall in the range $x \leqslant 1$.

13.9 BASIC EXAMPLES FOR THE STUDENT

B13.1 Evaluate the following indefinite integrals

(a) $\int (e^{3x} + e^{-3x})\, dx$ (f) $\int \frac{x}{\sqrt{1 + x^2}}\, dx$

(b) $\int \frac{dx}{(x + 3)}$ (g) $\int \frac{2x - 3}{4 - 3x + x^2}\, dx$

(c) $\int \frac{dx}{(3 - x)}$ (h) $\int \frac{dx}{x \log_e x}$

(d) $\int \sqrt{e^x}\, dx$ (i) $\int \frac{e^x - e^{-x}}{e^x + e^{-x}}\, dx$

(e) $\int x(x^2 - 3)^3\, dx$ (j) $\int x^2 e^{x^3}\, dx$

(k) $\int x e^x \, dx$

(l) $\int x e^{-x^2} \, dx$

(m) $\int \frac{e^{\sqrt{x}}}{\sqrt{x}} \, dx$

(n) $\int x^2 e^{-x} \, dx$

(o) $\int \frac{2x - 3}{\sqrt{4 - 3x + x^2}} \, dx$

(p) $\int (2x - 3)\exp(4 - 3x + x^2) \, dx$

(q) $\int \frac{dx}{x^2 - 9}$

(r) $\int \frac{dx}{x^2 - 5x + 6}$.

13.10 INTERMEDIATE EXAMPLES FOR THE STUDENT

I13.1 Evaluate the following integrals.

(a) $\int \frac{\exp \sqrt{x}}{\sqrt{x}} \, dx$

(b) $\int \frac{e^x + e^{-x}}{e^x - e^{-x}} \, dx$

(c) $\int \frac{1}{x \log_e x} \, dx$

(d) $\int \frac{\sqrt{\log_e x}}{x} \, dx$

(e) $\int \frac{(7 - x)}{x^2 - 14x + 9} \, dx$

(f) $\int \frac{7 - x}{\sqrt{x^2 - 14x + 9}} \, dx$

(g) $\int \frac{dx}{(x + 3)^6}$

(h) $\int \frac{x}{x^2 + 5} \, dx$

(i) $\int \frac{x}{\sqrt{x^2 + 5}} \, dx$

(j) $\int \frac{dx}{x^2 - 16}$

(k) $\int x e^x \, dx$

(l) $\int \log_e x \, dx$

(m) $\int x \log_e x \, dx$

(n) $\int \sqrt{3 - x} \, dx$

(o) $\int x e^{x^2} \, dx$.

I13.2 A random variable has a triangular probability distribution given by the probability density function

$p(x) = x - 3$ when $3 \leqslant x \leqslant 4$
$p(x) = 5 - x$ when $4 \leqslant x \leqslant 5$
$p(x) = 0$ otherwise.

Find

(i) the mean value of the variable

(ii) the probability that the variable will take a value between 3·5 and 4·0.

Sketch the probability density function.

I13.3 In a mathematical model involving two variable, x and y, it is known that y depends only on variable x, so that $y = y(x)$. At some point in a calculation involving this model, it is necessary to evaluate the integral

$$\int (2x - 5)\,\mathrm{d}x.$$

Give two possible interpretations of why this integral might arise and give two *possible* forms of $y = y(x)$ consistent with these interpretations. (There are many possible answers, but the two *simplest* interpretations are asked for.)

CHAPTER 14

Mathematical Models

14.1 THE PURPOSE OF MODELS

In Chapter 9 we examined the economic theory of the monopolist from a mathematical viewpoint, and at one stage introduced the equation below to represent the relationship between level of demand (q) and price level per unit (p):

$$p = a - bq.$$

Since this mathematical relationship is designed to simulate the relationship which exists between price level and demand level in the real world, we can describe it as a *model*, albeit a very simple one.

We subsequently used this particular equation, along with others, to build a more complete (mathematical) picture of the economic problem of the monopolist, and were eventually able to draw some conclusions as to the profitability of various production levels. This result had practical use in enabling us to identify the production level which maximized gross profit. Some general conclusions about mathematical models can be made from this particular example.

(i) In order to be able to make any use of the final conclusions of the full analysis, we have to be able to rely on the validity of the basic input to the model, such as the equation $p = a - bq$, in two ways. Firstly, it should follow the measured experimental data sufficiently closely to be a reasonable description of the actual situation. Secondly, we are likely to want to assume that this relationship will hold over a reasonable period of time. (We hope, in particular, that the values of the two *parameters* a and b will be fairly stable.)

(ii) Often, for a given set of data, we have a choice of curve types which can be fitted. This was shown in the chapter on curve-sketching (Chapter 10) when four curves were fitted to a small quantity of data on demand level versus price level and the cubic was shown to be a bad choice because of its behaviour outside the range of the direct experimental measurements. In this case, where we have used $p = a - bq$, the emphasis has been on mathematical simplicity, an advantage when mathematical complexity might make the overall problem virtually intractable, or where the sensitivity of the results to the *exact* choice of input curve is not too great.

(iii) The relationship (or model)

$$p = a - bq$$

is said to be *empirical* since the form of the relationship was found from experimental data. Following an inspection of the data, an appropriate mathematical relationship was sought which closely simulated the actual measurements.

(iv) Since there may be a wide choice of curve types from which to select when building a model, it must be remembered that these input relationships are *assumptions*, and have *not been proved*. The output from our economic monopoly problem, for example, cannot be better than the quality of the assumptions we build into it, of which $p = a - bq$ was one (i.e. rubbish in: rubbish out).

(v) Besides the *empirical* relationship $p = a - bq$, we also introduced some *prescriptive* relationships such as

$$\Pi = \text{TR} - \text{TC} \quad \text{(gross profit = total revenue − total costs)}$$

and

$$\text{TR} = \text{pq} \quad \text{(total revenue = price per unit} \times \text{demand level).}$$

These two relationships actually *define* Π and TR, and so can be considered as *prescriptions* for these two quantities. In general, complex mathematical models will involve a mixture of both empirical (experimental) and prescriptive (defining) relations.

We will now give two further examples of models: the first on stock control involving differential calculus, and the second on decay rates involving integral calculus.

14.2 A BASIC STOCK CONTROL PROBLEM

A manufacturer produces a fluorescent pink, and fairly lethal, sweet called popcorn in large batches, and his basic problem is to decide on how large each batch size should be. In order to develop a model of the situation, from which it is hoped a least-cost solution can be identified, the following *assumptions* and *definitions* are made.

(i) The *demand* for the popcorn can be treated as uniform and is known to be at a rate of D (tons per year).

(ii) When it makes a *batch*, the *lead-time* involved is negligible so that it can re-stock its warehouse virtually immediately. There is no *stock-out*.

(iii) There is a *fixed cost* involved each time the popcorn machine is *set going* to produce, even before any popcorn is actually produced. This is known to be S (£ per batch).

(iv) There is a warehousing (or *inventory*) cost involved in just storing popcorn. This includes such things as fire, rodent, and radioactivity control as well as the economic *opportunity cost* of having considerable capital tied up in the form of popcorn. This carrying cost is defined in two ways:

carrying cost as a proportion of purchase price, per year $= i$

or

carrying cost per ton of popcorn per year $= I$.

If C is the cost of popcorn per ton, then $I = iC$.

(v) The batch size produced is Q, and the most economic batch size is Q^* (both in tons).

(vi) The total acquisition cost (holding + purchasing but not production costs) is denoted by A, and the value of A at the *economic batch quantity* EBQ (sometimes EOQ—economic order quantity) is denoted by A^*.

(vii) The time interval between batches is T, and between economic batch quantities this becomes T^* (both in years).

The solution as to the most economic batch size to produce is not obvious, since there are two types of cost involved, each pulling in opposite directions. Frequent production of small batches will result in low warehousing/inventory/carrying costs but high set-up costs. On the other hand, infrequent production of large batches will reverse this, to give high warehousing/inventory/carrying costs (as the average stock level will be high) but low set-up costs (as set-up is infrequent).

We can represent the actual stock-level held by the saw-tooth graph, Figure 14.1. Note carefully the following points:

(i) the vertical leading edge—this represents *instant production and delivery* (one of our simplifying assumptions).

(ii) the uniform slope of the trailing edge—this represents the *uniform demand rate* (another simplifying assumption).

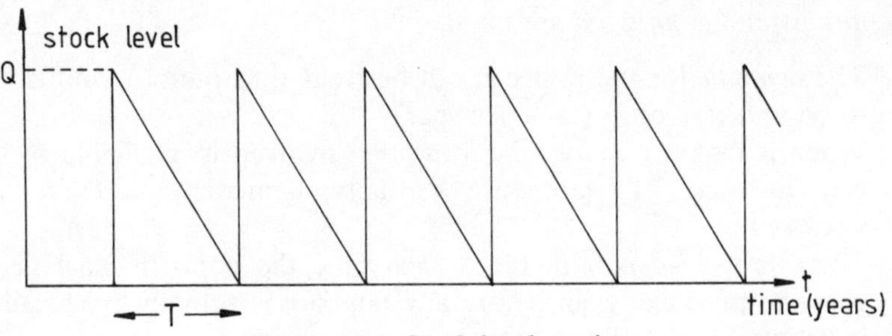

Figure 14.1 Stock level vs. time

(iii) the *repetition* of the pattern. Once the optimum value for Q and T have been found (Q^* and T^*), then this pattern will be repeated.

The first stage of the problem is to find an expression for the total acquisition cost A.

Annual Set-up Costs (ASC)

The total annual demand is D, and the size of each batch is Q, so the number of batches per year $= D/Q$. Each batch costs S to set up,

$$\therefore \text{Total annual set-up costs (ASC)} = \frac{SD}{Q}.$$

Annual Carrying Costs (ACC)

The average stock held during the year $= \frac{1}{2}Q$. This should be clear from the form of the saw-tooth graph, since the height of each 'tooth' is Q. Strictly, this could be found from the integral formula for the average

$$\bar{q} = \frac{1}{T}\int_0^T q \, \mathrm{d}t$$

where q represents the (variable) stock level, but for such a simple form of stock level, the area under the curve for q can be seen directly

$$\bar{q} = \frac{1}{T}\int_0^T q \, \mathrm{d}t = \frac{1}{T} \times \frac{1}{2} QT = \tfrac{1}{2}Q.$$

Now, the cost of popcorn is C (per ton), so that the average value of the stock $= \frac{1}{2}QC$

$$\therefore \text{Annual carrying cost (ACC)} = \tfrac{1}{2}QCi = \tfrac{1}{2}QI$$

$\therefore$ Total annual acquisition cost is given by

$$A = \frac{SD}{Q} + \tfrac{1}{2}IQ$$

which is of the general form $A = a/Q + bQ$. If this is sketched (Figure 14.2) it is seen to be the sum of a hyperbola and a straight line, with a minimum value at (Q^*, A^*). This minimum can be located using differential calculus in the usual way. We indicate the stationary values by means of an asterisk (*).

$$A = \frac{SD}{Q} + \tfrac{1}{2}IQ$$

$$\therefore \frac{\mathrm{d}A}{\mathrm{d}Q} = -\frac{SD}{Q^2} + \tfrac{1}{2}I = 0 \text{ at the minimum.}$$

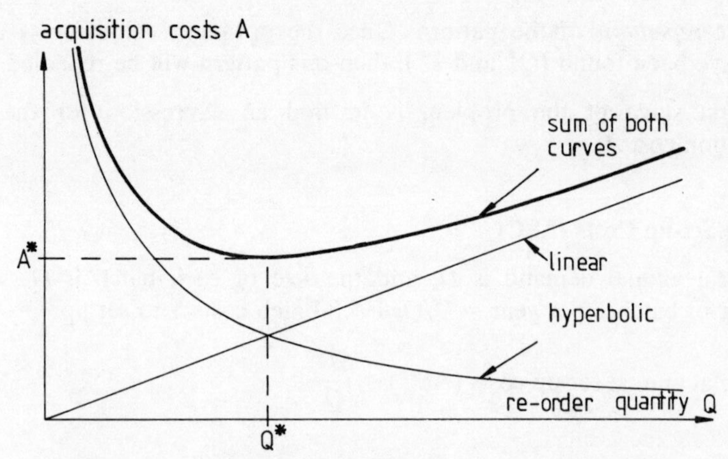

Figure 14.2 Total annual acquisition costs as a function of batch size

At the minimum,

$$\frac{1}{2}I = \frac{SD}{Q^{*2}},$$

$$\therefore Q^{*2} = \frac{2SD}{I}$$

$$\therefore Q^* = \sqrt{\frac{2SD}{I}}$$

$$\therefore A^* = \frac{SD}{Q^*} + \frac{1}{2}IQ^* = SD\sqrt{\frac{I}{2SD}} + \frac{1}{2}I\sqrt{\frac{2SD}{I}}$$

$$\therefore A^* = \sqrt{\frac{ISD}{2}} + \sqrt{\frac{ISD}{2}} = \sqrt{2ISD}\ .$$

At the minimum value of the acquisition cost, therefore, the contributions from the annual set-up costs and the annual carrying costs are equal (at $\sqrt{ISD/2}$), which justifies the way Figure 14.2 has been drawn.

It is not necessarily the case that these contributions will always be equal, particularly in more complicated models. However, since

Annual Acquisition Costs	=	Annual Carrying Costs	+	Annual Set-Up Costs
A	=	ACC	+	ASC

and minimizing A

$$\frac{dA}{dQ} = \frac{d(ACC)}{dQ} + \frac{d(ASC)}{dQ} = 0$$

$$\therefore \frac{d(ACC)}{dq} = -\frac{d(ASC)}{dQ}.$$

So that it is generally true that the *rate* of increase of carrying costs equals the *rate* of decrease of set-up costs, with respect to changes in the batch size, when A is minimized.

We also know that the demand rate is $D = Q/T = Q^*/T^*$

$$\therefore T^* = \frac{Q^*}{D} = \frac{1}{D}\sqrt{\frac{2SD}{I}} = \sqrt{\frac{2S}{ID}}.$$

Conclusions With the assumptions we have made above, the economic batch quantity is $\sqrt{2SD/I}$ tons:

which should be ordered at intervals of $\sqrt{2S/ID}$ years

when the total annual acquisition costs will be $\sqrt{2ISD}$ £/year.

Example

Suppose the demand level for popcorn is 500 tons/year; the set-up cost per batch is £1000; the carrying cost is 20% of stock held, per year; and the cost of popcorn is £2000 per ton. Find Q^*, T^*, and A^* under the same assumptions as the above model.

$$D = 500,\ S = 1000,\ i = 0{\cdot}20,\ C = 2000$$

and

$$I = iC = 0{\cdot}2 \times 2000 = 400$$

$$\therefore Q^* = \sqrt{2SD/I} = \sqrt{2 \times 10^3.5 \times 10^2/(4 \times 10^2)} = \sqrt{25 \times 10^2} = 50 \text{ tons}$$

$$T^* = \sqrt{2S/ID} = \sqrt{2 \times 10^3/(4 \times 10^2.5 \times 10^2)} = \sqrt{10^{-2}}$$
$$= \tfrac{1}{10} = 0{\cdot}1 \text{ years}$$

$$A^* = \sqrt{2ISD} = \sqrt{2.4 \times 10^2.10^3.5 \times 10^2} = \sqrt{4 \times 10^8}$$
$$= 2 \times 10^4 \text{ £ per year.}$$

Hence 10 batches per year should be produced, of 50 tons each, total value £1m, and total acquisition cost of £20 000 per year.

14.3 POSSIBLE REFINEMENTS TO THE BASIC STOCK CONTROL MODEL

The value of the results which emerged from the basic stock control problem depend almost entirely on the validity of the assumptions which were built into the model. More sophisticated models have been developed which attempt to use more realistic assumptions, but which necessarily are also more complex. Refinements which have been incorporated include the following

(i) The assumption of zero production lead time (or instantaneous batch production) is replaced by a uniform rate of delivery or production (Figure 14.3(a)).

(ii) Uniform demand is replaced by a fluctuating demand rate with a known (and usually steady) *average* demand rate. Allowance for fluctuations can then be made by the addition of a *safety stock* which absorbs occasional excesses of demand to prevent a stock-out with a known degree of confidence (Figure 14.3(b)).

(iii) Stock-outs can be allowed in the model, and a cost is usually associated with a stock-out in much the same way as a carrying cost. In this case, orders are accumulated until a new batch arrives (Figure 14.3(c)).

(iv) So called *price-breaks* may be introduced. In these cases a financial incentive of some type may be offered for ordering large batches. If the batch size attracting the offer is greater than Q^* then there *may be* a case for taking this larger batch size because of the savings involved (Figure 14.3(d)).

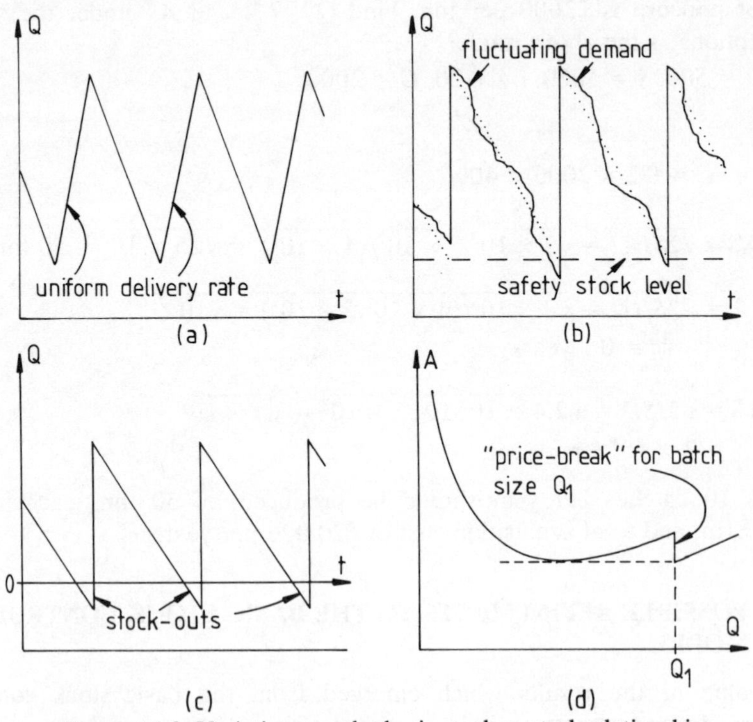

Figure 14.3 Variations on the basic stock control relationships

14.4 A BASIC DECAY RATE PROBLEM

A large number of apples held in storage are found to include a number of decaying fruit. This number is also known to be increasing and the spread is such that the rate of increase is proportional to the number of healthy apples remaining. It is found experimentally that the rate of increase of decay equals one tenth of the number of healthy apples, where time is measured in months. There is a total of 100 000 healthy apples to begin with. Find an expression which describes the number of decayed apples at any particular point in time.

This type of situation occurs in many disguised forms, such as radioactivity decay rates and memory decay rates. A general statement of this form of decay is that the *rate of movement* into a new state (of decay) *is proportional to the number still able to make that transition* (number still healthy).

Solution

Let x be the number of decayed apples at time t, and A the total number of apples (in this case, 100 000). The problem is to find a relation between x and t. t is measured in months.

The rate of increase in the number of decayed apples is $\mathrm{d}x/\mathrm{d}t$

The number of healthy apples = Total − Number decayed

$$= A - x.$$

The basic information given is that

$$\frac{\mathrm{d}x}{\mathrm{d}t} = \frac{1}{10}(A - x)$$

$$\therefore \int \frac{\mathrm{d}x}{A - x} = \frac{1}{10}\int \mathrm{d}t$$ (The integral signs are introduced here because we cannot isolate $\mathrm{d}x$ and $\mathrm{d}t$ otherwise.)

$$\therefore -\log_e(A - x) = \frac{1}{10}t + c$$

$$\therefore \log_e(A - x) = -\frac{1}{10}t - c$$

$$\therefore \exp -(c + t/10) = A - x$$

$$\therefore e^{-c}\, e^{-t/10} = A - x$$

but e^{-c} is another constant, say B

$$\therefore \underline{Be^{-t/10} = A - x}$$

which is the general relationship between x and t.

However, when $t = 0$, $x = 0$ (no decay at the beginning)

$$\therefore Be^0 = B = A \quad \text{so that} \quad Ae^{-t/10} = A - x$$

$$\therefore x = A(1 - e^{-t/10}).$$

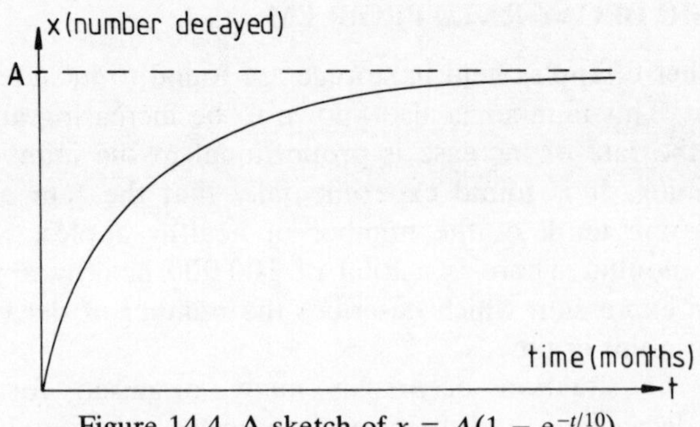

Figure 14.4 A sketch of $x = A(1 - e^{-t/10})$

We know that $A = 10^5$

$$\therefore x = \underline{10^5 (1 - e^{-t/10})} \quad \text{(Figure 14.4).}$$

From the figure it can seeen that the number of decayed apples approaches the total number possible, A, but in theory it takes an infinite length of time to actually reach A. To estimate the sort of time period during which a substantial number will have decayed (an example of a *characteristic time*) we can put

$$x = A\left(1 - \frac{1}{e}\right) = 0{\cdot}632A$$

when $t/10 = 1$ or $t = 10$. The figure 10 appears at this stage because we used the fraction 1/10 as the proportion which decayed monthly. Hence 63·2% decayed in the first ten months.

In general, if $x = A(1 - e^{-t/\tau})$ then the characteristic time of decay is τ.

The *half-life*, h, is the time taken for any population to be reduced by a factor of 0·5, that is when

$$1 - e^{-h/\tau} = 0.5$$

$$\therefore e^{-h/\tau} = 1 - 0{\cdot}5 = 0{\cdot}5$$

$$\therefore -\frac{h}{\tau} = \log_e 0{\cdot}5 = \bar{1}{\cdot}3068 = -0{\cdot}6932$$

$$\therefore \underline{h = 0{\cdot}693\tau}.$$

The half-life is therefore a fixed proportion (69·3%) of the characteristic time, and knowledge of one can give us the other.

From the basic relationship

$$x = 10^5 (1 - e^{-t/10})$$

the absolute number of decayed apples (x) and the rate of decay (dx/dt)

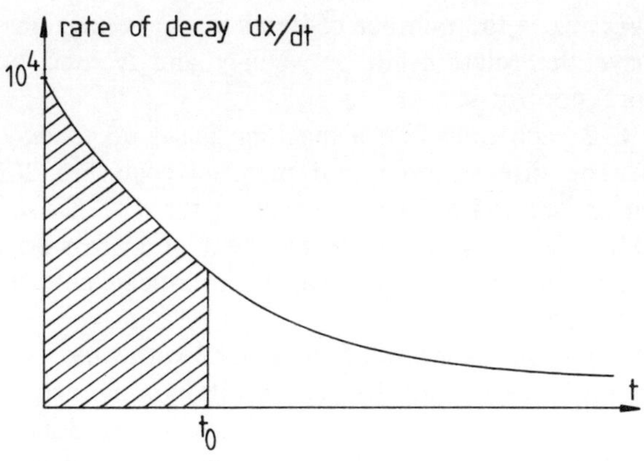

Figure 14.5 The rate of decay vs. time. The shaded area $= \int_0^{t_0} \frac{dx}{dt}\, dt$ = number of decayed apples at time t_0. The total area under the curve must equal A so that all apples have decayed as $t_0 \to \infty$

can be found for any value of t; for example,

$$\frac{dx}{dt} = 10^5\left(+ \frac{1}{10} e^{-t/10} \right) = 10^4 e^{-t/10}$$

(Figure 14.5); when $t = 0$, $dx/dt = 10^4$, so that the initial rate of decay is 10 000 apples per month (consistent with the rate equalling one tenth of the number of healthy apples, 100 000, at the start).

14.5 BASIC EXAMPLES FOR THE STUDENT

B14.1 The number of items manufactured per day (N) after t days from the beginning of a production run is given by

$$N = 100(1 - e^{-0.2t}).$$

(i) Sketch this function.
(ii) How many items will be produced per day 5 days from the start of a run?
(iii) How many days will pass before they are achieving 99 units per day at least?

B14.2 The number of items, N, in a jeweller's shop whose value exceeds x is given by

$$N = \frac{20\,000}{x^{3/2}} \quad \text{for } x > 1.$$

(i) How many are worth more than £100?
(ii) How many are worth between £1 and £4?

(iii) If $y(x)\,\mathrm{d}x$ is the number of items in the value range x to $x + \mathrm{d}x$, derive the relationship between y and N, and hence find an expression for y.

B14.3 It costs £12 each time that a machine is set up to make a batch of brushes. The direct labour and material costs are 20p per brush. The cost of keeping a finished brush in the warehouse for a year is 5p. 100 000 brushes are made per year and are sold at a uniform rate. A new batch of brushes can be produced almost immediately when required.

(i) Sketch the variation of stock level with time.
(ii) Find an expression for A, the total annual acquisition cost.
(iii) Find the economic batch quantity (EBQ), the interval between batches, and the minimum annual total acquisition cost.

14.6 INTERMEDIATE EXAMPLES FOR THE STUDENT

I14.1 A local authority is responsible for refuse disposal, and operates a system whereby rubbish is accumulated at a collection site, and then incinerated when the pile of rubbish is high enough. The cost of each incineration is £100 and this must be balanced against the costs of refuse storage. Because of health and fire precautions which are necessary while the pile is accumulating, it is found that overheads of storage amount to £10 for each 250 tons of rubbish per year. Altogether 80 000 tons of rubbish are dealt with each year (and it arrives at a virtually uniform rate throughout the year). In order to minimize its costs, how often should the authority incinerate the accumulated pile of rubbish; at what annual cost; and how large is each incineration?

I14.2 A car's petrol consumption increases with speed. At zero speed it consumes nothing; at 50 m.p.h. it consumes petrol at a rate of 40 miles per gallon, and its highest speed is 80 m.p.h. however high the petrol consumption rate. Fit an exponential type of curve to relate speed to petrol consumption.

I14.3 Bacteria cells reproduce simply by dividing into two, a procedure which they are observed to carry out with perfect regularity, every minute. Starting with one cell, derive a formula to describe the total number of bacteria after t hours. Roughly how many would you expect after 24 hours? (Assume that none die in this period.)

CHAPTER 15

Lagrange Multipliers

15.1 INTRODUCTION

Many business-type problems or decisions are concerned with identifying possible choices, and then selecting the optimum in accordance with some well-defined criterion of *excellence*. So, linear programming, for example (Chapter 5), is a technique used in determining the best use of resources, under known constraints, where *best* is defined specifically in terms of maximum profits or minimum costs or whatever. The typical stock-control problem, discussed in the previous chapter, is concerned with searching for the optimum re-ordering cycle (or equivalent) such that inventory costs are minimized.

This chapter is concerned with similar sorts of problems, and will illustrate ways in which we can maximize (or minimize) the values of functions when certain constraints are imposed. In particular we shall be introducing a technique called the method of *Lagrange multipliers*.

15.2 PARTIAL DERIVATIVES: A REMINDER

In Section 8.5 we introduced the technique of *partial differentiation* when we considered functions of more than one variable. For example, the (dependent) variable f where

$$f(x, y) = 2x^2 + 5y^2$$

is a function of two independent variables x and y. $f(x, y)$ can be illustrated geometrically by treating it as a *height* above the x–y plane and for the function above we have a bowl-like shape, which is parabolic in cross-section (Figure 15.1). To find the minimum (or maximum) of such a function we can differentiate and set the derivative equal to zero in the usual way. However since *two* independent variables are involved, we must use the *partial derivatives*.

$$f = 2x^2 + 5y^2$$

$$\therefore \frac{\partial f}{\partial x} = 2.2x + 0 = 4x = 0 \text{ at minimum}$$

$$\therefore \frac{\partial f}{\partial y} = 5.2y + 0 = 10y = 0 \text{ at minimum.}$$

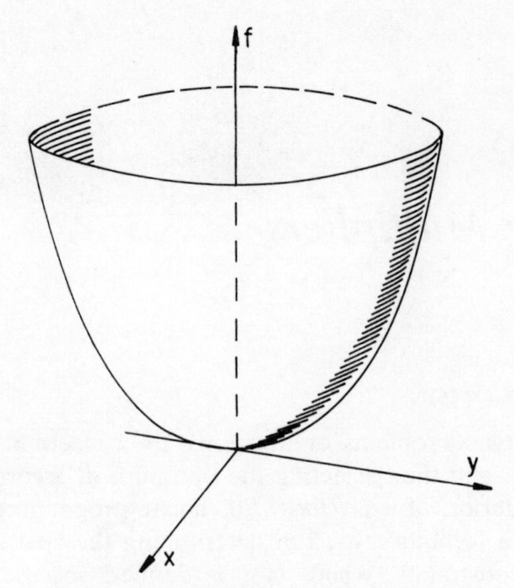

Figure 15.1 The function $f(x, y) = ax^2 + by^2$, $a, b > 0$

Hence $f = x = y = 0$ at the minimum value of the function, which should be intuitively obvious from the sketch in Figure 15.1 In finding the minimum point, it should be noted that both partial derivatives were set equal to zero. In words, this means that we have found a point which is a minumum in both the x-direction and the y-direction.

15.3 SADDLE POINTS

In the example just illustrated, it should be noted we have not actually proved that the point (0, 0) is a minimum, only that it has a zero gradient. The second derivatives need to be evaluated before we can be certain that

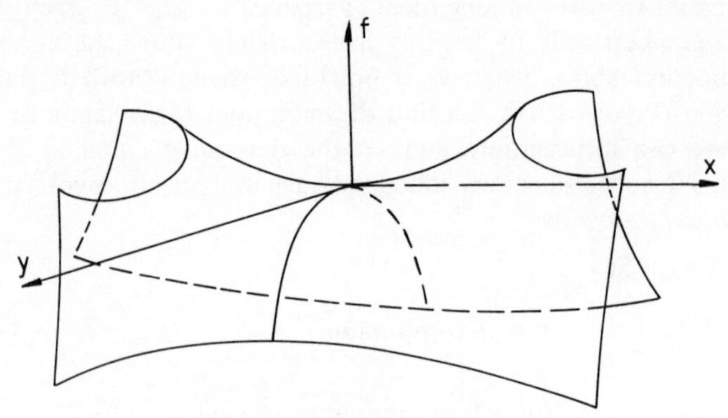

Figure 15.2 A saddle-point at (0, 0)

the point in question is actually a minimum. (The interpretation as a minimum is however correct in this case.)

For the curve

$$f(x, y) = -2x^2 - 5y^2$$

(upside-down version of the former) then (0, 0) is a *maximum*, which should be intuitively obvious (the upside-down bowl). However, the curve

$$f(x, y) = 2x^2 - 5y^2$$

also has a zero gradient at (0, 0) without there being a true maximum or minimum. A *saddle-point* occurs because

$\dfrac{\partial f}{\partial x} = 0$ is a *minimum* in the x-direction

$\dfrac{\partial f}{\partial y} = 0$ is a *maximum* in the y-direction.

This is illustrated in Figure 15.2.

15.4 THE INTRODUCTION OF CONSTRAINTS

Suppose that 100 metres of fencing are available to enclose a rectangular field, and we wish to know the size of the largest field which can be enclosed (Figure 15.3).

If the dimensions of the rectangle enclosed are x metres by y metres then the area of the field, f, is given by

$$f = xy.$$

The total length of fencing is given by

$$2x + 2y = 100 \text{ (metres)}.$$

The problem has now become one of maximizing the function $f(x, y)$, but

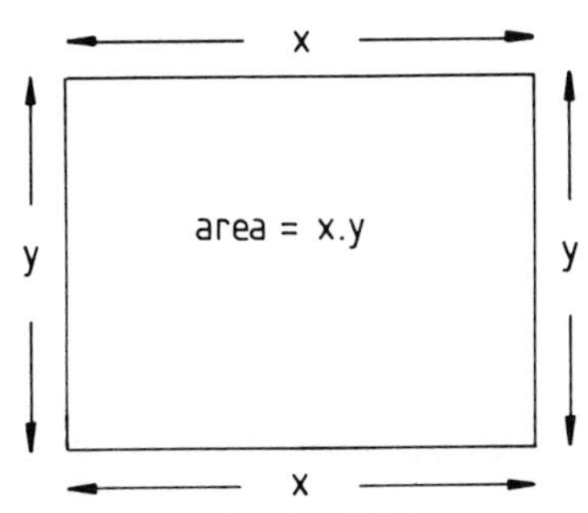

Figure 15.3 The field x metres by y metres

subject to a *constraint* in the form of a relationship between the variables x and y.

Two methods of solution will now be demonstrated.

15.5 SOLUTION BY SUBSTITUTION

The function representing the area of the field, f, is apparently a function of two variables (x and y). However, since a relationship exists between x and y we can, by substituting for either one of them, reduce f to a function of one variable only.

$$2x + 2y = 100 \qquad \therefore x + y = 50 \qquad \therefore y = 50 - x$$

so that $f = xy = x(50 - x) = 50x - x^2$ (Figure 15.4). This can then be maximized in the usual way

$$\frac{df}{dx} = 50 - 2x = 0 \text{ at max.}$$

$$\therefore x = 25, \qquad y = 25 \text{ (i.e. a square field).}$$

Hence the maximum field area is

$$f = 25.25 = 625 \text{ sq. m.}$$

In principle, substitution can always be used, but in practice the necessary substitution may be so difficult that the attempt has to be abandoned. We must use another method, such as that of *Lagrange multipliers*.

15.6 SOLUTION BY THE METHOD OF LAGRANGE MULTIPLIERS

We shall use the same example as in Section 15.5 to illustrate the method, before introducing more complex examples where the method really proves invaluable.

We first rewrite the constraint in the form $g(x, y) = 0$

$$\therefore g(x, y) = x + y - 50 = 0.$$

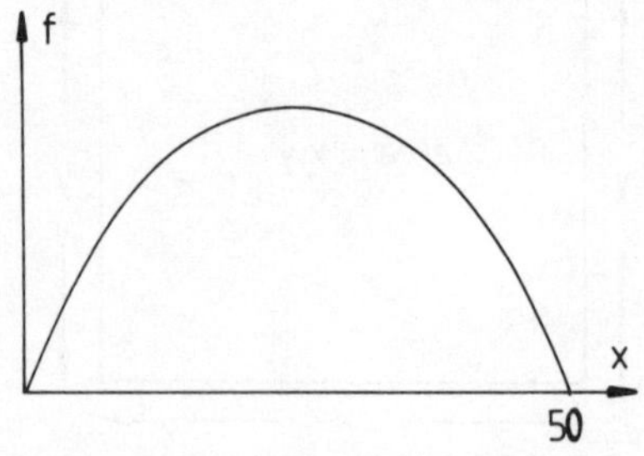

Figure 15.4 Field area vs. length of one side

We then define a new function $F(x, y)$ by

$$F(x, y) = f(x, y) - \lambda g(x, y) \quad \text{(general case)}$$
$$= xy - \lambda(x + y - 50). \quad \text{(this example)}$$

The parameter λ is called the *Lagrange multiplier* (which we treat as a constant in effect). We then proceed to find the maximum (or minimum) value of F as though there were no restrictions on x and y.

$$F = xy - \lambda x - \lambda y + 50\lambda$$

$$\therefore \frac{\partial F}{\partial x} = y - \lambda = 0 \qquad \therefore y = \lambda$$

$$\frac{\partial F}{\partial y} = x - \lambda = 0 \qquad x = \lambda.$$

Hence $x = y = \lambda$ and $x + y = 50$ (the original constraint)

$$\therefore x = y = 25, \quad \text{and} \quad f = xy = 625 \text{ sq. m.}$$

as before.

The method works because the function $g(x, y)$ is always arranged to be zero so that finding a stationary point of $f(x, y)$ subject to $g(x, y) = 0$ is equivalent to finding a stationary point of $F(x, y)$. Strictly, we also need to check that we have found a maximum value of f, rather than a minimum or a saddle point.

The technique can also be extended to problems involving several variables and several constraints. For example, suppose that we wish to maximize the function $f(x, y, z)$ subject to $g(x, y, z) = 0$ and $h(x, y, z) = 0$. In this case we instead maximize the function F, where

$$F(x, y, z) = f(x, y, z) - \lambda g(x, y, z) - \mu h(x, y, z)$$
$$(\text{or} \quad F = f - \lambda g - \mu h)$$

where λ and μ are the Lagrange multipliers. To find the maximum we must solve the five simultaneous equations in x, y, z, λ, and μ:

$$\frac{\partial F}{\partial x} = 0 \qquad \frac{\partial F}{\partial y} = 0 \qquad \frac{\partial F}{\partial z} = 0 \qquad g(x, y, z) = h(x, y, z) = 0.$$

15.7 EXAMPLES OF THE METHOD OF LAGRANGE MULTIPLIERS

(i) Find the maximum volume of a closed rectangular box which can be made from 150 sq. metres of material.

Firstly, we let the sides of the box be represented by x, y, and z. The volume f is given by $f = xyz$.

The limitation on the quantity of material we can use is therefore

$$2xy + 2yz + 2xz = 150$$

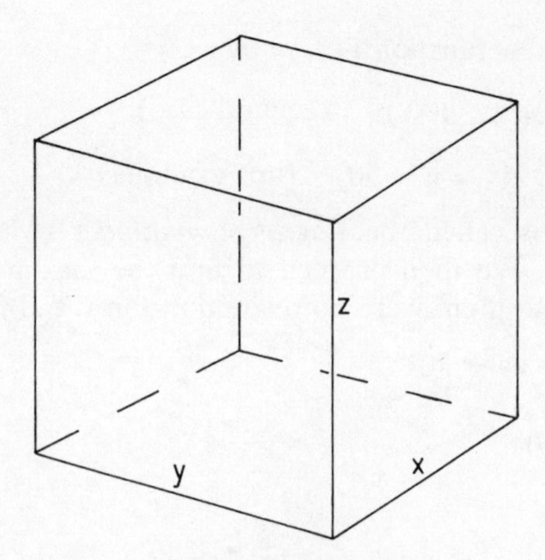

Figure 15.5 Rectangular box, x by y by z metres

since the expression on the left-hand side is the total surface area of the box. So,

$$f(x, y, z) = xyz \qquad g(x, y, z) = xy + yz + xz - 75$$

$$\therefore F(x, y, z) = xyz - \lambda(xy + yz + xz - 75)$$
$$= xyz - \lambda xy - \lambda yz - \lambda xz + 75\lambda$$

$$\left.\begin{aligned} \therefore \frac{\partial F}{\partial x} &= yz - \lambda y - \lambda z = 0 \\ \frac{\partial F}{\partial y} &= xz - \lambda x - \lambda z = 0 \\ \frac{\partial F}{\partial z} &= xy - \lambda x - \lambda y = 0 \end{aligned}\right\} \text{at the stationary value of } F \text{ (and } f\text{).}$$

We are then left with the problem of solving four simultaneous equations in four unknowns, x, y, z, and λ.

$$xy + yz + xz = 75 \qquad \text{(a)}$$
$$yz - \lambda y - \lambda z = 0 \qquad \text{(b)}$$
$$xz - \lambda x - \lambda z = 0 \qquad \text{(c)}$$
$$xy - \lambda x - \lambda y = 0 \qquad \text{(d)}$$

One neat trick we could use at this stage to solve these equations is to exploit the complete symmetry between x, y, and z. If we swapped *any* pair (x and y say) throughout all four equations we would end up with the same four equations. This can only occur if $x = y = z$, and so

$$xy + yz + xz = x^2 + x^2 + x^2 = 3x^2 = 75$$
$$\therefore x^2 = 25 \qquad x = y = z = 5.$$

However, this method works only for rather special cases.

From (b) and (d)

$$\lambda y = yz - \lambda z = xy - \lambda x$$
$$\therefore z(y - \lambda) = x(y - \lambda)$$
$$\therefore \underline{z = x} \qquad (\text{or } y - \lambda = 0).$$

From (c) and (d)

$$\lambda x = xz - \lambda z = xy - \lambda y$$
$$\therefore z(x - \lambda) = y(x - \lambda)$$
$$\therefore \underline{z = y} \qquad (\text{or } x - \lambda = 0).$$

Hence,

$$x = y = z \qquad (\text{or } x = y = \lambda)$$
$$\therefore xy + yz + xz = x^2 + x^2 + x^2 = 3x^2 = 75$$

and

$$\underline{x = y = z = 5.}$$

(We can now show that

$$\lambda = \frac{yz}{y + z} = \frac{25}{10} = 2{\cdot}5 \neq x \text{ or } y,$$

so that $x = y = z$ was the correct choice.)

Hence the largest volume is given by

$$\underline{f = xyz = 5.5.5 = 125 \text{ cu. m.}}$$

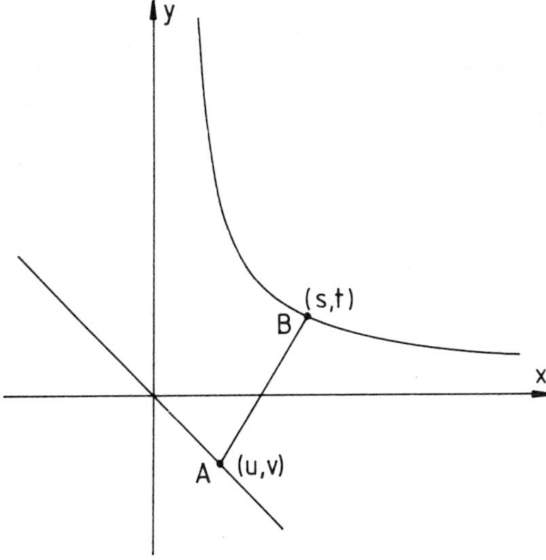

Figure 15.6 Distance AB between line and hyperbola

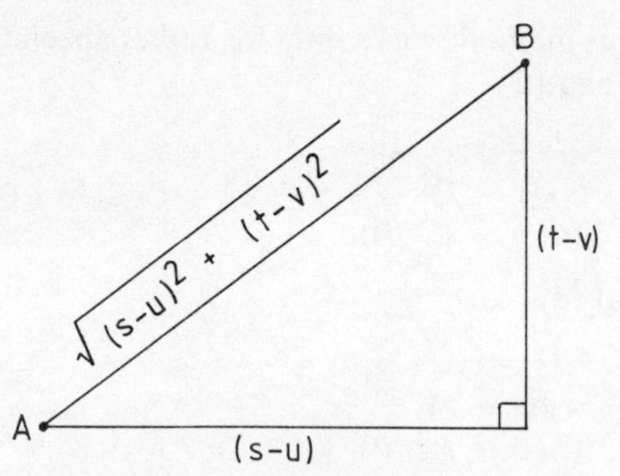

Figure 15.7 The relationship between s, u, t, and v

(ii) Find the minimum distance between the line $x + y = 0$ and the hyperbola $xy = 4$ (positive quadrant only).

Firstly, we imagine a line (AB) drawn between any point on the line (A) and any point on the hyperbola (B) (Figure 15.6). We let the point A have the coordinates (u, v), and the point B (s, t).

The length AB is given by L where

$$L = \sqrt{(s-u)^2 + (t-v)^2}$$

(Figure 15.7) which we wish to minimize subject to

$$u + v = 0 \text{ (line)}$$
$$st = 4 \text{ (hyperbola)}.$$

To simplify the analysis, we shall minimize L^2 rather than L, to remove the square root. The minimum value of L gives the minimum value of L^2, so the result will be the same. Hence

$$f = (s-u)^2 + (t-v)^2$$
$$g = u + v$$
$$h = st - 4$$
$$\therefore \; F = (s-u)^2 + (t-v)^2 - \lambda(u+v) - \mu(st-4)$$
$$\therefore \frac{\partial F}{\partial s} = 2(s-u) - \mu t = 0 \quad \therefore (s-u) = \tfrac{1}{2}\mu t$$
$$\therefore \frac{\partial F}{\partial t} = 2(t-v) - \mu s = 0 \quad \therefore (t-v) = \tfrac{1}{2}\mu s$$
$$\left.\begin{aligned} \frac{\partial F}{\partial u} &= -2(s-u) - \lambda = 0 \\ \frac{\partial F}{\partial v} &= -2(t-v) - \lambda = 0 \end{aligned}\right\} \therefore \frac{-\lambda}{2} = (s-u) = (t-v)$$
$$\therefore \; s - u = t - v = \tfrac{1}{2}\mu t = \tfrac{1}{2}\mu s$$
$$\therefore \; t = s \quad \text{and} \quad u = v.$$

But

$$u + v = 0 \quad \therefore u = v = 0$$

and

$$st = 4 \qquad \therefore s^2 = t^2 = 4 \quad \text{and} \quad s = t = 2.$$

Hence, the minimum value of f is given by

$$f = (2 - 0)^2 + (2 - 0)^2 = 8$$

$$\therefore L = \sqrt{8} = 2\sqrt{2}.$$

Hence the minimum distance between the line and the hyperbola is $2\sqrt{2}$. In fact, this particular problem can be solved much more simply from geometrical considerations (in Figure 15.6). The symmetry of x and y throughout the whole problem (meaning that exchanging x and y in no way affects the problem definition, as in example (i)) gives us immediately:

$$x = y \quad \text{and} \quad xy = 4 \text{ (at B)}$$
$$x + y = 0 \text{ (at A)}.$$

The distance AB is therefore L where

$$L = \sqrt{2^2 + 2^2} = \sqrt{8} \quad \text{as before.}$$

The method of Lagrange multipliers is more generally useful, of course, and can be used when such alternative arguments are not available. All the examples chosen have been used as simple illustrations of the technique.

15.8 THE SIMILARITY WITH LINEAR PROGRAMMING

Consider the field fencing problem once again (see Section 15.4). If we express the length of fencing as a constraint, then we have

$$xy = f \quad \text{and} \quad x + y \leqslant 50.$$

In this case, therefore, the function f is the *objective function*.

The problem is to maximize f subject to the constraint given by the inequality. Figure 15.8 shows an immediate similarity with linear programming problems. The main difference is that the objective function which we are trying to maximize (f) is nonlinear. The shaded area is the feasible region given by $x + y \leqslant 50$; the curves are all hyperbolae of the form $xy = f$ with various values of f.

The point P, at (25, 25), is where the largest field area (f) is obtained within the constraints of fencing available ($x + y \leqslant 50$).

This particular example is obviously suitable for demonstrating the analogy with linear programming, since it only has two variables (and a linear constraint). This makes a simple graphical representation possible.

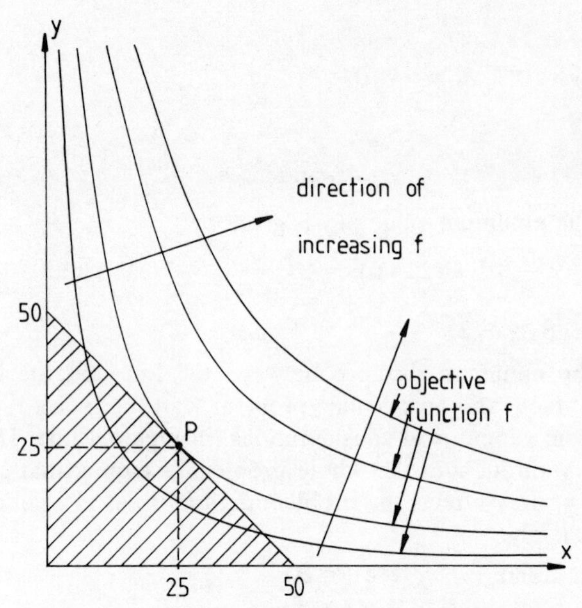

Figure 15.8 The field problem illustrated to resemble a linear programming example

The previous example (ii) requires 4 dimensions for graphical illustration as 4 variables (s, t, u, v) are involved.

15.9 BASIC EXAMPLES FOR THE STUDENT

B15.1 Find the maximum value of the function $f = x + y$ subject to the constraint $x^2 + y^2 = 9$.

What geometrical interpretation could you put on the question?

B15.2 Find the minimum distance between the circle $x^2 + y^2 = 1$ and the straight line $x + y = 4$.

15.10 INTERMEDIATE EXAMPLES FOR THE STUDENT

I15.1 It is required to enclose a rectangular field so that the area is 100 sq. metres. What is the minimun length of fencing which can be used?

I15.2 An economy size box of cornflakes has a volume of 12 000 c.c. (cubic centimetres). What is the minimum area of cardboard which can be used for an economy box? What are its dimensions? (Assume a rectangular box.)

I15.3 Find the minimum distance between the line $x + y = 1$ and the curve $xy = 4$.

More Advanced Examples for Part Four

A4.1 Differentiate the following functions with respect to x.

(a) $y= =x^3(1 + x^2)$

(b) $y = \dfrac{e^x + 1}{e^x}$

(c) $y = \log_e(x^2 + 5x + 6)$

(d) $y = x^3e^{3x}$

(e) $y = \sqrt{x + \log_e x}$

(f) $y = \log_e 3x^4 + \log_e 5$

(g) $y = \log_e\sqrt{x + 1}$

(h) $y = \left(\dfrac{a + x}{a - x}\right)\sqrt{\dfrac{b - x}{b + x}}$.

A4.2 Evaluate the following integrals

(a) $\displaystyle\int xe^{x^2}\,dx$

(b) $\displaystyle\int \frac{x}{\sqrt{1 - x^2}}\,dx$

(c) $\displaystyle\int \frac{dx}{a^2 - x^2}$

(d) $\displaystyle\int \frac{e^x + e^{-x}}{e^x - e^{-x}}\,dx$

(e) $\displaystyle\int \frac{dx}{x \log_e x}$

(f) $\displaystyle\int \frac{x^2 - 1}{\sqrt{x^3 - 3x + 5}}\,dx$

(g) $\displaystyle\int_0^1 (1 - e^{-x})\,dx$

(h) $\displaystyle\int_0^\infty xe^{-x^2}\,dx$

A4.3 Sketch the following curves, showing as many features of interest as you can.

(a) $y = x^3 - 7x^2 + 36$
(b) $y = x^4 - 16x^2$
(c) $xy = 2(x + y)$
(d) $y = \dfrac{1}{e^x + 1}$.

A4.4 A Member of Parliament wishes to rent a flat in or near London, and decides to select the area purely on the basis of cost. The advantage of the city centre is the proximity to his work, with low travelling costs, but the rent levels are very high. The outer suburbs have the reverse advantages, with lower rent levels, but with the higher costs involved in daily commuting.

From a study of the housing market, he realizes that the annual cost of renting a flat can be reasonably represented by the function R,

$$R = 15000e^{-x/5} \qquad \text{(in £ per year)},$$

where x is the distance (in miles) from the city centre. Also, the annual cost of commuting is £300 for each mile he lives away from the centre.

Show on a sketch that the total annual cost of renting and

travelling has a minimum value at a particular distance from the centre and find this distance.

What is the minimum total annual cost of travelling and renting according to this model?

If travelling costs are expected to increase rapidly, while rent levels are expected to remain stable, how would this affect the distance he should live from the city centre?

If the rate of inflation is assumed to affect rent and travelling costs in the same way, how would this affect your original answer?

Another Member of Parliament uses the same estimate for travelling costs (£300 per mile from centre, per year) but uses the following formula for rent levels:

$$R = \frac{11000}{x + 1} + 3000.$$

Compare this with the former model.

A4.5 A manufacturer produces cylindrical cans for use in the food industry. A tin can actually uses a cheap metal (such as steel) covered with a thin layer of (expensive) protective coating, such as tin. In order to use as little tin as possible, find the appropriate ratio (height/diameter) for the cans in order to minimize the surface area, for a given volume. (For a circle of diameter d, the circumference is πd, and the area $\pi d^2/4$. The volume of a cylinder is given by the product: Base Area × Vertical Height.)

A4.6 A cone-shaped vessel is to be made from the sector of a circle (Figure A) by bending the material into the form shown in Figure B. The radius of the original circle is 1 unit, so that the length of the sloping surface of the cone is also 1 unit. Using the notation of Figure B, the volume of a cone is given by

$$V = \tfrac{1}{3}\pi r^2 h.$$

Find the value of r and h so that the volume of the cone is a maximum. Hence find the angle a (Figure A) to achieve this maximum volume.

A4.7 Turnup Ltd produce small farm tractors, and are analysing their sales figures and pricing policies, etc., to determine the production level (per year) which will yield maximum profits.

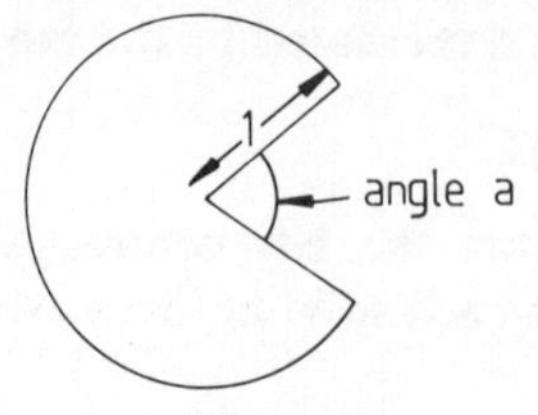

Figure A

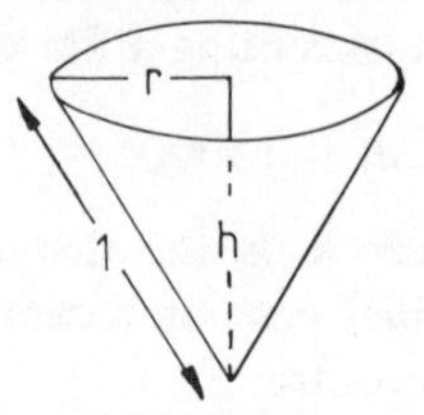

Figure B

They find that the price (p) vs. demand level (q) relationship is given by

$$p + 3q = 5000 \text{ (£)}.$$

The marginal cost of production (MC), per unit tractor, is given by

$$\text{MC} = 4200 - 4q \quad \text{(£/unit)}$$

where the production level is assumed to be the same as the demand level (q). Their fixed costs amount to £70 000 p.a. Find

(i) The production level at the break-even point.
(ii) The production level for the maximum profit.
(iii) The marginal costs, marginal revenue, profit, and elasticity of demand at the level of production which maximizes profit (found in (ii)).

What does the form of the relationship MC(q) imply?

A4.8 A paper mill manufactures various grades of paper and card, all of which are subsequently used in another section of the company for the production of packaging. One of their products is a high quality white paper used solely for surfacing lower quality cardboard in preparation for printing appropriate brand designs, etc. The total annual demand for this white paper is 400 tons, which is produced on one machine in batches. The cost of setting the machine up for each batch is £1000, and of storing the finished product is £80 per ton per year. The cost of raw material for this paper is £500 per ton of finished paper. What is the most economic batch size?

The company has been offered an alternative source of raw material, at £490 per ton, but it must buy a minimum at each purchase: enough for 400 tons of finished paper. Since the raw material must be used immediately, this would mean one batch (at most) being produced each year. What choice would you make, and why?

PART FIVE

Introduction

In Part Five we return to some of the material introduced in Part Three, concerned with linear relationships. Matrix algebra is developed from first principles within the context of sets of simultaneous linear equations, and the basic rules underlying matrix algebra explained. The technique side is further extended in Chapter 17 with a particular method, the Gauss–Jordan, for finding the inverse of a matrix. This is used subsequently in Chapter 19 when the simplex method for solving linear programming problems is introduced. Chapter 19 thus draws together the problems tackled in Chapter 5 with the methods of Chapters 16 and 17. Chapter 18 looks at two further types of model which employ matrix methods. Markov chains and the idea of transition probabilities are introduced in the first part with some very simple illustrations of their use (again, the pitfalls as well as the advantages are stressed); and the basic structure of Economic Input–Output models is explained in the second part. As throughout the book, the examples are over-simplified but serve to illustrate the basic ideas. It is felt that expertise with more complicated examples can be built on this basic understanding.

CHAPTER 16

Matrix Algebra

16.1 INTRODUCTION

We introduced the problem of simultaneous linear equations in Section 4.6, and showed that in simple cases (such as two equations in two variables) we can find a solution either graphically or by straightforward algebraic manipulation. However, problems sometimes arise where the number of equations and the number of variables rises far above two. (Real applications of linear programming may involve perhaps 1000 variables and a similar number of equations.) In principle, n independent equations in n variables can be solved by the same manual methods, but:

(i) Unless the solution is carried out in a systematic way, one is likely to end up with even more equations, no solution, and a desperate urge to become a poet.
(ii) If a systematic method of solution can be found, then it may be amenable to solution by a computer.

Matrix algebra allows the manipulation of simultaneous equations in a systematic and concise manner. So, as far as this chapter is concerned, matrix algebra is concerned essentially with sets of *linear equations*.

16.2 DEVELOPING A MATRIX TECHNOLOGY

We shall demonstrate the development of matrix algebra from basic principles. The first few examples may seem puzzling, because the techniques introduced may seem rather more complicated than the conventional method or form: bringing in a sledgehammer to crack a nut, so to speak. This is inevitable when the aim is to demonstrate the method as simply as possible, rather than to show applications.

First, consider the linear expression

$$5x - 3y \qquad \text{(a)}$$

which consists of 2 coefficients (5 and -3) and 2 variables (x and y). One might also consider (and compare) two other expressions, one with the same variables and different coefficients (b) and one with the same coefficients and different variables (c), as below:

$$1x + 2y \tag{b}$$

$$5u - 3v. \tag{c}$$

As written, the similarities between (a), (b), and (c) are not particularly emphasized, and the full form is not very concise (5, -3, x, and y have all been repeated). So, to try and overcome these objections, the next idea is to separate the coefficients from the variables in some way. For (a), one could write $(5 \quad -3)(x \quad y)$ for example, and introduce necessary conventions for restoring it to its normal form. Certainly, if one also writes $(5 \quad -3)(u \quad v)$ for (c), the similarity has improved, but there is no neat way of condensing (to save re-writing $(5 \quad -3)$) in this format. Instead, try

$$(5 \quad -3)\begin{pmatrix} x \\ y \end{pmatrix}.$$

For this jumble of brackets and numbers to mean anything, *conventions* regarding its interpretation are needed. In particular, how to 'put it back together again', so that

$$(5 \quad -3)\begin{pmatrix} x \\ y \end{pmatrix} = (5x - 3y). \tag{a}$$

Notice that we have taken each *element* from the row in the first bracket and multiplied it by the corresponding element from the *column* in the second bracket (to obtain $5x$ and $-3y$) and then added (to obtain $5x - 3y$). Using this *multiplication* convention, we can therefore write (b) as

$$(1 \quad 2)\begin{pmatrix} x \\ y \end{pmatrix} = (1x + 2y). \tag{b}$$

The brackets $(5 \quad -3)$, $(1 \quad 2)$ are simple examples of what we call a *matrix*: an array of numbers in a bracket. Because of the simple form these examples take, they are called *row matrices* (or *row vectors*). Similarly $\begin{pmatrix} x \\ y \end{pmatrix}$ is a *column matrix* (or *column vector*). The word *vector* is only used when the matrix takes the form of a single row or column.

The *dimensions* of a matrix are given by the number of rows and columns, *in that order*, so that

$(5 \quad -3)$ is a *one by two* matrix (i.e. (1×2))

and

$\begin{pmatrix} x \\ y \end{pmatrix}$ is a *two by one* matrix (i.e. (2×1)).

We can now represent both (a) and (b) simultaneously by stacking the row matrices thus

$$\begin{matrix}(5 & -3)\\(1 & 2)\end{matrix}\begin{pmatrix}x\\y\end{pmatrix} = \begin{matrix}(5x-3y)\\(1x+2y)\end{matrix}$$

but we usually condense this to the form

$$\begin{pmatrix}5 & -3\\1 & 2\end{pmatrix}\begin{pmatrix}x\\y\end{pmatrix} = \begin{pmatrix}5x-3y\\1x+2y\end{pmatrix}. \qquad \text{(d)}$$

The first bracket is a (2×2) matrix, and the RHS of the equation has become a (2×1) matrix (one element is $5x - 3y$, and the other is $1x + 2y$). Also notice how the *multiplication rule* has developed. Elements from the *row* in the first matrix are multiplied by corresponding elements from the *column* in the second matrix (to obtain $5x$ and $-3y$; or $1x$ and $2y$) and then added (to obtain $5x - 3y$; or $1x + 2y$).

Now, instead of introducing expression (b), expression (c) could have been introduced instead:

$$(5 \quad -3)\begin{pmatrix}x\\y\end{pmatrix} = (5x - 3y)$$

$$(5 \quad -3)\begin{pmatrix}u\\v\end{pmatrix} = (5u - 3v) \qquad \text{(c)}$$

which could be combined thus:

$$(5 \quad -3)\begin{pmatrix}x & u\\y & v\end{pmatrix} = (5x - 3y \quad 5u - 3v). \qquad \text{(e)}$$

Notice that the result (RHS of equation) is a (1×2) matrix. We can also generate a fourth expression $(1u + 2v)$ as well as combining (d) and (e) by now writing

$$\begin{pmatrix}5 & -3\\1 & 2\end{pmatrix}\begin{pmatrix}x & u\\y & v\end{pmatrix} = \begin{pmatrix}5x-3y & 5u-3v\\1x+2y & 1u+2v\end{pmatrix}$$

Dimensions: (2×2) (2×2) (2×2).

The LHS is a neat way of writing down all the information contained on the RHS, but without any duplication. Notice that no symbol is repeated on the LHS, whilst all coefficients and variables appear twice on the RHS. Also notice that in the (2×2) matrix on the RHS, the element in the *second row* and *first column* (say) is obtained from the *second row* of the first matrix, and the *first column* of the second matrix, on the LHS. (This sequence *Row–Column* (or RC) runs throughout the conventions of matrix algebra.)

One further point can be shown here which will prove useful in the next section. A numerical factor applied to a matrix can be taken inside the matrix if it is applied to all elements; for example,

$$2\begin{pmatrix}5 & -3\\1 & 2\end{pmatrix}\begin{pmatrix}x\\y\end{pmatrix} = \begin{pmatrix}10x - 6y\\2x + 4y\end{pmatrix} = \begin{pmatrix}10 & -6\\2 & 4\end{pmatrix}\begin{pmatrix}x\\y\end{pmatrix} \quad \text{(c.f.(d))}$$

so that

$$2\begin{pmatrix}5 & -3\\1 & 2\end{pmatrix} = \begin{pmatrix}10 & -6\\2 & 4\end{pmatrix}.$$

16.3 LINEAR EQUATIONS

Suppose we wish to solve the following two simultaneous equations by matrix methods

$$a = 5x - 3y = 1$$
$$b = x + 2y = 8.$$

(a and b are introduced for convenience. Their role becomes clearer below.)

Using conventional methods, we would find x by eliminating y and find y by eliminating x in the following way:

$$2a = 10x - 6y = 2.1 \qquad = 2$$
$$3b = 3x + 6y = 3.8 \qquad = 24$$

$$\therefore\ 2a + 3b \quad = 13x \quad = 2.1 + 3.8 \quad = 26$$

$$\therefore \frac{2}{13}a + \frac{3}{13}b = x \qquad = \frac{2}{13}.1 + \frac{3}{13}.8 = \frac{26}{13} = 2$$

$$\therefore \underline{x = 2}\,.$$

Likewise

$$a = 5x - 3y = 1.1 \qquad = 1$$
$$5b = 5x + 10y = 5.8 \qquad = 40$$

$$\therefore 5b - a \quad = \quad 13y = 5.8 - 1.1 \quad = 39$$

$$\therefore \frac{5}{13}b - \frac{1}{13}a = \qquad y = \frac{5}{13}.8 - \frac{1}{13}.1 = \frac{39}{13} = 3$$

$$\therefore \underline{y = 3}\,.$$

If we now put this procedure into matrix form, we have

$$\frac{2}{13}a + \frac{3}{13}b = \frac{2}{13}.1 + \frac{3}{13}.8$$

and

$$-\frac{1}{13}a + \frac{5}{13}b = -\frac{1}{13}.1 + \frac{5}{13}.8$$

$$\therefore \frac{1}{13}\begin{pmatrix} 2 & 3 \\ -1 & 5 \end{pmatrix}\begin{pmatrix} a \\ b \end{pmatrix} = \frac{1}{13}\begin{pmatrix} 2 & 3 \\ -1 & 5 \end{pmatrix}\begin{pmatrix} 1 \\ 8 \end{pmatrix} \quad \text{(c.f.(d))}$$

$$\frac{1}{13}\begin{pmatrix} 2 & 3 \\ -1 & 5 \end{pmatrix}\begin{pmatrix} 5x - 3y \\ 1x + 2y \end{pmatrix} = \frac{1}{13}\begin{pmatrix} 2 & 3 \\ -1 & 5 \end{pmatrix}\begin{pmatrix} 1 \\ 8 \end{pmatrix}$$

$$\frac{1}{13}\begin{pmatrix} 2 & 3 \\ -1 & 5 \end{pmatrix}\begin{pmatrix} 5 & -3 \\ 1 & 2 \end{pmatrix}\begin{pmatrix} x \\ y \end{pmatrix} = \frac{1}{13}\begin{pmatrix} 2 & 3 \\ -1 & 5 \end{pmatrix}\begin{pmatrix} 1 \\ 8 \end{pmatrix} \qquad \text{(f)}$$

RHS Using the rules of matrix multiplication already derived we find

$$\frac{1}{13}\begin{pmatrix} 2 & 3 \\ -1 & 5 \end{pmatrix}\begin{pmatrix} 1 \\ 8 \end{pmatrix} = \frac{1}{13}\begin{pmatrix} 2.1 + 3.8 \\ -1.1 + 5.8 \end{pmatrix} = \frac{1}{13}\begin{pmatrix} 2 + 24 \\ -1 + 40 \end{pmatrix} = \frac{1}{13}\begin{pmatrix} 26 \\ 39 \end{pmatrix} = \begin{pmatrix} 2 \\ 3 \end{pmatrix}.$$

(This has, in effect, reversed the process we undertook, to reach equation (f).)

LHS Using the same rules, we can evaluate the left-hand side in two ways

$$\begin{aligned}
&\frac{1}{13}\begin{pmatrix} 2 & 3 \\ -1 & 5 \end{pmatrix}\begin{pmatrix} 5 & -3 \\ 1 & 2 \end{pmatrix}\begin{pmatrix} x \\ y \end{pmatrix} \\
&= \frac{1}{13}\begin{pmatrix} 2 & 3 \\ -1 & 5 \end{pmatrix}\begin{pmatrix} 5x - 3y \\ 1x + 2y \end{pmatrix} \\
&= \frac{1}{13}\begin{pmatrix} 2(5x - 3y) + 3(1x + 2y) \\ -1(5x - 3y) + 5(1x + 2y) \end{pmatrix} \\
&= \frac{1}{13}\begin{pmatrix} 10x - 6y + 3x + 6y \\ -5x + 3y + 5x + 10y \end{pmatrix} \\
&= \frac{1}{13}\begin{pmatrix} 13x \\ 13y \end{pmatrix} = \begin{pmatrix} x \\ y \end{pmatrix}.
\end{aligned}$$

$$\begin{aligned}
&\frac{1}{13}\begin{pmatrix} 2 & 3 \\ -1 & 5 \end{pmatrix}\begin{pmatrix} 5 & -3 \\ 1 & 2 \end{pmatrix}\begin{pmatrix} x \\ y \end{pmatrix} \\
&= \frac{1}{13}\begin{pmatrix} 2.5 + 3.1 & -2.3 + 3.2 \\ -1.5 + 5.1 & 1.3 + 5.2 \end{pmatrix}\begin{pmatrix} x \\ y \end{pmatrix} \\
&= \frac{1}{13}\begin{pmatrix} 10 + 3 & -6 + 6 \\ -5 + 5 & 3 + 10 \end{pmatrix}\begin{pmatrix} x \\ y \end{pmatrix} \\
&= \frac{1}{13}\begin{pmatrix} 13 & 0 \\ 0 & 13 \end{pmatrix}\begin{pmatrix} x \\ y \end{pmatrix} \\
&= \begin{pmatrix} 1 & 0 \\ 0 & 1 \end{pmatrix}\begin{pmatrix} x \\ y \end{pmatrix} = \begin{pmatrix} x \\ y \end{pmatrix}.
\end{aligned}$$

If we now equate the LHS and the RHS of equation (f) we find

$$\begin{pmatrix} x \\ y \end{pmatrix} = \begin{pmatrix} 2 \\ 3 \end{pmatrix} \quad \text{or} \quad \begin{aligned} x &= 2 \\ y &= 3. \end{aligned}$$

Summary

Starting with the original simultaneous equations, we have moved through the following stages

$$\left.\begin{aligned} 5x - 3y &= 1 \\ x + 2y &= 8 \end{aligned}\right\}\text{original form}$$

$$\left.\begin{pmatrix} 5 & -3 \\ 1 & 2 \end{pmatrix}\begin{pmatrix} x \\ y \end{pmatrix} = \begin{pmatrix} 1 \\ 8 \end{pmatrix}\right\}\text{matrix form}$$

$$\left.\frac{1}{13}\begin{pmatrix} 2 & 3 \\ -1 & 5 \end{pmatrix}\begin{pmatrix} 5 & -3 \\ 1 & 2 \end{pmatrix}\begin{pmatrix} x \\ y \end{pmatrix} = \frac{1}{13}\begin{pmatrix} 2 & 3 \\ -1 & 5 \end{pmatrix}\begin{pmatrix} 1 \\ 8 \end{pmatrix}\right\}\begin{array}{l}\text{multiplying both sides} \\ \text{by the same matrix}\end{array}$$

$$\left.\begin{pmatrix} x \\ y \end{pmatrix} = \begin{pmatrix} 2 \\ 3 \end{pmatrix}\right\}\begin{array}{l}\text{condensing each side using the rules for matrix} \\ \text{multiplication}\end{array}$$

$$\left.\begin{aligned} x &= 2 \\ \underline{y} &\underline{= 3} \end{aligned}\right\}\text{to find the solution.}$$

The key step in the process was the multiplication by the matrix $\boldsymbol{B}$ where

$$\boldsymbol{B} = \frac{1}{13}\begin{pmatrix} 2 & 3 \\ -1 & 5 \end{pmatrix} = \begin{pmatrix} \frac{2}{13} & \frac{3}{13} \\ -\frac{1}{13} & \frac{5}{13} \end{pmatrix}.$$

Now, on the assumption that we can always find the equivalent of the matrix $\boldsymbol{B}$, the method of solution is very simple. We simply multiply the matrix form of the original simultaneous equations by $\boldsymbol{B}$ on both sides, and we immediately obtain the solution. The core of the problem is now to find the matrix $\boldsymbol{B}$ in the general case. The big clue to the nature of $\boldsymbol{B}$ is to be found from the matrix product of $\boldsymbol{B}$ and the matrix of original coefficients ($\boldsymbol{A}$ say); that is,

$$\underbrace{\frac{1}{13}\begin{pmatrix} 2 & 3 \\ -1 & 5 \end{pmatrix}\begin{pmatrix} 5 & -3 \\ 1 & 2 \end{pmatrix}}_{\boldsymbol{B} \times \boldsymbol{A}} = \frac{1}{13}\begin{pmatrix} 13 & 0 \\ 0 & 13 \end{pmatrix} = \begin{pmatrix} 1 & 0 \\ 0 & 1 \end{pmatrix}$$

$$\boldsymbol{B} \times \boldsymbol{A} \qquad = \boldsymbol{I}.$$

The product is called the *unit* or *identity* matrix $\boldsymbol{I}$ and it leaves another matrix unchanged in a product (similar to the number 1 in multiplication):

$$\boldsymbol{IA} = \boldsymbol{AI} = \boldsymbol{A}.$$

In fact, because both $\boldsymbol{B}$ and $\boldsymbol{A}$ are also square matrices, (2×2), each is the *inverse* of the other. In matrix notation

$$\boldsymbol{B} = \boldsymbol{A}^{-1} \quad \text{or} \quad \boldsymbol{A} = \boldsymbol{B}^{-1}$$

and

$$\boldsymbol{AA}^{-1} = \boldsymbol{A}^{-1}\boldsymbol{A} = \boldsymbol{I}.$$

So, if we use the following shorthand for the matrices shown, the solution of the simultaneous equations can be represented in a very simple fashion. Let

$$A = \begin{pmatrix} 5 & -3 \\ 1 & 2 \end{pmatrix} \quad B = A^{-1} = \frac{1}{13}\begin{pmatrix} 2 & 3 \\ -1 & 5 \end{pmatrix} \quad X = \begin{pmatrix} x \\ y \end{pmatrix} \quad C = \begin{pmatrix} 1 \\ 8 \end{pmatrix}.$$

then

$$\begin{aligned} AX &= C && \text{(original equations)} \\ A^{-1}AX &= A^{-1}C && \text{(multiply both sides by the inverse of } A\text{)} \\ IX &= A^{-1}C && \text{(unit matrix appears on LHS)} \\ \therefore X &= A^{-1}C && \text{(solution).} \end{aligned}$$

Methods for finding the inverse matrix and for solving sets of linear equations will be developed in the next chapter. We now want to consolidate the material of this chapter by defining more formally the properties of matrices.

16.4 FORMAL PROPERTIES OF MATRICES

1. A **matrix** is a rectangular array of numbers, usually enclosed in brackets, and denoted by capital letters. The size of a matrix is defined by the number of rows and columns; for example,

$$A = \begin{pmatrix} 3 & 5 & 4 \\ 2 & 0 & -1 \end{pmatrix}$$ (2×3) matrix with 6 **elements**.

2. Elements are usually denoted by small letters, and the element in the ith row and the jth column can be denoted by suffixes thus: a_{ij}; for example,

$$A = \begin{pmatrix} a_{11} & a_{12} & a_{13} \\ a_{21} & a_{22} & a_{23} \end{pmatrix}.$$

3. The **zero matrix 0** has all elements equal to zero.

4. A **square matrix** has equal numbers of rows and columns.

5. The **unit** or **identity matrix**, I, is square, with all elements zero except on the principal diagonal, where all elements equal one; for example,

$$I = \begin{pmatrix} 1 & 0 & 0 \\ 0 & 1 & 0 \\ 0 & 0 & 1 \end{pmatrix}$$ is a unit matrix of *order* 3.

6. If A and B are square matrices of the same order (size) such that $AB = BA = I$ then B is called the *inverse* of A (and vice versa), and may be

written as A^{-1}, so that

$$AA^{-1} = A^{-1}A = I.$$

7. Two matrices A and B of the *same size* are *equal* if corresponding elements in the two matrices are all equal, so that $a_{ij} = b_{ij}$ for all i and j; for example,

$$\begin{pmatrix} 2 & 3 \\ 0 & -1 \end{pmatrix} = \begin{pmatrix} a & b \\ c & d \end{pmatrix} \quad \text{implies that } a = 2, b = 3, c = 0, d = -1$$

8. If a matrix A is *multiplied* by a numerical factor (or *scalar*) then all the elements are multiplied by the factor; for example,

$$3\begin{pmatrix} 1 & 2 \\ -3 & 0 \end{pmatrix} = \begin{pmatrix} 3 & 6 \\ -9 & 0 \end{pmatrix}.$$

9. Two matrices A and B may be *added* (or *subtracted*) if they are of the *same size*. Corresponding elements add (or subtract); for example,

$$\begin{pmatrix} 1 & 3 & 2 \\ -2 & 5 & 0 \end{pmatrix} + \begin{pmatrix} 0 & 4 & 7 \\ 3 & -2 & 1 \end{pmatrix} = \begin{pmatrix} 1 & 7 & 9 \\ 1 & 3 & 1 \end{pmatrix}.$$

10. Rules 8 and 9 above, when combined, allow us to take a *linear combination* of any number of matrices as long as they are all of the *same size*. Again, the corresponding elements combine according to the particular linear combination; for example,

$$\text{if } A = 2B - 3C + D \quad \text{then} \quad a_{ij} = 2b_{ij} - 3c_{ij} + d_{ij}.$$

11. The *transpose* of a matrix A, written as A', is the matrix formed by interchanging rows and columns of matrix A. Unless A is square, therefore, the dimensions of A' are reversed compared with A; for example,

$$A = \begin{pmatrix} 1 & 3 & 4 \\ 0 & -2 & 6 \end{pmatrix} \quad A' = \begin{pmatrix} 1 & 0 \\ 3 & -2 \\ 4 & 6 \end{pmatrix} \quad \begin{matrix} A \text{ is 2 by 3} \\ A' \text{ is 3 by 2.} \end{matrix}$$

12. A matrix A, of dimensions $(m \times n)$, may be multiplied by another matrix B, of dimensions $(n \times p)$, to give a matrix C of dimensions $(m \times p)$.

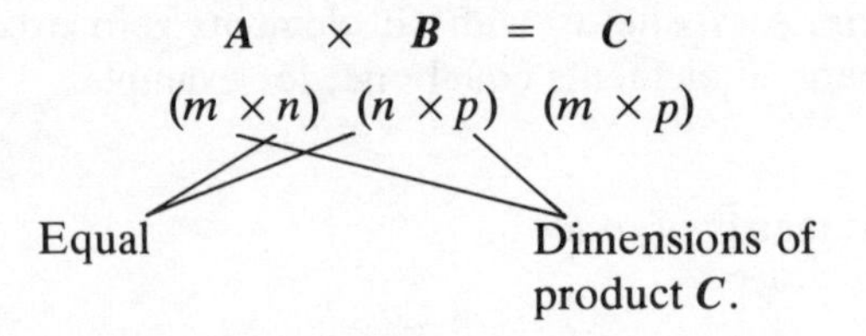

Notice that the number of *columns in the first matrix* (n) must equal the number of *rows in the second matrix* (n). This means that unless $m = p$ as well, the product $B \times A$ is not possible.

13. Even if $m = n = p$, so that both products $\boldsymbol{AB}$ and $\boldsymbol{BA}$ are possible, and they have the same dimensions, in general we find $\boldsymbol{AB} \neq \boldsymbol{BA}$. If $\boldsymbol{AB} = \boldsymbol{BA}$ for particular cases, then $\boldsymbol{A}$ and $\boldsymbol{B}$ are said to *commute*. For example,

$$\boldsymbol{A} = \begin{pmatrix} 1 & 2 \\ 0 & 3 \end{pmatrix} \quad \boldsymbol{B} = \begin{pmatrix} 4 & -1 \\ 5 & 0 \end{pmatrix} \quad \boldsymbol{AB} = \begin{pmatrix} 14 & -1 \\ 15 & 0 \end{pmatrix} \quad \boldsymbol{BA} = \begin{pmatrix} 4 & 5 \\ 5 & 10 \end{pmatrix}$$

(so that $\boldsymbol{A}$ and $\boldsymbol{B}$ do *not* commute in this case).

14. The rule for *matrix multiplication* seems curious, but if we refer back to Section 16.2 it should be clearer how the method is built up; for example,

$$\boldsymbol{A} = \begin{pmatrix} 1 & 2 & 3 \\ 4 & 5 & 6 \end{pmatrix} \quad \boldsymbol{B} = \begin{pmatrix} 7 & 10 \\ 8 & 11 \\ 9 & 12 \end{pmatrix} \quad \begin{array}{l} \boldsymbol{A} \text{ is } (2 \times 3) \\ \boldsymbol{B} \text{ is } (3 \times 2). \end{array}$$

Because of the dimensions of $\boldsymbol{A}$ and $\boldsymbol{B}$, both $\boldsymbol{AB}$ and $\boldsymbol{BA}$ are possible:

$$\boldsymbol{C} = \boldsymbol{AB} \text{ will have dimensions } (2 \times \not{3})(\not{3} \times 2) \quad \text{i.e. } (2 \times 2)$$

$$\boldsymbol{D} = \boldsymbol{BA} \text{ will have dimensions } (3 \times \not{2})(\not{2} \times 3) \quad \text{i.e. } (3 \times 3)$$

$$\boldsymbol{C} = \begin{pmatrix} 1 & 2 & 3 \\ \hline 4 & 5 & 6 \end{pmatrix} \left(\begin{array}{c|c} 7 & 10 \\ 8 & 11 \\ 9 & 12 \end{array}\right)$$

$$= \begin{pmatrix} 1.7 + 2.8 + 3.9 & 1.10 + 2.11 + 3.12 \\ 4.7 + 5.8 + 6.9 & 4.10 + 5.11 + 6.12 \end{pmatrix} = \begin{pmatrix} 50 & 68 \\ 122 & 167 \end{pmatrix}$$

$$\boldsymbol{D} = \begin{pmatrix} 7 & 10 \\ \hline 8 & 11 \\ 9 & 12 \end{pmatrix} \left(\begin{array}{c|c|c} 1 & 2 & 3 \\ 4 & 5 & 6 \end{array}\right)$$

$$= \begin{pmatrix} 7.1 + 10.4 & 7.2 + 10.5 & 7.3 + 10.6 \\ 8.1 + 11.4 & 8.2 + 11.5 & 8.3 + 11.6 \\ 9.1 + 12.4 & 9.2 + 12.5 & 9.3 + 12.6 \end{pmatrix} = \begin{pmatrix} 47 & 64 & 81 \\ 52 & 71 & 90 \\ 57 & 78 & 99 \end{pmatrix}.$$

15. As long as the dimensions of the matrices are appropriate we can form a product from three or more matrices; for example,

$$\boldsymbol{ABC} = (\boldsymbol{AB})\boldsymbol{C} = \boldsymbol{A}(\boldsymbol{BC}) \qquad \textit{associative}$$

but

$$\boldsymbol{ABC} \neq \boldsymbol{BAC} \neq \boldsymbol{BCA} \text{ etc.} \qquad \textit{non-commutative}.$$

If

$$\boldsymbol{A} \text{ is } (m \times n), \boldsymbol{B} \text{ is } (n \times p), \text{ and } \boldsymbol{C} \text{ is } (p \times q)$$

then

$$\boldsymbol{ABC} \text{ is } (m \times q).$$

16. The product of the appropriate unit matrix and any matrix $\boldsymbol{A}$ leaves $\boldsymbol{A}$ unchanged. (*Appropriate* means the right size.)

$$\boldsymbol{IA} = \boldsymbol{A} \quad \text{or} \quad \boldsymbol{AI} = \boldsymbol{A}.$$

16.5 BASIC EXAMPLES FOR THE STUDENT

B16.1 Put the following expressions in matrix form

(a) $5x + 3y$ (c) $2x - y$ and

(b) $-x + y$ $3x + 4y$.

B16.2 Put the following sets of simultaneous equations into matrix form

(a)
$$\begin{aligned} 2x + 3y &= 6 \\ 4x - y &= 2 \end{aligned}$$

(b)
$$\begin{aligned} x - y + 4z &= 16 \\ 3x + 2z &= 3 \\ 5x + 4y - z &= 7. \end{aligned}$$

B16.3 Suppose

$$\boldsymbol{A} = \begin{pmatrix} 2 & 0 \\ 1 & -3 \end{pmatrix} \quad \boldsymbol{B} = \begin{pmatrix} 1 & 3 & 0 \\ 4 & -1 & 5 \end{pmatrix} \quad \boldsymbol{C} = \begin{pmatrix} -3 & 0 \\ -1 & 2 \end{pmatrix}$$

(a) What values do a_{21} and b_{32} take?

(b) Write down the matrix $3\boldsymbol{A}$ in full.

(c) Find $(\boldsymbol{A} + \boldsymbol{C})$ and $(\boldsymbol{A} + \boldsymbol{B})$.

(d) Find $\boldsymbol{AB}$, $\boldsymbol{AC}$, $\boldsymbol{CA}$, and $\boldsymbol{BA}$.

(e) What is the inverse matrix of $\boldsymbol{A}$?

(f) Solve the following simultaneous equations using matrix algebra

$$\begin{aligned} 2x \quad &= 8 \\ x - 3y &= -2. \end{aligned}$$

16.6 INTERMEDIATE EXAMPLES FOR THE STUDENT

I16.1 Check by multiplication that the inverse of matrix $\boldsymbol{A}$ is matrix $\boldsymbol{B}$:

$$\boldsymbol{A} = \begin{pmatrix} 1 & 0 & 2 \\ -1 & 3 & 2 \\ 0 & -2 & 1 \end{pmatrix} \quad \text{and} \quad \boldsymbol{B} = \frac{1}{11}\begin{pmatrix} 7 & -4 & -6 \\ 1 & 1 & -4 \\ 2 & 2 & 3 \end{pmatrix}.$$

Use the inverse to solve the following simultaneous equations by matrix algebra

$$\begin{aligned} x \quad + 2z &= 6 \\ -x + 3y + 2z &= 1 \\ -2y + z &= -1. \end{aligned}$$

I16.2
$$\boldsymbol{A} = \begin{pmatrix} 1 & 3 & 0 \\ 0 & 4 & -1 \\ 2 & 0 & -3 \end{pmatrix} \qquad \boldsymbol{C} = \begin{pmatrix} 3 \\ 0 \\ 1 \end{pmatrix}.$$

Find (i) $\boldsymbol{AC}$
(ii) $\boldsymbol{CA}$
(iii) $\boldsymbol{C'A}$
(iv) $\boldsymbol{C'A'}$.

What relationship exists between $\boldsymbol{AC}$ and $\boldsymbol{C'A'}$?

I16.3 Suppose that the inverse of the matrix $\boldsymbol{A}$ is given by $\boldsymbol{B}$, where

$$\boldsymbol{A} = \begin{pmatrix} 3 & 2 \\ 1 & -1 \end{pmatrix} \qquad \boldsymbol{B} = \begin{pmatrix} a & b \\ c & d \end{pmatrix}.$$

Use the definition of the inverse to find values for a, b, c, and d and hence find $\boldsymbol{B}$.

CHAPTER 17

The Gauss–Jordan Technique

17.1 INTRODUCTION

The intention in this chapter is to develop a standard matrix technique for finding

(i) solutions to sets of simultaneous linear equations, and
(ii) the inverse of a square matrix.

The method to be demonstrated is known as the Gauss–Jordan technique, and in principle is no different from the usual manual method of solving simultaneous equations by systematically eliminating variables one by one. The usefulness (and tediousness!) of the method rests in its very routine and systematic rules, which make the method ideal for translating large-scale problems into a form soluble by computer.

A standard set of rules for undertaking a type of calculation, such as the Gauss–Jordan method, is generally called an *algorithm*.

17.2 SIMULTANEOUS EQUATIONS

From the previous chapter, it will be recalled that a set of two simultaneous equations in two variables can be solved in the following way, using matrix algebra:

$$\left.\begin{aligned} 5x - 3y &= 1 \\ x + 2y &= 8 \end{aligned}\right\}\text{original form}$$

$$\begin{pmatrix} 5 & -3 \\ 1 & 2 \end{pmatrix}\begin{pmatrix} x \\ y \end{pmatrix} = \begin{pmatrix} 1 \\ 8 \end{pmatrix} \quad \Big\}\text{matrix form} \quad \Big\{ \boldsymbol{AX} = \boldsymbol{C}.$$

Multiply both sides by the appropriate (inverse) matrix

$$\frac{1}{13}\begin{pmatrix} 2 & 3 \\ -1 & 5 \end{pmatrix}\begin{pmatrix} 5 & -3 \\ 1 & 2 \end{pmatrix}\begin{pmatrix} x \\ y \end{pmatrix} = \frac{1}{13}\begin{pmatrix} 2 & 3 \\ -1 & 5 \end{pmatrix}\begin{pmatrix} 1 \\ 8 \end{pmatrix} \qquad \boldsymbol{A}^{-1}\boldsymbol{AX} = \boldsymbol{A}^{-1}\boldsymbol{C}$$

and then, since the product $\boldsymbol{A}^{-1}\boldsymbol{A}$ is the unit matrix $\boldsymbol{I}$,

$$\frac{1}{13}\begin{pmatrix}13 & 0\\ 0 & 13\end{pmatrix}\begin{pmatrix}x\\ y\end{pmatrix} = \begin{pmatrix}1 & 0\\ 0 & 1\end{pmatrix}\begin{pmatrix}x\\ y\end{pmatrix} = \frac{1}{13}\begin{pmatrix}2 & 3\\ -1 & 5\end{pmatrix}\begin{pmatrix}1\\ 8\end{pmatrix} \qquad \boldsymbol{IX} = \boldsymbol{A}^{-1}\boldsymbol{C}.$$

We finally obtain

$$\begin{pmatrix}x\\ y\end{pmatrix} = \frac{1}{13}\begin{pmatrix}26\\ 39\end{pmatrix} = \begin{pmatrix}2\\ 3\end{pmatrix} \qquad \boldsymbol{X} = \boldsymbol{A}^{-1}\boldsymbol{C} = \boldsymbol{D} \text{ (say).}$$

The solution is obtained when we have the final form

$$\boldsymbol{X} = \boldsymbol{A}^{-1}\boldsymbol{C} = \boldsymbol{D} \qquad \text{(a)}$$

because this essentially lists the values of x, y (z, etc.) in matrix form

$$\begin{pmatrix}x\\ y\end{pmatrix} = \begin{pmatrix}2\\ 3\end{pmatrix} \quad \text{or} \quad x = 2 \text{ and } y = 3.$$

It is worth emphasizing, therefore, that we wish to reach an equation of a type similar to (a) above, and in particular, *to find* $\boldsymbol{D}$.

Two methods will be demonstrated (both of which will employ the Gauss–Jordan (GJ) technique).

(i) If one inspects the stages in the above example, it will be seen that the whole process could be cut down to the following basic steps:

$$\boldsymbol{AX} = \boldsymbol{C}$$

$$\boldsymbol{IX} = \boldsymbol{D}.$$

That is, if we can convert $\boldsymbol{A}$ to $\boldsymbol{I}$, then $\boldsymbol{C}$ is converted to $\boldsymbol{D}$. The first demonstrations of the GJ technique will be doing just this.

(ii) If the *inverse* of $\boldsymbol{A}$ could be found, then $\boldsymbol{D}$ can be obtained directly by putting

$$\boldsymbol{D} = \boldsymbol{A}^{-1}\boldsymbol{C}$$

since $\boldsymbol{C}$ is already known.

In the above example

$$\boldsymbol{C} = \begin{pmatrix}1\\ 8\end{pmatrix} \qquad \boldsymbol{A}^{-1} = \frac{1}{13}\begin{pmatrix}2 & 3\\ -1 & 5\end{pmatrix}$$

so that

$$\boldsymbol{D} = \boldsymbol{A}^{-1}\boldsymbol{C} = \frac{1}{13}\begin{pmatrix}2 & 3\\ -1 & 5\end{pmatrix}\begin{pmatrix}1\\ 8\end{pmatrix} = \begin{pmatrix}2\\ 3\end{pmatrix}.$$

This second method is actually lengthier; but if we had also to solve

$$\begin{aligned} 5x - 3y &= 17 \\ x + 2y &= 6 \end{aligned} \qquad \text{(notice that the } \textit{coefficients} \text{ have not changed)}$$

then

$$AX = \begin{pmatrix} 17 \\ 6 \end{pmatrix}$$

$$A^{-1}AX = \frac{1}{13}\begin{pmatrix} 2 & 3 \\ -1 & 5 \end{pmatrix}\begin{pmatrix} 17 \\ 6 \end{pmatrix} = \frac{1}{13}\begin{pmatrix} 52 \\ 15 \end{pmatrix} = \begin{pmatrix} 4 \\ 1 \end{pmatrix}$$

and we have very quickly found the new solutions, $x = 4$, $y = 1$. We now develop the GJ technique in full, for each of the methods suggested above.

17.3 THE GAUSS–JORDAN TECHNIQUE (i)

We shall demonstrate the technique by applying it to both the matrix notation and the usual algebraic form of the equations to show the equivalence more clearly.

We wish to solve the simultaneous equations below (i.e. find x, y and z)

Algebraic form / *Matrix form*

$$\begin{aligned} 4x - y + 3z &= 11 \\ 2x \quad + z &= 5 \\ -3x + y + 2z &= 5 \end{aligned} \qquad \begin{pmatrix} 4 & -1 & 3 \\ 2 & 0 & 2 \\ -3 & 1 & 2 \end{pmatrix}\begin{pmatrix} x \\ y \\ z \end{pmatrix} = \begin{pmatrix} 11 \\ 5 \\ 5 \end{pmatrix}.$$

Step 1 Reduce the *first* coefficient in the *first* equation to 1 by dividing the whole equation by 4.

$$\begin{aligned} x - \frac{1}{4}y + \frac{3}{4}z &= \frac{11}{4} \\ 2x \quad + z &= 5 \\ -3x + y + 2z &= 5 \end{aligned} \qquad \left|\begin{array}{ccc|c} 1 & -\frac{1}{4} & \frac{3}{4} & \frac{11}{4} \\ 2 & 0 & 1 & 5 \\ -3 & 1 & 2 & 5 \end{array}\right|.$$

An abbreviated notation is being used for the matrix version, since it is not essential to involve x, y, and z in the changes which are made.

Step 2 Eliminate the first variable (x) in all equations except the first, by subtracting suitable multiples of this first equation from the others. The first equation itself should remain unchanged.

$$\begin{aligned} x - \frac{1}{4}y + \frac{3}{4}z &= \frac{11}{4} \\ 0 + \frac{1}{2}y - \frac{1}{2}z &= -\frac{1}{2} \\ 0 + \frac{1}{4}y + \frac{17}{4}z &= \frac{53}{4} \end{aligned} \qquad \left|\begin{array}{ccc|c} 1 & -\frac{1}{4} & \frac{3}{4} & \frac{11}{4} \\ 0 & \frac{1}{2} & -\frac{1}{2} & -\frac{1}{2} \\ 0 & \frac{1}{4} & \frac{17}{4} & \frac{53}{4} \end{array}\right| \quad \begin{array}{l} \\ \text{Subtracted } 2 \times \text{eqn 1} \\ \text{Added } 3 \times \text{eqn 1.} \end{array}$$

Step 3 We now produce a coefficient of 1 in the next position down the principal diagonal. So, the second equation is multiplied by 2 to give a coefficient of 1 for the variable y. The other two equations remain unchanged.

$$
\begin{array}{l}
x - \frac{1}{4}y + \frac{3}{4}z = \frac{11}{4} \\
0 + \ \ y - \ \ z = -1 \\
0 + \frac{1}{4}y + \frac{17}{4}z = \frac{53}{4}
\end{array}
\quad
\left|\begin{array}{ccc|c}
1 & -\frac{1}{4} & \frac{3}{4} & \frac{11}{4} \\
0 & 1 & -1 & -1 \\
0 & \frac{1}{4} & \frac{17}{4} & \frac{53}{4}
\end{array}\right|
\quad \begin{array}{l} \\ \text{Multiplied eqn 2 by 2.} \\ \\ \end{array}
$$

Step 4 (Similar to step 2.) Eliminate the second variable (y) in all equations except the second, by subtracting suitable multiples of this second equation from the others. The second equation itself should remain unchanged.

$$
\begin{array}{l}
x + 0 + \frac{1}{2}z = \frac{10}{4} \\
0 + y - \ \ z = -1 \\
0 + 0 + \frac{18}{4}z = \frac{54}{4}
\end{array}
\quad
\left|\begin{array}{ccc|c}
1 & 0 & \frac{1}{2} & \frac{10}{4} \\
0 & 1 & -1 & -1 \\
0 & 0 & \frac{18}{4} & \frac{54}{4}
\end{array}\right|
\quad \begin{array}{l} \text{Added } \frac{1}{4} \text{ of eqn 2} \\ \\ \text{Subtracted } \frac{1}{4} \text{ of eqn 2.} \end{array}
$$

Step 5 (Similar to steps 1 and 3.) We now produce a coefficient of 1 in the next position down the principal diagonal. So, the third equation is multiplied by 4/18 to give a coefficient of 1 for the variable z. The other two equations remain unchanged.

$$
\begin{array}{l}
x + 0 + \frac{1}{2}z = \frac{10}{4} \\
0 + y - \ \ z = -1 \\
0 + 0 + \ \ z = \ \ 3
\end{array}
\quad
\left|\begin{array}{ccc|c}
1 & 0 & \frac{1}{2} & \frac{10}{4} \\
0 & 1 & -1 & -1 \\
0 & 0 & 1 & 3
\end{array}\right|
\quad \begin{array}{l} \\ \\ \text{Multiplied eqn 3 by 4/18.} \end{array}
$$

Step 6 (Similar to steps 2 and 4.) Eliminate the third variable (z) in all equations except the third, by subtracting suitable multiples of this third equation from the other two. The third equation itself should remain unchanged.

$$
\begin{array}{l}
x + 0 + 0 = \frac{4}{4} = 1 \\
0 + y + 0 = 2 \\
0 + 0 + z = 3
\end{array}
\quad
\left|\begin{array}{ccc|c}
1 & 0 & 0 & 1 \\
0 & 1 & 0 & 2 \\
0 & 0 & 1 & 3
\end{array}\right|
\quad \begin{array}{l} \text{Subtracted } \frac{1}{2} \text{ of eqn 3} \\ \text{Added eqn 3.} \\ \\ \end{array}
$$

This is the desired result. Rewriting in the full matrix form we have

$$
\begin{pmatrix} 1 & 0 & 0 \\ 0 & 1 & 0 \\ 0 & 0 & 1 \end{pmatrix}
\begin{pmatrix} x \\ y \\ z \end{pmatrix} =
\begin{pmatrix} 1 \\ 2 \\ 3 \end{pmatrix}
\quad \text{or} \quad
\begin{pmatrix} x \\ y \\ z \end{pmatrix} =
\begin{pmatrix} 1 \\ 2 \\ 3 \end{pmatrix}
$$

$\therefore \underline{x = 1, y = 2, z = 3.}$

Notice that we have carried out the intention indicated in paragraph (i) above:

$$AX = C \quad \text{has become} \quad IX = D.$$

The sequence of operations in the GJ technique is important. The matrix A is converted to I by developing the pattern in the order below

$$\begin{pmatrix} * & * & * \\ * & * & * \\ * & * & * \end{pmatrix} \to \begin{pmatrix} 1 & * & * \\ * & * & * \\ * & * & * \end{pmatrix} \to \begin{pmatrix} 1 & * & * \\ 0 & * & * \\ 0 & * & * \end{pmatrix} \to \begin{pmatrix} 1 & * & * \\ 0 & 1 & * \\ 0 & * & * \end{pmatrix}$$

$$\to \begin{pmatrix} 1 & 0 & * \\ 0 & 1 & * \\ 0 & 0 & * \end{pmatrix} \to \begin{pmatrix} 1 & 0 & * \\ 0 & 1 & * \\ 0 & 0 & 1 \end{pmatrix} \to \begin{pmatrix} 1 & 0 & 0 \\ 0 & 1 & 0 \\ 0 & 0 & 1 \end{pmatrix}.$$

The diagonal row of 'ones' are created by *multiplication/division of the row*. The 'zeros' are created by *adding/subtracting multiples of the row with the 'one' most recently created*. This sequence is important, and it should be noticed that once created, the zeros and ones are not subsequently disturbed.

Strictly speaking it is not essential that the 'ones' should appear in a diagonal, or that they are developed in this particular order (though there must be only a single 'one' in each row or column). We shall keep to this order throughout this section, however, to avoid any confusion.

17.4 THE GAUSS–JORDAN TECHNIQUE (ii) (INVERSE)

A similar technique can be developed to find the inverse of any square matrix.

By the definition of the inverse

$$\left.\begin{aligned} &A^{-1}A = AA^{-1} = I \\ \text{but} \\ &IA^{-1} = A^{-1}. \end{aligned}\right\} \qquad \text{(b)}$$

So, by converting A into I, we convert I into A^{-1}.

Again, we illustrate this with an example. Suppose we wish to find the inverse of A where

$$A = \begin{pmatrix} 4 & -1 & 3 \\ 2 & 0 & 1 \\ -3 & 1 & 2 \end{pmatrix} \quad A^{-1} = \begin{pmatrix} a & b & c \\ d & e & f \\ g & h & i \end{pmatrix}.$$

The two equations (b) can be written:

$$\begin{pmatrix} 4 & -1 & 3 \\ 2 & 0 & 1 \\ -3 & 1 & 2 \end{pmatrix}\begin{pmatrix} a & b & c \\ d & e & f \\ g & h & i \end{pmatrix} = \begin{pmatrix} 1 & 0 & 0 \\ 0 & 1 & 0 \\ 0 & 0 & 1 \end{pmatrix} \quad \text{or} \quad \boldsymbol{AA}^{-1} = \boldsymbol{I}$$

$$\begin{pmatrix} 1 & 0 & 0 \\ 0 & 1 & 0 \\ 0 & 0 & 1 \end{pmatrix}\begin{pmatrix} a & b & c \\ d & e & f \\ g & h & i \end{pmatrix} = \begin{pmatrix} a & b & c \\ d & e & f \\ g & h & i \end{pmatrix} \quad \text{or} \quad \boldsymbol{IA}^{-1} = \boldsymbol{A}^{-1}.$$

By converting $\boldsymbol{A}$ to $\boldsymbol{I}$ (LHS), then $\boldsymbol{I}$ is converted to $\boldsymbol{A}^{-1}$ (RHS). We use the same GJ method.

Step 1

$$\left|\begin{array}{ccc|ccc} 4 & -1 & 3 & 1 & 0 & 0 \\ 2 & 0 & 1 & 0 & 1 & 0 \\ -3 & 1 & 2 & 0 & 0 & 1 \end{array}\right|$$

$$\left|\begin{array}{ccc|ccc} 1 & -\frac{1}{4} & \frac{3}{4} & \frac{1}{4} & 0 & 0 \\ 2 & 0 & 1 & 0 & 1 & 0 \\ -3 & 1 & 2 & 0 & 0 & 1 \end{array}\right| \quad \begin{array}{l} \text{Divided eqn 1 by 4} \\ \\ \\ \end{array}$$

$$\left|\begin{array}{ccc|ccc} 1 & -\frac{1}{4} & \frac{3}{4} & \frac{1}{4} & 0 & 0 \\ 0 & \frac{1}{2} & -\frac{1}{2} & -\frac{1}{2} & 1 & 0 \\ 0 & \frac{1}{4} & \frac{17}{4} & \frac{3}{4} & 0 & 1 \end{array}\right| \quad \begin{array}{l} \\ \text{Subtracted } 2 \times \text{eqn 1} \\ \text{Added } 3 \times \text{eqn 1} \end{array}$$

$$\left|\begin{array}{ccc|ccc} 1 & -\frac{1}{4} & \frac{3}{4} & \frac{1}{4} & 0 & 0 \\ 0 & 1 & -1 & -1 & 2 & 0 \\ 0 & \frac{1}{4} & \frac{17}{4} & \frac{3}{4} & 0 & 1 \end{array}\right| \quad \begin{array}{l} \\ \text{Multiplied eqn 2 by 2} \\ \\ \end{array}$$

$$\left|\begin{array}{ccc|ccc} 1 & 0 & \frac{1}{2} & 0 & \frac{1}{2} & 0 \\ 0 & 1 & -1 & -1 & 2 & 0 \\ 0 & 0 & \frac{18}{4} & 1 & -\frac{1}{2} & 1 \end{array}\right| \quad \begin{array}{l} \text{Added } \frac{1}{4} \text{ of eqn 2} \\ \\ \text{Subtracted } \frac{1}{4} \text{ of eqn 2} \end{array}$$

$$\left|\begin{array}{ccc|ccc} 1 & 0 & \frac{1}{2} & 0 & \frac{1}{2} & 0 \\ 0 & 1 & -1 & -1 & 2 & 0 \\ 0 & 0 & 1 & \frac{2}{9} & -\frac{1}{9} & \frac{2}{9} \end{array}\right| \quad \begin{array}{l} \\ \\ \text{Multiplied by 4/18} \end{array}$$

$$\left|\begin{array}{ccc|ccc|}1 & 0 & 0 & -\frac{1}{9} & \frac{5}{9} & -\frac{1}{9} \\ 0 & 1 & 0 & -\frac{7}{9} & \frac{17}{9} & \frac{2}{9} \\ 0 & 0 & 1 & \frac{2}{9} & -\frac{1}{9} & \frac{2}{9}\end{array}\right. \quad \begin{array}{l}\text{Subtracted } \tfrac{1}{2} \text{ of eqn 3} \\ \text{Added eqn 3.} \\ \\ \end{array}$$

In full, this reads

$$\begin{pmatrix}1 & 0 & 0 \\ 0 & 1 & 0 \\ 0 & 0 & 1\end{pmatrix}\begin{pmatrix}a & b & c \\ d & e & f \\ g & h & i\end{pmatrix} = \frac{1}{9}\begin{pmatrix}-1 & 5 & -1 \\ -7 & 17 & 2 \\ 2 & -1 & 2\end{pmatrix} \quad \text{or} \quad \boldsymbol{I}\boldsymbol{A}^{-1} = \boldsymbol{A}^{-1}.$$

Hence the inverse. $\boldsymbol{A}^{-1}$, is given by the matrix on the RHS. This can be checked by direct multiplication:

$$\boldsymbol{A}\boldsymbol{A}^{-1} = \frac{1}{9}\begin{pmatrix}4 & -1 & 3 \\ 2 & 0 & 1 \\ -3 & 1 & 2\end{pmatrix}\begin{pmatrix}-1 & 5 & -1 \\ -7 & 17 & 2 \\ 2 & -1 & 2\end{pmatrix} = \frac{1}{9}\begin{pmatrix}9 & 0 & 0 \\ 0 & 9 & 0 \\ 0 & 0 & 9\end{pmatrix} = \begin{pmatrix}1 & 0 & 0 \\ 0 & 1 & 0 \\ 0 & 0 & 1\end{pmatrix} = \boldsymbol{I}.$$

Similarly, one can show that $\boldsymbol{A}^{-1}\boldsymbol{A} = \boldsymbol{I}$.

17.5 USE OF THE INVERSE

Solve the three simultaneous equations

$$\left.\begin{aligned}4x - y + 3z &= 1 \\ 2x \qquad + z &= -1 \\ -3x + y + 2z &= 12\end{aligned}\right\} \quad \text{notice that the coefficients on the LHS give the same matrix } \boldsymbol{A} \text{ as above}$$

$$\therefore \begin{pmatrix}4 & -1 \\ 2 & 0 \\ -3 & 1\end{pmatrix}\begin{pmatrix}3 \\ 1 \\ 2\end{pmatrix}\begin{pmatrix}x \\ y \\ z\end{pmatrix} = \begin{pmatrix}1 \\ -1 \\ 12\end{pmatrix}$$

$$\therefore \frac{1}{9}\begin{pmatrix}-1 & 5 & -1 \\ -7 & 17 & 2 \\ 2 & -1 & 2\end{pmatrix}\begin{pmatrix}4 & -1 & 3 \\ 2 & 0 & 1 \\ -3 & 1 & 2\end{pmatrix}\begin{pmatrix}x \\ y \\ z\end{pmatrix} = \frac{1}{9}\begin{pmatrix}-1 & 5 & -1 \\ -7 & 17 & 2 \\ 2 & -1 & 2\end{pmatrix}\begin{pmatrix}1 \\ -1 \\ 12\end{pmatrix}$$

$$\left.\therefore \begin{pmatrix}x \\ y \\ z\end{pmatrix} = \frac{1}{9}\begin{pmatrix}-18 \\ 0 \\ 27\end{pmatrix} = \begin{pmatrix}-2 \\ 0 \\ 3\end{pmatrix}\right\} \quad \text{we have used the properties of the inverse on the LHS}$$

$$\therefore \underline{x = -2, y = 0, z = 3.}$$

17.6 BASIC EXAMPLES FOR THE STUDENT

B17.1 Use the GJ technique to solve the three simultaneous equations:

$$\begin{aligned} 2x - 2y + z &= -8 \\ 3x + y - z &= 0 \\ x + 4y + 3z &= 11. \end{aligned}$$

B17.2 Find the inverse of the matrix $\boldsymbol{A}$, and use it to solve the equation

$$\boldsymbol{AX} = \boldsymbol{C}$$

where

$$\boldsymbol{A} = \begin{pmatrix} 2 & -2 & 1 \\ 3 & 1 & -1 \\ 1 & 4 & 3 \end{pmatrix} \quad \boldsymbol{X} = \begin{pmatrix} x \\ y \\ z \end{pmatrix} \quad \boldsymbol{C} = \begin{pmatrix} 6 \\ 8 \\ 13 \end{pmatrix}.$$

17.7 INTERMEDIATE EXAMPLES FOR THE STUDENT

I17.1 Express the equations below in matrix notation, and solve using the Gauss–Jordan method.

$$\begin{aligned} x + y + z &= 1 \\ 2x - y + z &= -1 \\ x + 3y - z &= 7. \end{aligned}$$

I17.2 Derive the inverse of the matrix $\boldsymbol{A}$

$$\boldsymbol{A} = \begin{pmatrix} 3 & 1 & 0 \\ -1 & 2 & 1 \\ 4 & 2 & 0 \end{pmatrix}$$

and use it to solve the following equations

$$\begin{aligned} x + 3y \quad &= 5 \\ 2x - y + z &= 6 \\ 2x + 4y \quad &= 8. \end{aligned}$$

I17.3 If two square matrices $\boldsymbol{B}$ and $\boldsymbol{C}$ have the property $\boldsymbol{BC} = \boldsymbol{I}$, show that $\boldsymbol{CB} = \boldsymbol{I}$.

CHAPTER 18

Matrix Applications I

18.1 MATRICES AS NETWORKS

Many problems can be appreciated more easily when represented pictorially rather than in formal mathematical symbols. Graphs for example are often simpler to understand, and give an overall view of a relationship between two variables more readily, than the equivalent algebraic representation.

Since a matrix can always be represented by a network-type diagram, it may be worthwhile employing this equivalence if a problem can be readily conceptualized in network terms, rather than using matrix algebra immediately. For example, we could suppose that the network in Figure 18.1 represents a railway system, in which a_{13} denotes the number of trains per day travelling from Station 1 to Station 3 (sixteen in this case). The total number of trains running per day can then be shown either on the network directly, or tabulated as a matrix ($\boldsymbol{A}$):

$$\boldsymbol{A} = \begin{pmatrix} a_{11} & a_{12} & a_{13} \\ a_{21} & a_{22} & a_{23} \\ a_{31} & a_{32} & a_{33} \end{pmatrix} = \begin{pmatrix} 0 & 11 & 16 \\ 8 & 0 & 15 \\ 19 & 12 & 0 \end{pmatrix}.$$

We have not formally linked this matrix to any set of linear equations, however, so that the relevance of the usual rules of matrix algebra has not been demonstrated in this case. The remainder of this chapter will be concerned with two applications of matrices where this network picture is a useful aid to comprehension:

(i) transition probabilities (and Markov chains), and
(ii) input–output models in economics.

18.2 TRANSITION PROBABILITIES

Suppose that from a survey of voting patterns among the English electorate it is found that

(i) of thosc who voted Labour at an election, 40% will vote Labour at the next election.

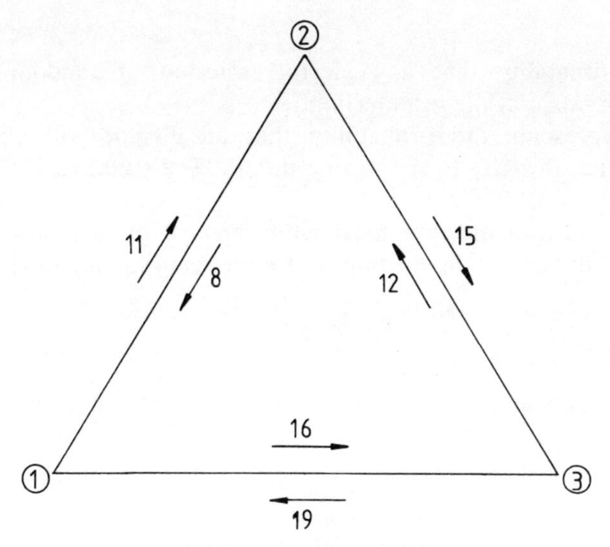

Figure 18.1 A network

(ii) of those who voted Conservative, 65% will do so again at the next election.
(iii) all other electors switch parties.

In the first election, 90% of the electorate voted Labour. How does this pattern change with successive elections? (Ignore problems connected with turnout of electors, and other parties etc.)

We can represent the problem in the following manner. Let

$x_L^{(n)}$ denote the probability that any elector, selected at random, will vote Labour at the nth election

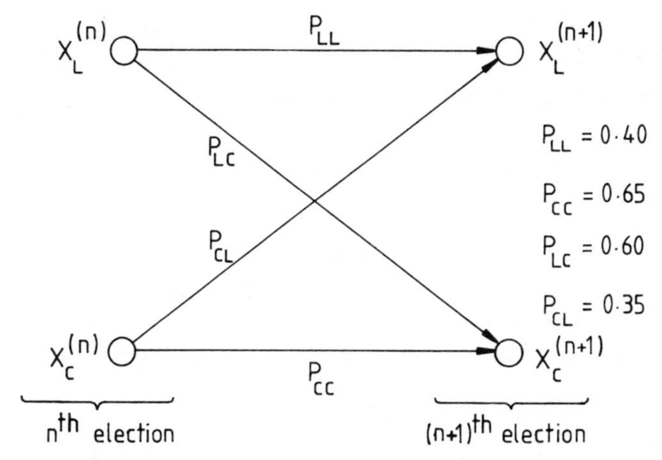

Figure 18.2 Voters' allegiances at two elections

and

$x_C^{(n)}$ the probability that any elector, selected at random, will vote Conservative at the nth election.

Also, p_{ij} represents the probability that an elector will change from supporting the ith party to supporting the jth. The situation is illustrated in Figure 18.2.

Voters at the nth election must either *change* or *maintain* their party allegiance at the $(n + 1)$th election, as we are ignoring any other option.

$$\therefore p_{LC} = 1 - 0{\cdot}40 = 0{\cdot}60$$

$$p_{CL} = 1 - 0{\cdot}65 = 0{\cdot}35.$$

At the $(n + 1)$th election, therefore, the probability that any elector will vote:

Labour is $x_L^{(n+1)} = \underbrace{x_L^{(n)} \times p_{LL}}_{\text{(previous Labour voters)}} + \underbrace{x_C^{(n)} \times p_{CL}}_{\text{(previous Conservative voters)}}$

Conservative is $x_C^{(n+1)} = \underbrace{x_L^{(n)} \times p_{LC}}_{\text{(previous Labour voters)}} + \underbrace{x_C^{(n)} \times p_{CC}}_{\text{(previous Conservative voters)}}$

$$\therefore \begin{pmatrix} x_L^{(n+1)} \\ x_C^{(n+1)} \end{pmatrix} = \begin{pmatrix} p_{LL} & p_{CL} \\ p_{LC} & p_{CC} \end{pmatrix} \begin{pmatrix} x_L^{(n)} \\ x_C^{(n)} \end{pmatrix}$$

or

$$(x_L^{(n+1)}\ x_C^{(n+1)}) = (x_L^{(n)}\ x_C^{(n)}) \begin{pmatrix} p_{LL} & p_{LC} \\ p_{CL} & p_{CC} \end{pmatrix}$$

or

$$\boldsymbol{X}^{(n+1)} = \boldsymbol{X}^{(n)}\boldsymbol{P}.$$

The matrix $\boldsymbol{P}$ is called the *transition probability matrix*; the elements p_{ij} are called the *transition probabilities*.

For the example given,

$$\boldsymbol{P} = \begin{pmatrix} 0{\cdot}40 & 0{\cdot}60 \\ 0{\cdot}35 & 0{\cdot}65 \end{pmatrix}.$$

(Notice that if *row* vectors are used for $\boldsymbol{X}$, then the *rows* of $\boldsymbol{P}$ add up to 1. If the column form of $\boldsymbol{X}$ is used, then the columns of $\boldsymbol{P}$ sum to 1. The row form is more usual, since if we generalize the suffix notation, putting L ≡ 1, C ≡ 2, then

$$\boldsymbol{P} = \begin{pmatrix} p_{11} & p_{12} \\ p_{21} & p_{22} \end{pmatrix}$$

is consistent with the usual matrix conventions.)

18.3 SUCCESSIVE OUTCOMES

In the first election ($n = 1$), 90% of the electorate voted Labour.

$$\therefore x_{L}^{(1)} = 0{\cdot}90 \quad x_{C}^{(1)} = 0{\cdot}10$$

$$\therefore (x_{L}^{(2)} \quad x_{C}^{(2)}) = (x_{L}^{(1)} \quad x_{C}^{(1)})\boldsymbol{P}$$

$$= (0{\cdot}90 \quad 0{\cdot}10)\begin{pmatrix} 0{\cdot}40 & 0{\cdot}60 \\ 0{\cdot}35 & 0{\cdot}65 \end{pmatrix}$$

$$= (0{\cdot}395 \quad 0{\cdot}605).$$

So, at the second election, 39·5% voted Labour and 60·5% voted Conservative.

At the third election,

$$(x_{L}^{(3)} \quad x_{C}^{(3)}) = (x_{L}^{(2)} \quad x_{C}^{(2)})\begin{pmatrix} 0{\cdot}40 & 0{\cdot}60 \\ 0{\cdot}35 & 0{\cdot}65 \end{pmatrix}$$

$$= (0{\cdot}395 \quad 0{\cdot}605)\begin{pmatrix} 0{\cdot}40 & 0{\cdot}60 \\ 0{\cdot}35 & 0{\cdot}65 \end{pmatrix}$$

$$= (0{\cdot}36975 \quad 0{\cdot}63025).$$

So, at the third election, 36·975% voted Labour and 63·025% voted Conservative.

To summarize, the results at the second election are given by

$$\boldsymbol{X}^{(2)} = \boldsymbol{X}^{(1)}\boldsymbol{P}$$

and at the third election by

$$\boldsymbol{X}^{(3)} = \boldsymbol{X}^{(2)}\boldsymbol{P} = \boldsymbol{X}^{(1)}\boldsymbol{P}\boldsymbol{P} = \boldsymbol{X}^{(1)}\boldsymbol{P}^{2}.$$

So we could in fact have calculated the results at the third election directly from those at the first by employing the two-election transition probability matrix, $\boldsymbol{P}^2$.

$$\boldsymbol{P}^2 = \begin{pmatrix} 0{\cdot}40 & 0{\cdot}60 \\ 0{\cdot}35 & 0{\cdot}65 \end{pmatrix}\begin{pmatrix} 0{\cdot}40 & 0{\cdot}60 \\ 0{\cdot}35 & 0{\cdot}65 \end{pmatrix} = \begin{pmatrix} 0{\cdot}37 & 0{\cdot}63 \\ 0{\cdot}3675 & 0{\cdot}6325 \end{pmatrix}.$$

This follows from the associative law of matrix multiplication:

$$(\boldsymbol{XP})\boldsymbol{P} = \boldsymbol{X}(\boldsymbol{PP}) = \boldsymbol{XP}^2.$$

18.4 EVENTUAL OUTCOME

It can be shown that if the matrix $\boldsymbol{P}$ is applied again and again, to derive the results of the fourth, fifth, sixth elections, etc., then an eventual *steady state* results. In other words, we find that the proportion of votes cast for each party eventually reaches a steady value.

Formally,

$$\lim_{n\to\infty} x_{\mathrm{L}}^{(n)} = \text{constant (L)}$$

$$\lim_{n\to\infty} x_{\mathrm{C}}^{(n)} = \text{constant (C).}$$

Rather than finding L and C by successive applications of $\boldsymbol{P}$ however (a long and tedious process, which simply continues the type of operation we were undertaking to find $x_{\mathrm{L}}^{(2)}$, and $x_{\mathrm{L}}^{(3)}$) a neat trick can be employed. If a steady state has been reached, then one more application of $\boldsymbol{P}$ cannot make any difference. So, the eventual parties' support (L and C) can be found from

$$(\mathrm{L} \quad \mathrm{C}) = (\mathrm{L} \quad \mathrm{C})\begin{pmatrix} 0{\cdot}40 & 0{\cdot}60 \\ 0{\cdot}35 & 0{\cdot}65 \end{pmatrix}$$

$$\therefore \mathrm{L} = 0{\cdot}40\mathrm{L} + 0{\cdot}35\mathrm{C}$$

$$\mathrm{C} = 0{\cdot}60\mathrm{L} + 0{\cdot}65\mathrm{C}$$

$$\therefore 0{\cdot}60\mathrm{L} = 0{\cdot}35\mathrm{C}, \quad \text{but } \mathrm{L} + \mathrm{C} = 1 \quad \text{(total support)}$$

$$\therefore \frac{0{\cdot}60}{0{\cdot}35}\mathrm{L} + \mathrm{L} = 1$$

$$\therefore \mathrm{L} = \frac{0{\cdot}35}{0{\cdot}95} = 0{\cdot}368421 \quad \text{and} \quad C = 0{\cdot}631579.$$

Eventually, therefore, 36·8421% of votes cast support Labour, and 63·1579% support Conservatives.

If we were to apply the original transition rules once more to this situation (i.e. 40% of Labour voters maintaining their party loyalty, and 65% of Conservatives likewise), the same number of votes change from Labour to Conservative as change from Conservative to Labour. The *overall* situation therefore remains steady. The proportion of the Labour vote at successive elections is shown in Figure 18.3 (solid line).

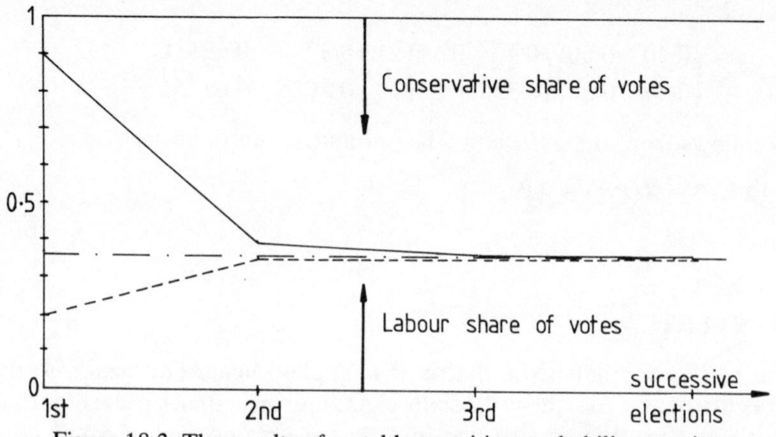

Figure 18.3 The results of a stable transition probability matrix

Notice that the eventual party balance is *independent* of the situation at the first election. To illustrate this, Figure 18.3 includes not only the case where Labour wins 90% of the votes at the first election (the case we calculated in detail above), but also the case where the Conservatives win 80% at the first election (broken line). By the third election the support for the two parties always divides into the approximate ratio 37:63.

18.5 ASSUMPTIONS IN THE MARKOV CHAIN

The last comment suggests that the model we have used for describing (and predicting?) the outcomes of successive elections is not very realistic. It is therefore worthwhile recalling the basic assumptions we made in the method so that its limitations are more clearly noted.

(i) Although only two parties were considered, the model could easily have been extended to include others by extending the dimensions of all the matrices involved.
(ii) Abstainers can also be involved in the process, by allocating them to a hypothetical 'Abstainers Anonymous' Party.
(iii) The outcome at any one election (the $(n + 1)$th say) depends only on two other matrices (or sets of data):
  (a) The outcome at the previous election (the nth), but no earlier election.
  (b) The transition probability matrix $\boldsymbol{P}$.
(iv) The matrix $\boldsymbol{P}$ is assumed to be *stable* over time. In other words, the pattern of switching parties never changes if the likelihood of each switch p_{ij} is expressed as a (fixed) probability.
(v) The voting pattern of any single elector cannot be determined (except as a probability). We have *not* made the assumption that the same 40% are always loyal to the Labour Party (say); only that (any) 40% of their supporters at the previous election remain so at the next.

A series of elections or *experiments* obeying this set of rules is known as a *Markov chain* or *Markov sequence*. The method of analysis is more realistic in those cases where an eventual stable distribution between possible states (or parties) can be shown to be justified. For example, eventual market shares of competing products may be amenable to this technique of analysis. Assumptions (iii) and (iv) are the worst faults in the use of this model for electoral voting patterns.

18.6 INPUT–OUTPUT MODELS

We begin by considering a very simplified model of a closed economy. By *closed* we imply total self-sufficiency so that all the goods and labour required to produce further goods and labour are available within the economy. Moreover, no surplus is produced so that all goods and services are consumed entirely within the home economy (e.g. an island economy with no international trade).

Next, we imagine that the total economic activity can be divided into its basic components, each producing a single *product*. The degree of disaggregation of products into basic types therefore matches the degree of disaggregation of activities. This could be a very broad, crude, and simple division into two major sectors (such as manufacturing and services) or could be a much more detailed division into a hundred or so sectors. For illustration purposes we shall take the former case.

Figure 18.4 shows the flow of goods and services in our two-sector economy, where the total outputs of the two goods G_1 and G_2 are x_1 and x_2 respectively, measured in some common unit (e.g. £p).

For *each unit* of G_2 produced, we assume that the quantity of product 1 required is a_{12}, and of product 2, a_{22}. For x_2 units of the second good, G_2, we therefore require an input:

$a_{12}x_2$ of good G_1

$a_{22}x_2$ of good G_2.

Similarly for an output x_1 of the good G_1.

The total requirements of the two sectors, u_1 and u_2, are therefore given by

$$\begin{aligned} u_1 &= a_{11}x_1 + a_{12}x_2 \\ u_2 &= a_{21}x_1 + a_{22}x_2 \end{aligned} \quad \text{or} \quad \begin{pmatrix} u_1 \\ u_2 \end{pmatrix} = \begin{pmatrix} a_{11} & a_{12} \\ a_{21} & a_{22} \end{pmatrix} \begin{pmatrix} x_1 \\ x_2 \end{pmatrix}$$

$$\therefore \boldsymbol{U} = \boldsymbol{AX} \quad \text{where } \boldsymbol{U} = \begin{pmatrix} u_1 \\ u_2 \end{pmatrix} \quad \text{etc.}$$

This gives us a simple relationship between the inputs to the total economic production process, $\boldsymbol{U}$, and the total production output, or

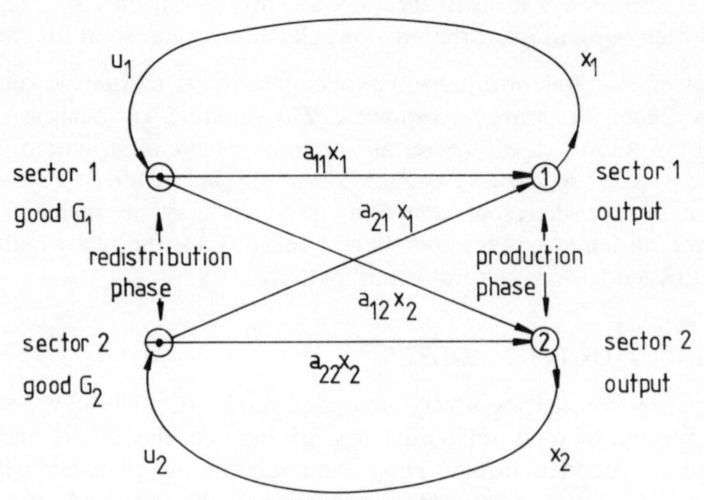

Figure 18.4 A simple two-sector closed economy

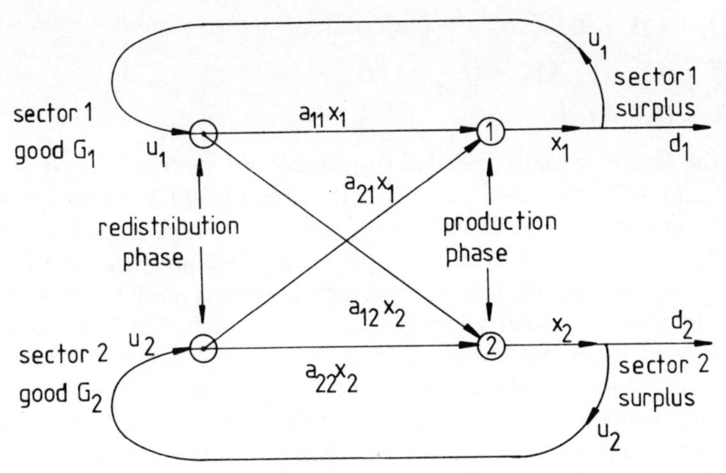

Figure 18.5 A simple two-sector open economy

intensity matrix (or vector) $\boldsymbol{X}$. The matrix $\boldsymbol{A}$ is called the *technology input–output matrix* for the system.

It may have been noticed from the flow network that apparently $u_1 = x_1$, etc., or $\boldsymbol{U} = \boldsymbol{X}$. This is in fact the case for a *closed two-sector system*, but we want to maintain the distinction between inputs ($\boldsymbol{U}$) and outputs ($\boldsymbol{X}$) as these are not equal with more complicated (and more realistic) assumptions. We now change our assumptions a little, in fact, and allow the production of a surplus, perhaps for export. Referring to Figure 18.5, we allow a surplus of good G_1 of amount d_1, and similarly for good G_2.

The relationship

$$\boldsymbol{U} = \boldsymbol{AX}$$

still holds, but it is now easy to see that the total output of sector 1, x_1, is split between the surplus d_1 and the contribution to the input u_1, and similarly for x_2:

$$\left.\begin{aligned} x_1 &= u_1 + d_1 \\ x_2 &= u_2 + d_2 \end{aligned}\right\} \quad \text{or } \boldsymbol{X} = \boldsymbol{U} + \boldsymbol{D} \text{ where } \boldsymbol{D} = \begin{pmatrix} d_1 \\ d_2 \end{pmatrix}.$$

The economic problem can now be stated as follows. If we have a given input–output matrix $\boldsymbol{A}$ (largely determined by the state of technology, etc.), and a given *demand matrix* (or vector) $\boldsymbol{D}$ (concerned with the desired level of surplus or exports), then what production outputs ($\boldsymbol{X}$) should we aim for?

The solution can be most readily found using basic matrix algebra.

$$\boldsymbol{X} = \boldsymbol{U} + \boldsymbol{D} \quad \text{and} \quad \boldsymbol{U} = \boldsymbol{AX}$$

$$\therefore \boldsymbol{X} = \boldsymbol{AX} + \boldsymbol{D}$$

$$\therefore \boldsymbol{IX} - \boldsymbol{AX} = \boldsymbol{D}$$

$$\therefore (\boldsymbol{I} - \boldsymbol{A})\boldsymbol{X} = \boldsymbol{D}$$
$$\therefore (\boldsymbol{I} - \boldsymbol{A})^{-1}(\boldsymbol{I} - \boldsymbol{A})\boldsymbol{X} = (\boldsymbol{I} - \boldsymbol{A})^{-1}\boldsymbol{D}$$
$$\therefore \boldsymbol{X} = (\boldsymbol{I} - \boldsymbol{A})^{-1}\boldsymbol{D}.$$

This is the intensity matrix needed to produce the demand matrix $\boldsymbol{D}$. For a given $\boldsymbol{D}$ and $\boldsymbol{A}$ it is in principle always possible to find $\boldsymbol{X}$. However, not all the elements of $\boldsymbol{X}$ need be positive; and $x_j < 0$ implies that insufficient good of type j are produced by the home economy to satisfy home demand. Imports are therefore necessary if both $\boldsymbol{A}$ and $\boldsymbol{D}$ are not to be altered. In principle, such input–output models of national economies (or even the world economy) can be built up, but the larger the number of economic divisions introduced, the greater the size of the matrices involved (i.e. more equations with more variables). For very large models, appropriate computer methods are essential for rapid solution.

18.7 BASIC EXAMPLES FOR THE STUDENT

B18.1 A gambler makes a decision each day whether or not to place a bet on a horse race. Express by means of a row vector in each case the following events:
(a) He does place a bet.
(b) He doesn't place a bet.
(c) He is 40% certain to place a bet.
His pattern of behaviour is consistent: if he bets one day, he is only 10% likely to place a bet on the following day. If he fails to place a bet one day, he is 80% certain to place one on the following day. If he definitely places a bet on Monday then
(d) What is he likely to do on Wednesday?
(e) What eventual pattern of behaviour will emerge?
How would the form of your calculation be altered if he also placed bets on greyhounds, but never on the same day as on horses?

B18.2 The input–output matrix for an economy is given by

$$\boldsymbol{A} = \begin{pmatrix} 0{\cdot}2 & 0{\cdot}4 \\ 0{\cdot}2 & 0{\cdot}4 \end{pmatrix}$$

Show on a diagram what this implies. If it is intended that 10 units of good G_1 and 20 units of good G_2 should be made available for export, write down the demand matrix $\boldsymbol{D}$, and determine the necessary output of each type of good. (N.B. The inverse of the matrix $\boldsymbol{B}$ where

$$\boldsymbol{B} = \begin{pmatrix} a & b \\ c & d \end{pmatrix} \quad \text{is} \quad \boldsymbol{B}^{-1} = \frac{1}{(ad - bc)}\begin{pmatrix} d & -b \\ -c & a \end{pmatrix}.)$$

18.8 INTERMEDIATE EXAMPLES FOR THE STUDENT

I18.1 A student tries not to be late for lectures too often. If he is late one day, he is 90% certain to be on time the next. If he is on time, then there is a 30% chance he will be late the next day.

(a) Express the transition probabilities in a 2×2 matrix.

(b) Express the fact that he is late (on Monday, say) as a row vector.

(c) Find the probability that he is late on Wednesday.

(d) How often is he late in the long run?

I18.2 Three brands of margarine, A, B, and C share a market. The transition probability matrix for customers changing brands is $\boldsymbol{P}$ where

$$\boldsymbol{P} = \begin{pmatrix} 0{\cdot}8 & 0{\cdot}15 & 0{\cdot}05 \\ 0{\cdot}2 & 0{\cdot}7 & 0{\cdot}1 \\ 0{\cdot}2 & 0{\cdot}15 & 0{\cdot}65 \end{pmatrix}.$$

What proportion of the market will each brand eventually hold?

I18.3 Suppose one has a situation similar to that in question I18.2, with three brands, except that no matter what the market share the transition probability matrix leaves it unchanged at any stage. What is the new transition probability matrix?

I18.4 Suppose that the input–output matrix for a two-sector economy is $\boldsymbol{A}$ and the demand matrix is $\boldsymbol{D}$, where

$$\boldsymbol{A} = \begin{pmatrix} 0{\cdot}1 & 0{\cdot}2 \\ 0{\cdot}4 & 0{\cdot}3 \end{pmatrix} \quad \boldsymbol{D} = \begin{pmatrix} 7 \\ 3 \end{pmatrix}.$$

Find the intensity matrix $\boldsymbol{X}$. What does it mean? (N.B. The inverse of the matrix $\boldsymbol{B}$ where

$$\boldsymbol{B} = \begin{pmatrix} a & b \\ c & d \end{pmatrix} \quad \text{is} \quad \boldsymbol{B}^{-1} = \frac{1}{(ad - bc)} \begin{pmatrix} d & -b \\ -c & a \end{pmatrix}.)$$

CHAPTER 19

Matrix Applications II

19.1 LINEAR PROGRAMMING: A REMINDER

In Chapter 5 we introduced the technique of linear programming as a method of solving a large class of optimization problems. At that stage the problems involved the maximization (or minimization) of a linear *objective function*, subject to a number of linear constraints; and because we restricted ourselves to two variables the problems could all be represented geometrically by a simple diagram in two dimensions.

In Section 5.3 we gave the following example:

Resources of a farmer:	Cash, limited to £5000
	Land, limited to 100 acres
Alternative crops:	Wheat, costing £40/acre, profit £30/acre
	Barley, costing £60/acre, profit £40/acre
Questions:	How many acres of wheat (x)
	How many acres of barley (y)
	for maximum profits?

The problem then reduced to one of maximizing the objective function which we shall denote generally by the symbol f.

$$f = 30x + 40y$$

subject to the four constraints

$$\begin{aligned} x + y &\leqslant 100 \qquad & x &\geqslant 0 \\ 40x + 60y &\leqslant 5000 \qquad & y &\geqslant 0. \end{aligned}$$

Figure 19.1 shows the *feasible region* (shaded) and the *objective function*, f (series of parallel lines), which when maximized gives us the most profitable mix of wheat and barley which can be achieved with the resources available. This is at the point A, at a vertex of the feasible region, where $x = 50$ and $y = 50$.

A full explanation of why point A should yield the maximum profit is given in Section 5.3.

It was pointed out in Section 5.5 that if the farmer had introduced another cereal as a possible crop, then another variable would have had to

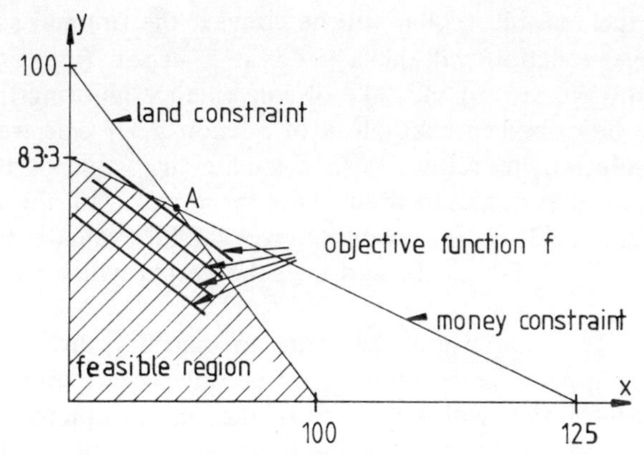

Figure 19.1 The linear programming example. (Not to scale)

have been introduced (z, say, for the acreage of maize). Geometrical interpretation in two dimensions is then no longer possible; a three-dimensional model would be needed with the feasible region being represented by a *volume*, and the constraints and objective function by *planes*. If further variables are introduced, then further dimensions are needed and the problem can no longer be represented in pictorial or geometrical terms. More general methods of solution are now required, and we will now give a brief outline of an important algebraic technique, the *simplex method*.

19.2 THE BASIC IDEA OF THE SIMPLEX METHOD

For a series of linear constraints the feasible region, such as that illustrated above, will always be *convex* in shape. Simply, though rather loosely put, the feasible region will have a series of outward pointing corners, but no 'dents' or concave boundaries. The two-dimensional feasible regions illustrated in Figures 19.2(a) and (b) satisfy this description *convex*, but not 19.2(c).

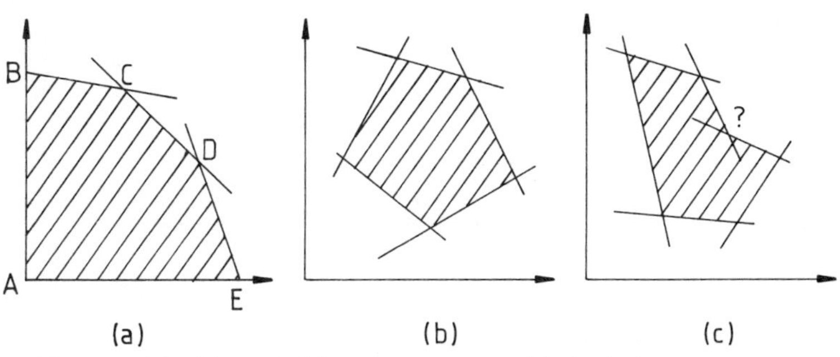

Figure 19.2 (a) and (b) convex areas; (c) includes a concave region (marked?)

Given that the feasible region will be convex, the optimal solution for a linear objective function will always be at a *vertex* (or corner) of the feasible region. (We are for the sake of convenience ignoring the *rare* cases of degeneracy described in example 4 of Section 5.3.) One way of finding this optimal solution, therefore, is to calculate the value of the objective function, f, at each vertex, and then select the vertex with the largest value for f. In Figure 19.2(a), for example, we would evaluate the objective function f at A, B, C, D, and E, and take the vertex which gave the largest value for f.

If instead of two dimensions we were to consider three or more, the same *principles* apply. The feasible region is convex (in a multi-dimensional sense), and the solution will be at one of the (multi-dimensional) corners. Again, the basic idea is to evaluate the objective function, f, at each vertex in turn, and to identify the largest such value. The simplex method is one which attempts to make our selection of vertices (and, particularly, the *order* in which we evaluate them) reasonably systematic and efficient.

19.3 VERTICES OF CONVEX REGIONS

The simplex method relies on the evaluation of the objective function f at successive vertices of the feasible region. In this section the problem of identifying the vertices of a convex region in algebraic rather than geometric terms will be examined. We shall continue to use the example of the farmer and his choice of crops (as in Sections 5.3 and 19.1) which enables us to show both the geometric and the algebraic versions; the basic

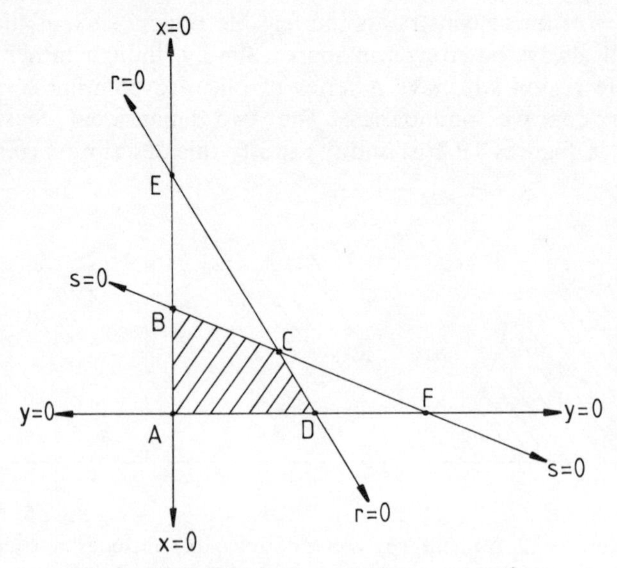

Figure 19.3 The four constraints as equations

method is still valid for 3 or more dimensions, however, when the geometric model cannot be illustrated.

The constraints were written (see Section 19.1)

$$x + y \leqslant 100 \qquad 40x + 60y \leqslant 5000 \qquad x \geqslant 0 \qquad y \geqslant 0.$$

We now introduce a *slack variable* into each of the principal constraints so as to convert the inequalities to equalities. Therefore

$$x + y + r = 100 \qquad 40x + 60y + s = 5000$$
$$x \geqslant 0 \qquad y \geqslant 0 \qquad r \geqslant 0 \qquad s \geqslant 0.$$

One immediate advantage of this change is that the *boundary* of the feasible region representing the land constraint $x + y = 100$ can instead be denoted by the simpler equation $r = 0$. In fact, from Figure 19.3, we can now denote all the relevant lines by similar equations.

The four constraints can now be represented by the four equations

$$x = 0 \qquad y = 0 \qquad r = 0 \qquad s = 0.$$

Also, each intersection of the constraints (labelled A to F) can now be identified by two simple equations. For example,

A occurs at $x = 0, y = 0$
C occurs at $r = 0, s = 0$, etc.

It follows that movement from vertex A to B to C can be represented by the succession

$$\begin{matrix} x = 0 \\ y = 0 \end{matrix} \text{ to } \begin{matrix} x = 0 \\ s = 0 \end{matrix} \text{ to } \begin{matrix} r = 0 \\ s = 0. \end{matrix}$$

The simplex method involves such sequences.

19.4 ALGEBRA OF THE SIMPLEX METHOD

The linear programming problem now becomes:

Maximize $f = 30x + 40y$
Subject to $x + y + r = 100 \qquad 40x + 60y + s = 5000$
$x \geqslant 0 \qquad y \geqslant 0 \qquad r \geqslant 0 \qquad s \geqslant 0.$

We shall concern ourselves exclusively with such *maximization* problems. L.P. problems involving minimization can be handled with little modification, but for the purposes of illustrating the principles of the method they will not be considered here (see Section 19.7).

In order to facilitate translation into matrix form, the principal equations can be written:

	Equation							
(1)	$x + y + r = 100$	1	1	1	0	0	100
(2)	$40x + 60y + s = 5000$	40	60	0	1	0	5000
(3)	$-30x - 40y + f = 0$	−30	−40	0	0	1	0

The initial matrix tableau, illustrated above, will be developed in detail later. We shall for the moment consider the geometric and conventional algebraic forms.

Notice that the *slack variables* can be readily identified in the matrix form by the columns including only 'zeros' and 'ones'. The objective function f is written in terms of two variables here (x and y) which, for the moment, are the *basic variables*. If we were to equate the basic variables to zero, we would obtain one of the possible *basic feasible solutions* which, though not necessarily the optimum, is one of the solutions corresponding to a vertex of the feasible region. At this stage, for example, if

$$x = 0 \quad \text{and} \quad y = 0 \quad \text{(basic variables)}$$

then

$$r = 100 \quad \text{and} \quad s = 5000 \quad \text{and} \quad f = 0$$

which corresponds to the vertex A in Figures 19.3 and 19.4.

The basic procedure, then, is to start at any one vertex of the feasible region (usually $x = 0$, $y = 0$, or point A in Figure 19.4), and to move to an *adjacent* vertex so as to increase the value of f.

This is continued until it is impossible to move to an adjacent vertex and at the same time increase f; any further movement would, in other words, reduce the value of f. When this stage is reached we have succeeded in identifying the optimal vertex, the one which represents the largest value of f within the feasible region.

As the simplex method proceeds, the label *basic* is applied to the pair of zero-valued variables corresponding to the vertex currently being considered; so that for example, as we proceed through the vertices A, B,

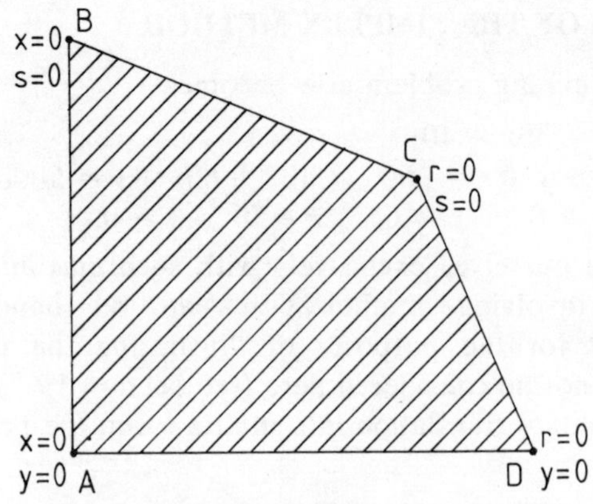

Figure 19.4 The feasible region with four vertices A, B, C, D identified by the appropriate zero-valued (basic) variables

and C in Figure 19.4, the pairs of basic variables are, successively

x and y
then x and s
then r and s.

At each stage, as we proceed from each vertex to one of its neighbours, one basic variable is dropped and another takes its place.

If we commence at A, $x = 0$, $y = 0$, and $f = 0$. To increase f we move to an adjacent vertex, B or D. The form of f suggests increasing y rather than x, because of the *larger coefficient* (40 rather than 30)

$$f = 30x + 40y.$$

However, we can only increase y within the limits set by all the constraints.

From (1) above $y_{max} = 100$ (if $x = r = 0$).

From (2) above $y_{max} = \frac{5000}{60} = 83{\cdot}\dot{3}$ (if $x = s = 0$).

Since both must be satisfied, we take the smaller, and have $y = 83{\cdot}\dot{3}$, $x = s = 0$, and this takes us to point B on Figure 19.4. We also find that $f = 30x + 40y = 3333{\cdot}\dot{3}$.

The next step is therefore to treat x and s as our basic variables (since we are now at B, $x = s = 0$), and y and r as our slack variables. In terms of the coefficients of our basic equations (and hence the matrix tableau), we therefore want the slack variables y and r to have only 'zeros' and 'ones' in their respective columns, and the function f to be expressed in terms of x and s. This can be achieved by using equation (2) above to eliminate the variable y from both (1) and (3), thus:

$$\text{Stage 1}\begin{cases} x + y + r = 100 & (1)\\ 40x + 60y + s = 5000 & (2)\\ -30x - 40y + f = 0 & (3)\end{cases}$$

$$\text{Stage 2}\begin{cases} x + y + r = 100 & (1)\\ \frac{40}{60}x + y + \frac{1}{60}s = \frac{5000}{60} & = (2) \div 60 = (4)\\ -30x - 40y + f = 0 & (3)\end{cases}$$

$$\text{Stage 3}\begin{cases} \frac{1}{3}x + r - \frac{1}{60}s = \frac{100}{6} & = (1) - (4) = (5)\\ \frac{2}{3}x + y + \frac{1}{60}s = \frac{500}{6} & (4)\\ -\frac{10}{3}x + \frac{2}{3}s + f = \frac{20000}{6} & = (3) + 40(4) = (6)\end{cases}$$

From the pattern of coefficients, it can be seen that y and r have taken on the form of slack variables (coefficients of 1 or 0), and f is expressed in terms of the new basic variables x and s. We now wish to increase the value of f further if possible, where

$$f = \frac{10}{3}x - \frac{2}{3}s + \frac{20000}{6} \quad (x \geqslant 0, s \geqslant 0).$$

It can be seen immediately that f can be increased by increasing the value of x, but not by increasing the value of s. Notice that, as in the first attempt at increasing f (when f was expressed in terms of x and y), we focus our attention on the largest negative coefficient in the last line (equation (3), or equation (6)) which is $-10/3$ in this example. This point will be useful to remember when we introduce the matrix notation.

We wish, therefore, to increase x as far as possible within the constraints

From (5) $x = 50 \quad (r = 0, s = 0)$

From (4) $x = 125 \quad (y = 0, s = 0)$.

Again, as both must be satisfied we take the smaller, so that $x = 50$, $r = 0$, $s = 0$ and we are now at point C on Figure 19.4. The new basic variables are r and s, and x and y become the slack variables. The set of equations (5), (4), and (6) must now be rearranged with this in mind

$$\text{Stage 3}\left\{\begin{array}{lr} \frac{1}{3}x + r - \frac{1}{60}s = \frac{100}{6} & (5) \\ \frac{2}{3}x + y + \frac{1}{60}s = \frac{500}{6} & (4) \\ -\frac{10}{3}x + \frac{2}{3}s + f = \frac{20000}{6}. & (6) \end{array}\right.$$

We use equation (5) to eliminate the variable x from (4) and (6).

$$\text{Stage 4}\left\{\begin{array}{lr} x + 3r - \frac{1}{20}s = 50 & = 3 \times (5) = (7) \\ \frac{2}{3}x + y + \frac{1}{60}s = \frac{500}{6} & (4) \\ -\frac{10}{3}x + \frac{2}{3}s + f = \frac{20000}{6} & (6) \end{array}\right.$$

$$\text{Stage 5}\left\{\begin{array}{lr} x + 3r - \frac{1}{20}s = 50 & (7) \\ y - 2r + \frac{1}{20}s = 50 & = (4) - \frac{2}{3}(7) = (8) \\ 10r + \frac{1}{2}s + f = 3500 & = (6) + \frac{10}{3}(7) = (9) \end{array}\right.$$

At point C, therefore, $r = s = 0$ and $x = 50$, $y = 50$, $f = 3500$. It is also important to notice that the form of f is now

$$f = 3500 - 10r - \tfrac{1}{2}s.$$

Since $r \geqslant 0$, $s \geqslant 0$ always, then f can only *decrease* in value if r and/or s are increased above zero. The maximum value of f therefore occurs when r and s both take their minimum values ($r = s = 0$), at point C. We have now therefore reached the highest value of f within the feasible region so that the solution to the problem is

$x = 50$ (acres of wheat)
$y = 50$ (acres of barley)
$f = 3500$ (£, maximum profit).

19.5 THE MATRIX FORMAT

The procedure outlined above is systematic, but at first sight the manipulation of the equations (1) to (9) seems confusing. If the whole process is put into a matrix format, rather similar to that employed for the Gauss–Jordan technique, the simplex method becomes very mechanical, and the operations become clearer, even if the underlying logic seems more distant. We shall start with the matrix tableau introduced early in Section 19.4, and follow through the whole process again, with appropriate comments on each stage. The actual numbers involved prove to be identical of course with the full algebraic format already used. The only exception is that the final column on the left-hand side of the equations, corresponding to the variable f, is omitted. Reference back will show that this column is not relevant to the main calculations.

$$\begin{array}{cccc|c} 1 & 1 & 1 & 0 & 100 \\ 40 & \textcircled{60} & 0 & 1 & 5000 \\ \hline -30 & -40 & 0 & 0 & 0 \end{array}$$

Original matrix tableau, with the f-column omitted.

Step 1 Identify the column with largest negative coefficient in the bottom row: here it is (-40) in the *second column*.

Step 2 Identify the row which provides the smallest positive ratio between the figure in the final column and (in this case) the second column; that is,

$$\frac{100}{1} \quad \text{or} \quad \frac{5000}{60}.$$

In this case it is the *second row*.

Step 3 We now use the Gauss–Jordan-type technique to reduce the number in the *second column*, and the *second row* (i.e. 60, which has been encircled) to 1, and to produce zeros elsewhere in the second column

$$\begin{array}{cccc|c} 1 & 1 & 1 & 0 & 100 \\ \frac{2}{3} & \textcircled{1} & 0 & \frac{1}{60} & \frac{5000}{60} \\ \hline -30 & -40 & 0 & 0 & 0 \end{array}$$

Second row has been divided by 60 to reduce the encircled figure to 1.

$$\begin{array}{cccc|c} \frac{1}{3} & 0 & 1 & -\frac{1}{60} & \frac{100}{6} \\ \frac{2}{3} & 1 & 0 & \frac{1}{60} & \frac{500}{6} \\ \hline -\frac{10}{3} & 0 & 0 & \frac{2}{3} & \frac{20000}{6} \end{array}$$

Row 1 − Row 2.

40 × Row 2 + Row 3.

Step 4 Identify the column with the largest negative coefficient in the bottom row (first column, with −10/3).

Step 5 Identify the row with the smallest positive ratio between the final and the first column; that is,

$$\frac{100}{6} \div \frac{1}{3} = 50 \quad \text{or} \quad \frac{500}{6} \div \frac{2}{3} = 125.$$

In this case, the first row.

Step 6 Again, we use the Gauss–Jordan-type technique to reduce the number in the *first column* and the *first row* (i.e. 1/3) to 1 and to produce zeros elsewhere in the first column.

$$\begin{array}{cccc|c} \textcircled{1} & 0 & 3 & -\frac{1}{20} & 50 \\ \frac{2}{3} & 1 & 0 & \frac{1}{60} & \frac{500}{6} \\ \hline -\frac{10}{3} & 0 & 0 & \frac{2}{3} & \frac{20000}{6} \end{array}$$

Row one has been multiplied by 3 to bring the encircled figure to 1.

$$\begin{array}{cccc|c} 1 & 0 & 3 & -\frac{1}{20} & 50 \\ 0 & 1 & -2 & \frac{1}{20} & 50 \\ \hline 0 & 0 & 10 & \frac{1}{2} & 3500 \end{array}$$

Row 2 $-\frac{2}{3}\times$ Row 1.

Row 3 $+\frac{10}{3}\times$ Row 1.

Step 7 Identify the column with the largest negative coefficient in the bottom row.

There is none (all are positive). Procedure need be taken no further.

If we expand this tableau back into its full algebraic form it reads

$$x + 3r - s/20 = 50$$

$$y - 2r + s/20 = 50$$

$$10r + s/2 + f = 3500 \quad \text{(remembering the omitted } f \text{ column)}$$

which is the same as the final version of the algebraic procedure covered earlier. We therefore see immediately that the final column of the simplex tableau gives us the required results:

$$x = 50 \qquad y = 50 \qquad f = 3500 \quad (\text{at } r = 0, s = 0).$$

19.6 MORE THAN TWO VARIABLES

The geometric model, moving from vertex to vertex, etc., is difficult or impossible to visualize in more than two dimensions. The simplex technique can readily be extended to 3 or more dimensions, simply by enlarging the tableaux appropriately, and by repeating the fairly mechanical procedure until, once more, the bottom row achieves only positive figures. We have not proved that the technique will always converge on the correct solution for more than two dimensions, and some complications can arise (such as continuous *cycling* round the correct solution when redundant constraints are involved). The technique usually works well in practice, however.

19.7 MINIMIZATION PROBLEMS

In minimization problems the objective function g is to be given its minimum value subject to the usual linear constraints. This can be achieved by putting $f = -g$ and then maximizing f. The only complication is that the origin ($x = 0$, $y = 0$) does not usually lie within the feasible region, so that initial effort is directed towards finding two basic variables whose origin does lie in the feasible region, and using this point to start from. The details of this procedure will not be developed here, since otherwise the broad principles of the technique remain the same.

19.8 BASIC EXAMPLES FOR THE STUDENT

B19.1 In a linear programming problem, involving two independent variables (x, y) and two constraints, the initial matrix tableau for solving by the simplex method is found to be:

$$\begin{array}{rrrr|r} 5 & 2 & 1 & 0 & 44 \\ 3 & 4 & 0 & 1 & 46 \\ \hline -3 & -2 & 0 & 0 & 0 \end{array}$$

Find the maximum value of the objective function (f) which can be achieved within the constraints, and the corresponding values for x and y.

B19.2 In a linear programming problem, the aim is to maximize the function f, while simultaneously satisfying the two constraints given below. Solve, using the simplex method:

$$f = 4x + 3y$$
$$x + 2y \leqslant 16$$
$$5x + 2y \leqslant 48.$$

19.9 INTERMEDIATE EXAMPLES FOR THE STUDENT

I19.1 The initial matrix tableau for solving a linear programming problem by the simplex method is given below. Find the solution, using the simplex technique.

$$\begin{array}{rrrr|r} 3 & 4 & 1 & 0 & 21 \\ 5 & 3 & 0 & 1 & 24 \\ \hline -5 & -4 & 0 & 0 & 0 \end{array}$$

I19.2 The objective function in a linear programming example is given by

$$f = x + y.$$

Solve the problem, where the function f is to be maximized subject to the constraints below, using the simplex method.

$$2x + 4y \leqslant 20$$
$$3x + 4y \leqslant 23$$
$$5x + 3y \leqslant 31.$$

More Advanced Examples for Part Five

A5.1 If two matrices $\boldsymbol{P}$ and $\boldsymbol{Q}$ are said to commute, what does this mean?

$\boldsymbol{A}$ and $\boldsymbol{B}$ are two square matrices of the same dimensions. Show that $\boldsymbol{A}$ and $\boldsymbol{B}$ will always commute if

$$\boldsymbol{A} = \boldsymbol{B} + c\boldsymbol{I}$$

where c is a constant and $\boldsymbol{I}$ is the unit matrix.

A5.2 For a certain process the transition matrix is given by

$$\boldsymbol{P} = \begin{pmatrix} 0{\cdot}2 & 0{\cdot}8 \\ 0{\cdot}6 & 0{\cdot}4 \end{pmatrix}$$

for the 1st, 4th, 7th, 10th, etc., transitions.

For the 2nd, 5th, 8th, 11th, etc., transitions, however, the appropriate transition matrix becomes

$$\boldsymbol{Q} = \begin{pmatrix} 0{\cdot}5 & 0{\cdot}5 \\ 0{\cdot}2 & 0{\cdot}8 \end{pmatrix}.$$

For the 3rd, 6th, 9th, 12th, etc., transitions, the transition matrix is now

$$\boldsymbol{R} = \begin{pmatrix} 0{\cdot}4 & 0{\cdot}6 \\ 0{\cdot}6 & 0{\cdot}4 \end{pmatrix}.$$

The transition matrix cycles therefore through the three forms, $\boldsymbol{PQR}$, $\boldsymbol{PQR}$, etc. If the system is definitely in the first state to begin with, so that its state can be defined by the vector $(1 \quad 0)$, find the probability of its being in each of the two states

(i) after the 3rd transition

(ii) in the long run.

A5.3 (a) Using any matrix technique you wish, solve the equation

$$\boldsymbol{AX} = \boldsymbol{C}$$

to find $\boldsymbol{X}$, and hence x, y, and z, where

$$\boldsymbol{A} = \begin{pmatrix} 3 & -2 & 1 \\ -1 & 3 & 4 \\ 3 & 0 & 2 \end{pmatrix} \quad \boldsymbol{X} = \begin{pmatrix} x \\ y \\ z \end{pmatrix} \quad \boldsymbol{C} = \begin{pmatrix} 5 \\ -17 \\ 0 \end{pmatrix}.$$

(b) If you had been given the same problem as in (a) above, but with $\boldsymbol{C}$ replaced by $\boldsymbol{D}$, where

$$\boldsymbol{D} = \begin{pmatrix} 2 \\ 6 \\ 5 \end{pmatrix}$$

why would it have been useful to know or to derive the inverse of $\boldsymbol{A}$?

PART SIX

Introduction

This last chapter looks at series, their sums, and the expansion of power series, for two main reasons. The major part is concerned with the development of the simpler arithmetic and geometric progressions, and their important use in financial calculations. The formulae underlying the basic finance tables are explained from first principles. The second aspect is the demonstration of series expansions for the purposes of approximation. When an exact statement of a problem appears intractable, the use of appropriate approximations can often give a surprisingly good result, with relatively little effort. So again, this section is aimed at further developing the reader's facility of manipulating and understanding formulae which at first may seem rather daunting.

CHAPTER 20

Series

20.1 INTRODUCTION

Many mathematical models assume, for the sake of convenience, that all relevant variables can be treated as continuous rather than discrete. Graphs of such functions are then continuous smooth curves (in two dimensions) or surfaces (in three) and do not include steps, breaks, or *discontinuities*. Taking an example used in an earlier chapter, the relation between total production costs and production level usually assumes a continuous output, rather than one consisting of a whole number of units (Figure 20.1(a)). In fact, even when a continuous function may be a legitimate description of a system, our measurements may be taken at equal intervals of time, providing us with a finite series of data from which the underlying continuous function may be inferred. Figure 20.1(b) shows measurements of the height of a plant (say) taken at regular intervals, from which the height vs. time relationship may be estimated.

These examples show two ways in which a *series* of quantities or numbers can arise, and in this chapter we shall be concerned with a few special types of series which have applications within a management context.

Firstly, however, consider the following series of numbers

$$1 \quad 3 \quad \frac{1}{2} \quad -\frac{1}{2} \quad -1 \quad 2 \quad 0$$

which might represent the air temperature at noon on successive days in January. There is no clear pattern of *progression* from one term to the next, and statistical information would be the most we could deduce from this series (e.g. *mean* and *variance* of the air temperature). We are therefore concerned principally with series where each term has a *fixed and well-understood relationship* with the previous one. Succeeding terms can then be found by repeated application of this relationship.

An example of a series which had such a well-defined progression was introduced into Chapter 18, in fact, when Markov chains were introduced. It will be recalled that in the example based on allegiance to political parties, successive election results were given by

$$\boldsymbol{X} \quad \boldsymbol{XP} \quad \boldsymbol{XP}^2 \quad \boldsymbol{XP}^3 \quad \boldsymbol{XP}^4 \quad \text{etc.}$$

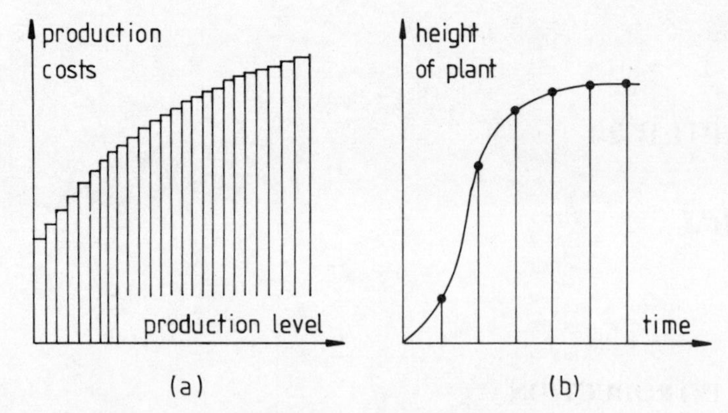

Figure 20.1 Examples of discrete measurements

which is a well-defined series with a superficial resemblance to a geometric progression (Section 20.4). In this chapter, however, we shall consider only series involving numbers rather than matrices.

20.2 ARITHMETIC PROGRESSION

Consider the following series of numbers, which form an *arithmetic progression* (A.P.)

10 000   9200   8400   7600   6800

They might represent the book value of a piece of equipment which is depreciating by a fixed sum per year (£800). It will be readily appreciated that the principal characteristic of an A.P. is a *common difference* between successive terms (−800 in this case). The general form of an A.P. is therefore

$$\underset{\text{1st term}}{a} \quad a+d \quad a+2d \quad a+3d \quad \ldots \quad \underset{n\text{th term}}{a+(n-1)d}$$

so that in the depreciation example above, $a = 10\,000$ and $d = -800$. The book value of the equipment at the beginning of the second year (or end of the first) is given by the second term $(a + d)$. Similarly, the book value at the beginning of the nth year (or end of the $(n - 1)$th) is given by the nth term $(a + (n - 1)d)$.

We can deduce from this the book value at any point in time, so that at the end of the ninth year, for example, $n = 10$ and

$$\begin{aligned} a + (n-1)d &= 10\,000 - (9)800 \\ &= 2800. \end{aligned}$$

Hence the book value at the end of 9 years is £2800. Figure 20.2 shows that an A.P. is equivalent to a linear relation if the variable is taken to be continuous. (This implies, for example, that the equipment is depreciating

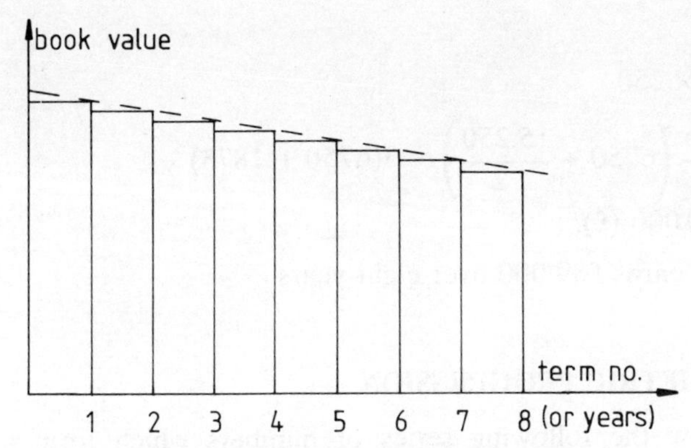

Figure 20.2 Graphical representation of an arithmetic progression, where $a > 0$, $d < 0$

continuously and that its book value is adjusted continuously as well, rather than just once per year.)

20.3 SUM OF AN ARITHMETIC PROGRESSION

The sum of the first n terms of an arithmetic progression is S_n where

$$S_n = a + (a + d) + \ldots + (a + (n - 2)d) + (a + (n - 1)d)$$

and in reverse order

$$S_n = (a + (n - 1)d) + (a + (n - 2)d) + \ldots + (a + d) + a.$$

Summing the two forms

$$2S_n = (2a + (n - 1)d) + (2a + (n - 1)d) + \ldots + (2a + (n - 1)d) + (2a + (n - 1)d)$$

$$= n(2a + (n - 1)d)$$

$$\therefore \quad S_n = \frac{n}{2}(2a + (n - 1)d).$$

Example

Find the total earnings over 8 years for a man whose salary is initially £6750 p.a. and which increases by £250 p.a. every six months thereafter.

For each six-month interval, he earns half his annual salary. So, we have the sum of 16 terms (16×6 months), with

$$a = \tfrac{1}{2} \times 6750$$

and

$$d = \tfrac{1}{2} \times 250$$

$$\therefore S_{16} = \frac{16}{2}\left(6750 + \frac{15.250}{2}\right) = 8(6750 + 1875)$$
$$= 69\,000\ (\pounds).$$

He therefore earns £69 000 over eight years.

20.4 GEOMETRIC PROGRESSION

Consider now the following series of numbers which form a *geometric progression* (G.P.)

10 000   9000   8100   7290   6561.

They might again represent the book value of depreciating equipment but where the depreciation is a fixed *percentage* of its initial value, each year (10% here). So, the principal characteristic of a G.P. is a *common ratio* between successive terms (0·9 in this example). The general form of a G.P. is therefore

$$\underset{\text{1st term}}{a} \quad ar \quad ar^2 \quad ar^3 \quad \ldots \quad \underset{n\text{th term}}{ar^{n-1}}$$

so that in the example, $a = 10\,000$ and $r = 0{\cdot}9$. At the end of the ninth year, using this G.P. model rather than the A.P. model, the book value is

$$ar^{n-1} = 10\,000 \times (0{\cdot}9)^{10-1} = 10\,000 \times (0{\cdot}9)^9$$
$$= 3874.$$

Hence the book value is £3874 at the end of 9 years. Figure 20.3 shows the connection between a G.P. and an *exponential* relation. Note that if r is negative, then the signs of successive terms alternate, so that two exponential curves are generated, in effect.

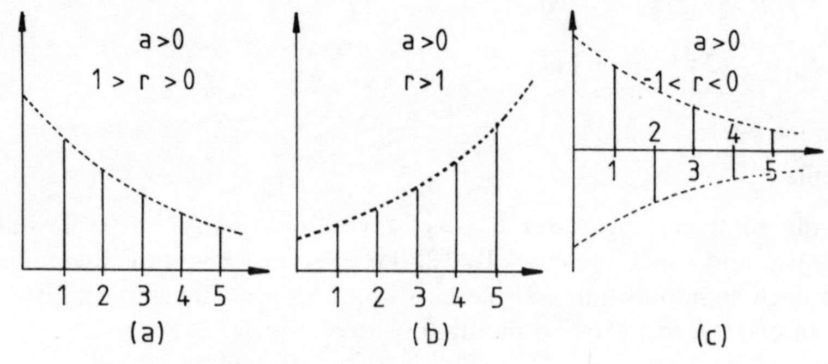

Figure 20.3 The geometric progression for various ratios, r

20.5 SUM OF A GEOMETRIC PROGRESSION

The sum of the first n terms of a geometric progression is S_n where

$$S_n = a + ar + ar^2 + \ldots + ar^{n-1}.$$

Now

$$rS_n = ar + ar^2 + ar^3 + \ldots + ar^{n-1} + ar^n.$$

On subtraction

$$S_n(1 - r) = a(1 - r^n)$$

$$\therefore S_n = a\frac{(1 - r^n)}{(1 - r)}.$$

For an *infinite series* (i.e. an infinite number of terms) then we find that

$$S_\infty \to \infty \quad \text{if} \quad r \geqslant 1$$

$$S_\infty \to \pm\infty \quad \text{if} \quad r \leqslant -1$$

and

$$S_\infty \to \frac{a}{1 - r} \quad \text{if } -1 < r < 1 \quad (\text{so that } r^n \to 0 \text{ as } n \to \infty).$$

The series illustrated in Figures 20.3(a) and (c) therefore give a finite sum for an infinite number of terms; while Figure 20.3(b), with ever-increasing terms, gives an infinite sum.

20.6 APPLICATIONS OF THE GEOMETRIC PROGRESSION

The properties of the geometric progression underlie many calculations in accounting problems, such as compound interest, discounting, annuities, mortgage loans, and sinking funds. Some examples are now given to illustrate the mathematical ideas behind each of these calculations.

20.6.1 Compound Interest

If a sum P (the *principal*) is invested at an annual rate of interest x, which is compounded annually, then the total sum invested at the *commencement* of each year is

P	$P(1+x)$	$P(1+x)^2$	$P(1+x)^3$	$\ldots$	$P(1+x)^{n-1}$.
1st year	2nd year				nth year

(Annual compounding implies that, each year, the interest gained during that year is also invested and begins to earn interest itself.)

The total sum therefore follows a geometric progression, with $a = P$ and $r = 1 + x$.

Suppose a sum of £1000 is invested at 10% compound interest. What final amount will have accumulated after 10 years?

$$a = 1000 \qquad r = 1 + 0{\cdot}1 = 1{\cdot}1 \qquad n = 11$$

$$\therefore \text{Final amount} = 1000(1{\cdot}1)^{10} = 1000(2{\cdot}594) = 2594\ (£)$$

(*Simple* interest, with no compounding, would have given a final amount of $1000 + 1000 \times 0{\cdot}1 \times 10 = 2000$ (£).)

20.6.2 Discounting

A sum a_n held n years from now has a present value a_0 which must be less than a_n. This is because a_0 could be invested now, for a period of n years, and would grow to a value a_n. This argument allows us to relate a_0 and a_n.

If a_0 is invested at a rate of interest x, after n years its value becomes a_n where

$$a_n = a_0(1 + x)^n = a_0 r^n$$

$$\therefore a_0 = \frac{a_n}{r^n} \qquad \text{the present value of } a_n.$$

For example, a sum of £300 in 3 years' time has a present value given by

$$a_0 = \frac{300}{r^3} = \frac{300}{(1{\cdot}1)^3} = £225.39$$

if a *discount rate* of 10% is assumed. Present-value tables enable this calculation to be achieved with minimum effort. The relevant section of present-value tables reads:

Years	...	9%	10%	11%	...
1		0·9174	0·9091	0·9009	
2		0·8417	0·8264	0·8116	Present value
3		0·7722	0·7513	0·7312	of 1 unit paid
4		0·7084	0·6830	0·6587	*n* years hence
5		0·6499	0·6209	0·5935	
.					
.					
.					

This gives the present value of *1 unit*, for appropriate choices of years and discount rate. In this case

$$a_0 = £300 \times 0{\cdot}7513 = £225.39.$$

20.6.3 Annuities

An annuity is a series of payments of equal value, a, made at the end of each year for n years. The present value of the annuity, A, is therefore the sum of the present values of all these payments, each of which is discounted back to the present time.

$$A = \underset{\substack{\uparrow \\ \textit{discounted} \\ \textit{1 year}}}{\frac{a}{r}} + \frac{a}{r^2} + \frac{a}{r^3} + \ldots + \underset{\substack{\uparrow \\ \textit{discounted} \\ n \text{ years}}}{\frac{a}{r^n}}$$

$$= \frac{1}{r^n}(ar^{n-1} + ar^{n-2} + \ldots + ar + a)$$

$$\therefore A = \frac{a}{r^n}\frac{(1 - r^n)}{(1 - r)}.$$

So, for example, the present value (or cost) of an annuity of £500 per annum, for 8 years, and assuming a compound rate of interest of 10%, is given by

$$A = \frac{500}{(1{\cdot}1)^8}\frac{(1 - (1{\cdot}1)^8)}{(1 - (1{\cdot}1))} = £2667.$$

Again, annuity tables (or cumulative present value factor tables), make this calculation straightforward. The relevant section of the table reads

Years	9%	10%	11%	
⋮				
7	5·033	4·868	4·712	Present value of
8	5·535	<u>5·335</u>	5·146	1 unit per year
9	5·995	5·759	5·537	for n years
⋮				

This gives the present value of *1 unit* each year for appropriate choices of years and discount rate. In this case

$$A = £500 \times 5{\cdot}335 = £2667.$$

20.6.4 Mortgage Loans

A building society (say) pays a loan of L, which is then repaid as n equal annual instalments of amount a. These repayments include both interest due on the loan outstanding, as well as repayment of the principal. The problem is to find the relation between L, a, n, and r. (Again $r = 1 + x$ where x is the rate of interest.) In fact the calculation is virtually identical to that of the annuity as the following reasoning shows. Instead of calling the original sum paid a loan from the building society to the borrower, imagine instead that the building society is purchasing an annuity of £a per annum for n years at a cost of £L from the borrower. The annuity formula can then be applied immediately to give

$$L = \frac{a}{r^n} \frac{(1 - r^n)}{(1 - r)}.$$

Again, tables can be used for these calculations. Suppose we wish to know the annual repayment on a loan of £10 000, over 30 years, assuming an interest rate of 10% again. Two methods are illustrated (using annuity tables, and capital recovery tables).

Annuity Tables

years	9%	10%	11%
⋮			
29	10·198	9·370	8·650
30	10·274	9·427	8·694
31	10·343	9·479	8·733
⋮			

Here, the present value of £1 each year for the next 30 years is £9·427. So, for a present value of £10 000 the annual payment would have to be larger, by a factor of

$$\frac{10\,000}{9{\cdot}427} = 1061.$$

Considering each annual payment to be a loan repayment, the annual repayment must be
£1061.

Capital Recovery Tables

years	9%	10%	11%
⋮			
29	0·0981	0·1067	0·1156
30	0·0973	0·1061	0·1150
31	0·0967	0·1055	0·1145
⋮			

Here, the annual repayment on a loan of £1 at 10% over the next 30 years is given directly as £0·1061. Hence the annual repayment on a loan 10 000 times larger is

$$10\,000 \times £0{\cdot}1061 = £1061.$$

(The capital recovery table is sometimes known as the Annual Equivalent Annuity Table.)

20.6.5 Sinking Funds

A company wishes to put aside an equal amount £a each year, so that after n years the total accumulated fund reaches a predetermined sum S for the purchase of new equipment.

The first instalment will be invested for n years altogether, so that its eventual contribution to the fund after n years is

$$a(1 + x)^n = ar^n.$$

Similarly, the second instalment will contribute an amount given by

$$ar^{n-1}.$$

So, the total fund collected will be given by

$$S = ar^n + ar^{n-1} + ar^{n-2} + \ldots + ar + a.$$

(The assumption has been made here that the last payment just completes the fund, and earns no interest.)

$$\therefore S = a\frac{(1 - r^n)}{(1 - r)}.$$

Future-value tables assist in these calculations. Suppose we wish to know the annual amount which must be invested to reach a fund of £10 000 in 10 years with a 10% rate of interest.

Years	9%	10%	11%	
⋮				
9	13·0210	13·5795	14·1640	Future value of
10	15·1929	15·9374	16·7220	1 unit per year
11	17·5603	18·5312	19·5614	for n years
⋮				

Hence £1 per year for 10 years at 10% will give a final fund of £15·9374.

The amount which must be invested annually to give a final fund of £10 000 must therefore be larger by a factor of

$$\frac{10\ 000}{15{\cdot}9374} = 627{\cdot}45.$$

Hence £627.45 must be invested in the sinking fund annually.

20.7 SERIES EXPANSION FOR METHODS OF APPROXIMATION

Series have another field of use, rather different in nature to those which have been discussed so far. Sometimes when a mathematical model of a system is built, exact solution is not possible because appropriate mathematical techniques are not available. In these cases some attempt at approximation is often useful. One particularly useful technique is to expand awkward terms as a series in such a way that many of the terms can be discarded altogether, and a reasonably good solution can be identified without resort to tedious or time-consuming numerical methods. As a simple example, suppose we wish to solve the equation

$$e^x = 1000x \tag{1}$$

The exponential function is a nuisance in such cases, and it could be expanded as an infinite series:

$$e^x = 1 + \frac{x}{1!} + \frac{x^2}{2!} + \frac{x^3}{3!} + \ldots$$

If the solution to equation (1) is a *small* value of x, then this series can be cut short, since high powers of x will be trivial in size. We might even retain just two terms, so

$$1 + x \cong e^x = 1000x$$

$$\therefore 1 \cong 999x \qquad x \cong 1/999 = 0{\cdot}001001.$$

The result is indeed a small value for x, which justifies our curtailment of the series expansion. Increased precision could be achieved by including further terms, though in this example a good result is achieved with only two.

20.8 THE BINOMIAL EXPANSION

A useful series expansion, which can often prove invaluable in methods involving approximation, is the binomial expansion. It allows us to expand expressions of the general form $(1 + x)^n$ as series, usually infinite.

The binomial theorem states that

$$(1 + x)^n = 1 + nx + n(n-1)\frac{x^2}{2!} + n(n-1)(n-2)\frac{x^3}{3!}$$

$$+ \ldots + n(n-1)(n-2)\cdots(n-r+1)\frac{x^r}{r!} + \ldots$$

The following conditions of validity apply.

(i) If n is a +ve integer, then the series will eventually terminate (to give a finite series), and will be valid for all values of x; for example

$$(1 + x)^3 = 1 + 3x + 3.2.\frac{x^2}{2!} + 3.2.1.\frac{x^3}{3!} + 0$$

$$= 1 + 3x + 3x^2 + x^3.$$

(ii) If n is −ve, or is a +ve or −ve fraction, then the series will not terminate (an infinite series), and will be valid only for values of x

which satisfy

$$-1 < x < 1 \quad \text{or} \quad |x| < 1.$$

For example,

$$\sqrt{5} = (1+4)^{1/2} = [4(1+\tfrac{1}{4})]^{1/2} = 4^{1/2}(1+\tfrac{1}{4})^{1/2}$$

$$= 2\left(1 + \frac{1}{2}\cdot\frac{1}{4} + \frac{1}{2}\left(-\frac{1}{2}\right)\frac{(1/4)^2}{2!} + \frac{1}{2}\left(-\frac{1}{2}\right)\left(-\frac{3}{2}\right)\frac{(1/4)^3}{3!} + \ldots\right)$$

$$= 2\left(1 + \frac{1}{8} - \frac{1}{128} + \frac{1}{1024} - \ldots\right)$$

$$\cong 2{\cdot}234$$

(In fact, $\sqrt{5} = 2{\cdot}236$.) Notice how, in order to satisfy the constraint on x, a factor of $4^{1/2}$ was taken outside the bracket which gave the necessary fractional form of x (viz. 1/4).

Also the smaller x, the more rapidly the series *converges*. With only four terms we have found $\sqrt{5}$ to within 0·1% precision, because succeeding terms in the series expansion become very small very quickly.

20.9 BASIC EXAMPLES FOR THE STUDENT

B20.1 Find the sums of the following series:

(a) $1 - (0{\cdot}9) + (0{\cdot}9)^2 - (0{\cdot}9)^3 + \ldots$ (∞ series)

(b) $1 + 3 + 5 + 7 + 9 + \ldots$ (n terms)

(c) $\dfrac{1}{1{\cdot}1} + \dfrac{1}{(1{\cdot}1)^2} + \dfrac{1}{(1{\cdot}1)^3} + \ldots$ (n terms)

B20.2 A man's salary commences at £3000 p.a. and increases by £150 p.a. each year. Find his mean annual salary over 20 years.

B20.3 What is the present value of the amounts below, discounted at the rate shown and over the period given?

(a) £5000 9% 5 years

(b) £1000 11% 2 years.

B20.4 What is the present value of an annuity of £1000 p.a. over 9 years, assuming a rate of interest of 9%?

B20.5 Calculate the annual mortgage repayments on a loan of £8000 over 30 years at interest rates of 9%, 10%, and 11%. What does your answer imply about the early repayments?

20.10 INTERMEDIATE EXAMPLES FOR THE STUDENT

I20.1 Find the sum of the first 10 terms of the following series:

(a) $1 + \dfrac{1}{2} + \dfrac{1}{4} + \dfrac{1}{8} + \ldots$

(b) $-9-7-5-3\ldots$

(c) $1+\dfrac{1}{0{\cdot}9}+\dfrac{1}{(0{\cdot}9)^2}+\dfrac{1}{(0{\cdot}9)^3}+\ldots$

I20.2 The accumulated value of investing £1000 for 8 years is £2000. What is the rate of interest? (Assume annual compounding.)

I20.3 What is the present value of an annuity of £400 p.a. for 20 years, assuming a rate of interest of 9%?

I20.4 What is the annual mortgage repayment on a loan of £10 000 over 3 years at 12%? Check your answer by direct calculation, year by year for the three years.

I20.5 Write down the first four terms in the expansion of

(a) $(1+2x)^{-1/3}$ (b) $(1-4x)^{3/4}$.

State in each case the restrictions on the value of x.

For questions I20.2, I20.3, and I20.4 you should give the mathematical formula appropriate to the question. However, you may use accounting tables (where appropriate) to find the results wanted.

More Advanced Examples for Part Six

A6.1 Expand the following expressions as series, using the binomial expansion, giving four terms in each expansion. For what range of x is each series valid?

(a) $\dfrac{1}{1-x}$

(b) $\sqrt{1-3x}$

(c) $\sqrt{1+x}$

Find the value of $\sqrt{10}$ to within 0·1%, using the binomial expansion.

A6.2 Sum the following infinite series:

(a) $1 - 0{\cdot}9 + 0{\cdot}81 - 0{\cdot}729 + \ldots$

(b) $1 - \dfrac{x}{1!} + \dfrac{x^2}{2!} - \dfrac{x^3}{3!} + \dfrac{x^4}{4!} - \ldots$

A6.3 Derive a formula (in terms of A, R, and N) from first principles for the present value of an annuity of £A per annum, the first payment of £A being made at the end of the first year, and the final (Nth) payment being made at the end of the Nth year. The rate of interest (or discount) is assumed to be X, and $R \equiv 1 + X$.

From tables, find the present value of an annuity of

(a) £5000 p.a. for 25 years, at 10%

(b) £3500 p.a. for 20 years, at 3%.

A6.4 Derive the formula which relates the value of the original loan, L, to the annual repayment, A, over a period of N years (where the annual repayment A is the same for all years, e.g. a mortgage loan), and to R, where $R \equiv 1 + X$ and X is the rate of interest. Assume that X remains steady throughout the repayment period.

From tables, find the annual repayment on

(a) a loan of £10 000 over 30 years, at 12%

(b) a loan of £15 000 over 3 years, at 10%.

From these two results, how could you estimate the annual repayment on a long term loan at a fairly high interest rate, very quickly and roughly?

Answers to Basic Examples

B 1.1 (a) $x < 300$ (b) $x \leqslant 10$ (c) $15 < x < 28$
(d) $96 \leqslant x \leqslant 150$ (e) $50 \leqslant x \leqslant 150$.

B 1.2 (a) 12 (b) 31 (c) 0.

B 1.3 (a) $x < 1$ (b) $z \leqslant y$ (c) $x \geqslant 1$.

B 1.4 0·2, 0·632, 0·00632, 63·2, 200, 6·32.

B 1.5 0·3, 3, 30, 0·2, 20.

B 1.6 (a) 3 (b) 2 (c) x^3y^6 (d) 1 (e) $x^{3/2}y^{9/4}$
(f) $x^{-3/2}y^{-5/2}$.

B 1.7 (a) 185 (b) 15·87 (c) 0·8262 (d) 9·636 (e) 0·6613
(f) 1.

B 1.8 (a) 1 (b) 21 (c) −5 (d) 106.

B 2.1 (a) {M, A, N, G, E, T} (b) {+2, −2} (c) {1, 2, 3, 4}
(d) {0, −1}.

B 2.2 (a) √ (b) × (c) √ (d) × (e) √ (f) ×.

B 2.3 (a) 20 (b) 30 (c) 2.

B 2.4 (a) (b) (c)

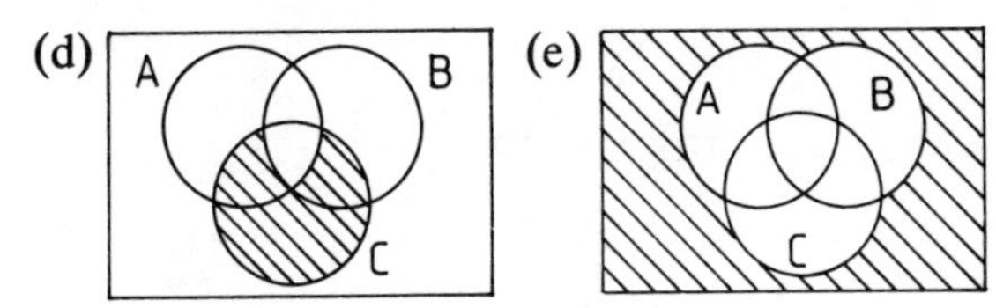

B 2.5

B 3.1

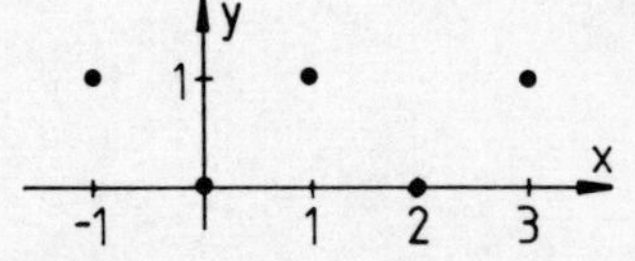

B 3.2 (a)

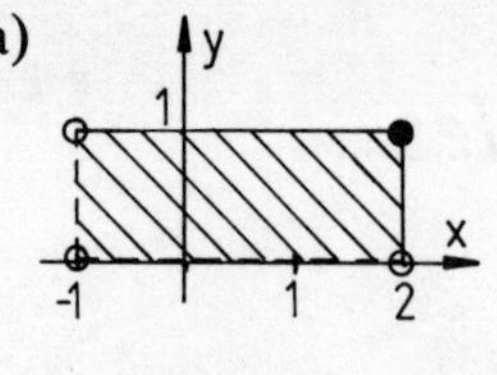

(b)

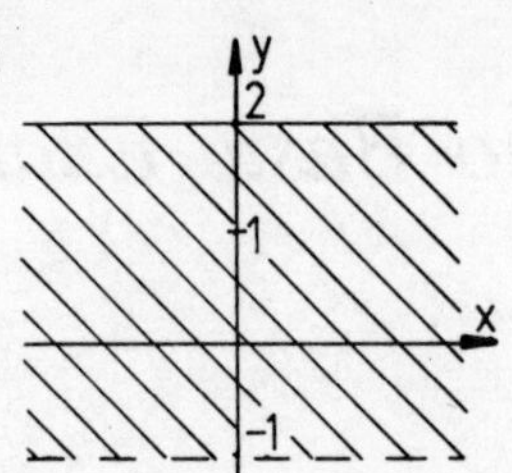

(c)

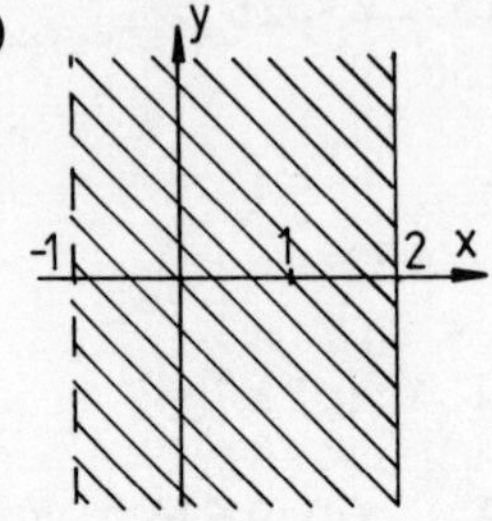

B 3.3 (a)

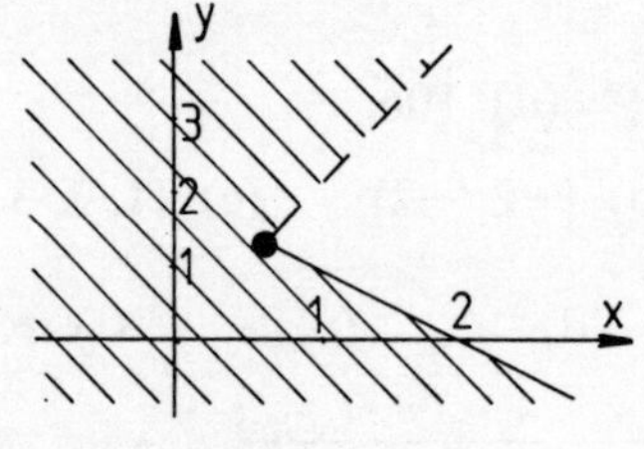

(b)

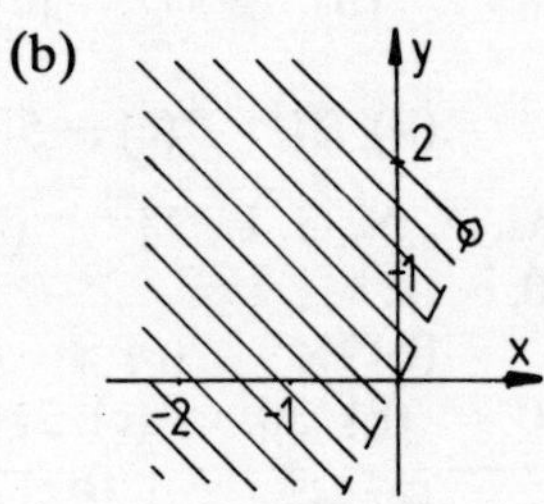

(c)

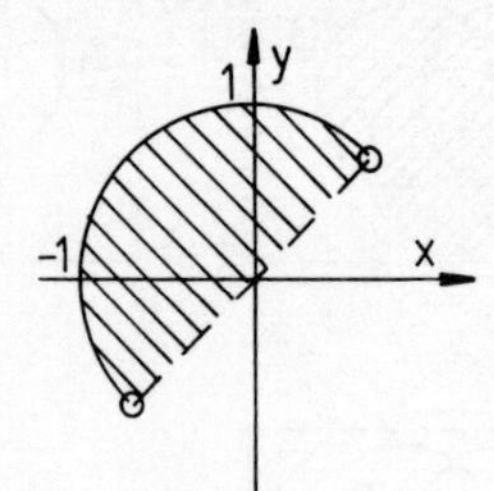

None are functions.

B 3.4 (a)

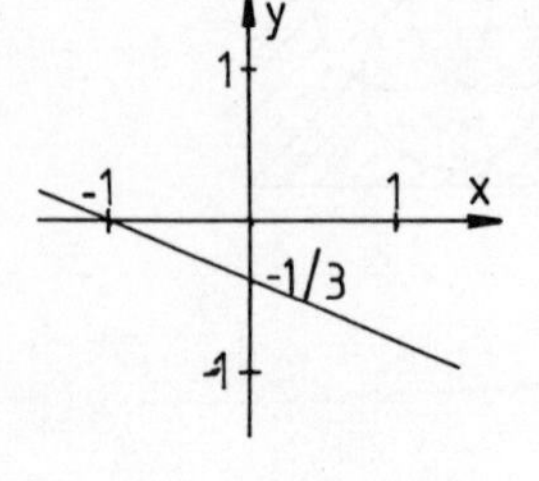

(b)

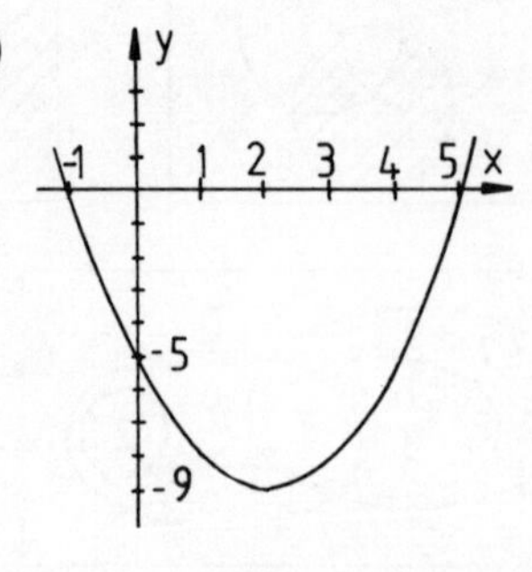

(c)

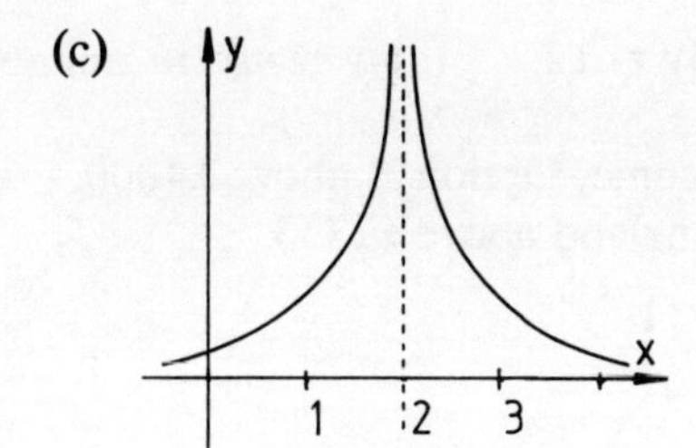

B 4.1 (a)

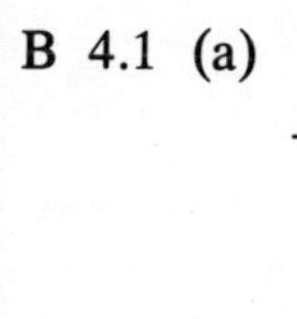

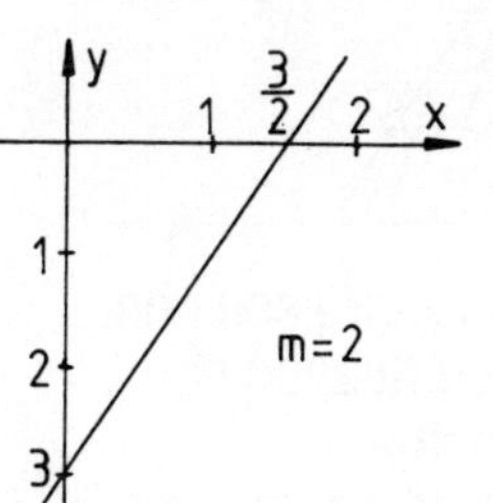

(b)

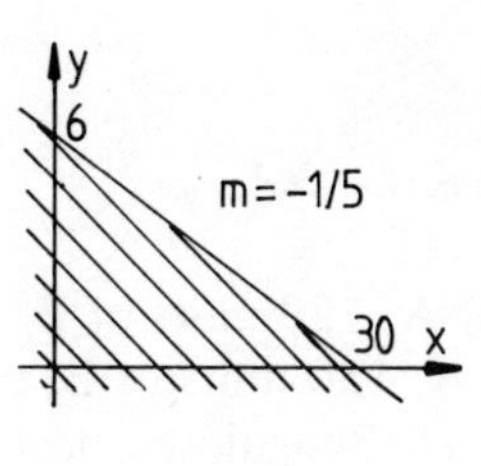

(c)

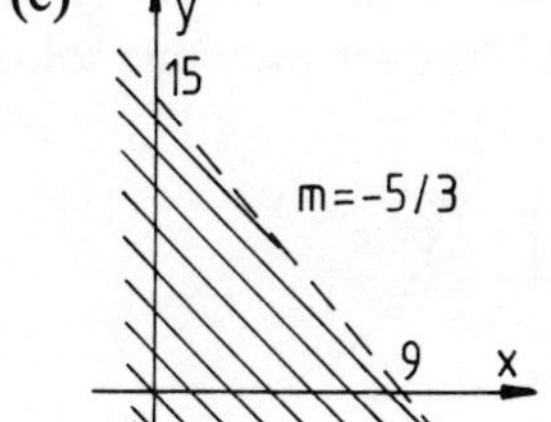

(d)

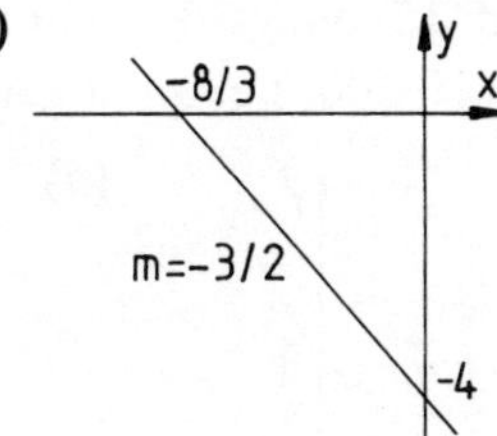

(e)

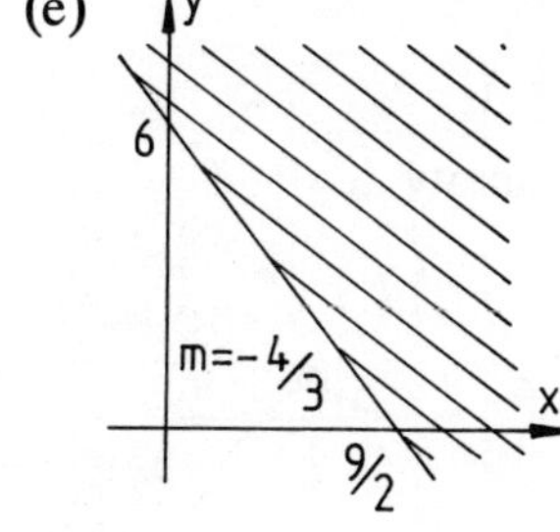

B 4.2

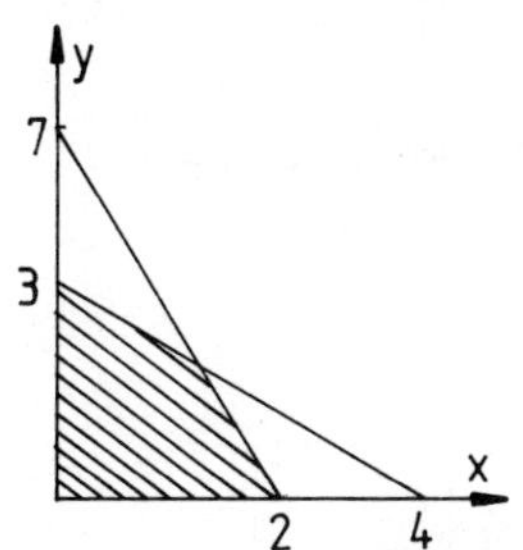

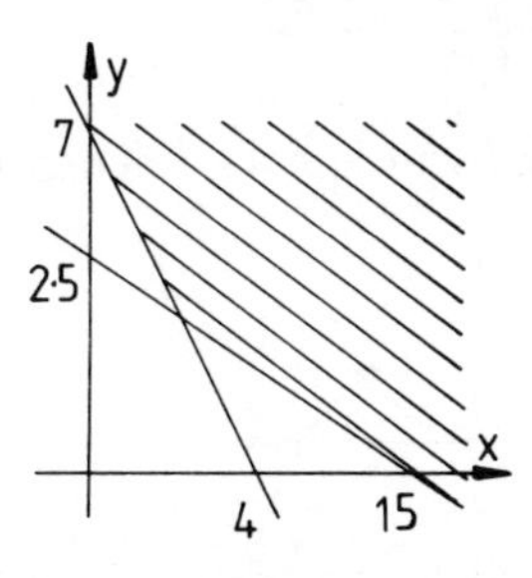

B 4.3 (a) $x = -2$, $y = 3$, $z = -1$ Intersection of 3 planes at one point.

(b) No solution. Two parallel lines with no intersection.

B 4.4 (a) $y + 3x = 0$ (b) $4x + 3y = 12$ (c) $y = mx$, m unknown
(d) $7y + x = 2$.

B 4.5 (a) Option A below £4000 income, Option B above £4000.
(b) Option A below £2353, England above £2353.

B 5.1 (a)

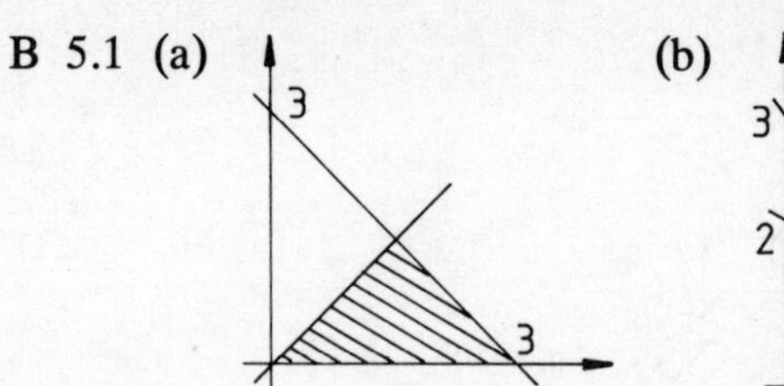

(b)

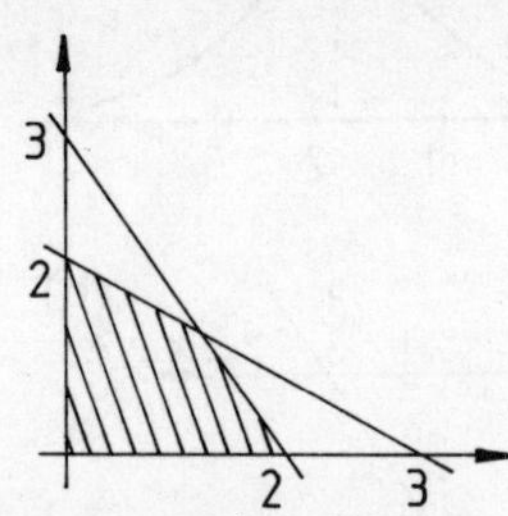

B 5.2 40 units of A, 120 units of B, total profit £800 000.

B 5.3 0 units of A, 150 units of B, total profit £900 000.

B 5.4 6 Houses, 16 Bungalows, profit £10 400.

If 5 units of land are purchased at a total cost of £200, then 8 houses and 15 bungalows can be built with a profit of £(10 700 − 200) = £10 500, i.e. a net gain of £100 over previous solution.

B 6.1 (a)

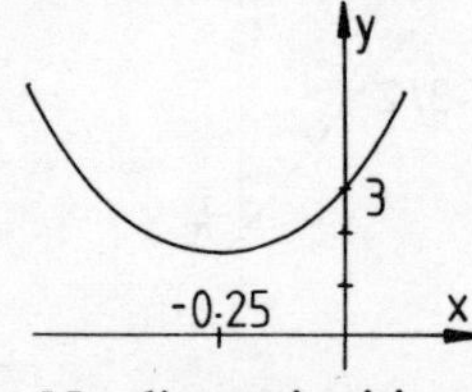

No discontinuities

(b)

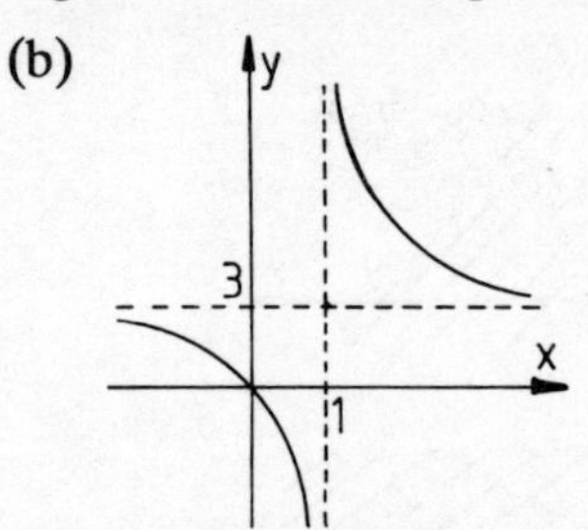

Discontinuity at x = 1

(c)

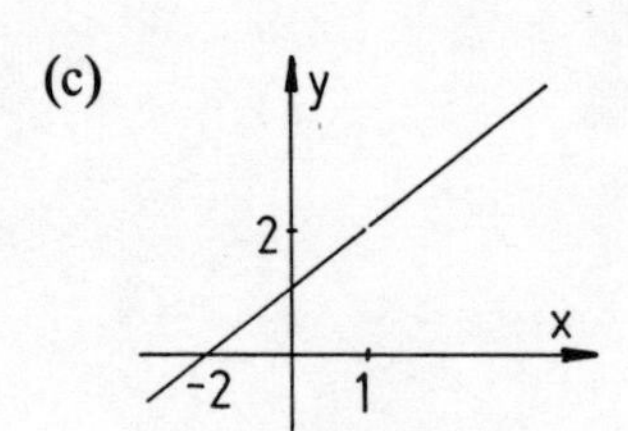

Discontinuity at x = 1

(d)

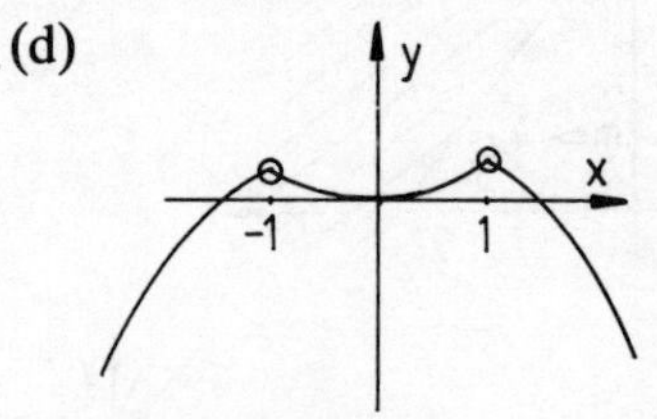

Discontinuities at x = ±1

(e)

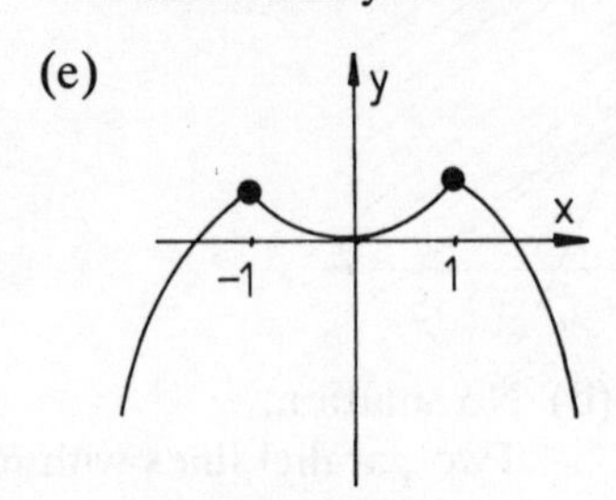

No discontinuities

B 6.2 (a) 4 (b) 2 (c) -4 (d) No limit (e) 0 (f) 0
(g) 4 (h) No limit (i) 1.

B 7.1 (a) $2x$ (b) $3x^2$ (c) $-2/x^3$ (d) $-3/x^4$.
B 7.2 (a) $4x, 4$ (b) $3 - 4/x^2 - 4/x^3, -5$ (c) $3ax^2 + 2bx + c$, $3a + 2b + c$ (d) $2(x + a), 2(1 + a)$ (e) $3/x, 3$
(f) $5e^x - 1/x + 15/x^2, 5e + 14$.

B 8.1 (a) $\log_e x + 1$ (b) $(\log_e x - 1)/(\log_e x)^2$ (c) $e^x(\log_e x + 1/x)$
(d) $\dfrac{e^x}{(\log_e x)^2}\left(\log_e x - \dfrac{1}{x}\right)$ (e) $2(x + 3)^2/x + 4(x + 3)\log_e x$
(f) $\log_e x$ (g) $2(x + 3)\log_e x((x + 3)/x + \log_e x)$
(h) $-3/(x + 1)^2$ (i) $-1/x$ (j) $4e^{4x}$ (k) $\dfrac{-x^2 + 2x + 11}{((x + 2)(x + 3))^2}$
(l) $\dfrac{3x^2 + 4}{x^3 + 4x}$ (m) 1 (n) $(2x - 2)\exp(x^2 - 2x + 1)$
(o) $(\exp(\log_e x))/x = 1$ (p) $9x^2/2\sqrt{3x^3 - 4}$ (q) $\dfrac{-(8x^3 + 6x)}{3(2x^4 + 3x^2 - 1)^{4/3}}$
(r) $\dfrac{-x}{(x^2 + 4)^{3/2}}$ (s) $-\dfrac{2}{5}\dfrac{(x - 1/x^3)}{(x^2 + 1/x^2)^{6/5}}$ (t) $\dfrac{1}{y}\dfrac{dy}{dx}$
(u) $\dfrac{3x^2 - 1}{2(x^3 - x)}$.
B 8.2 (a) $3x^2y^2$ (b) $1/x$ (c) $3(x + y)^2$ (d) $-2y/(x - y)^2$
(e) $\dfrac{x}{\sqrt{x^2 + y^2}}$ (f) $2(x + y)\exp(x^2 + 2xy + y^2)$ (g) 0
(h) $1/x$.

B 9.1 (a) $TR = 2500q - 0{\cdot}8q^2$ (b) $TC = 0{\cdot}2q^2 + 2000q + 22500$
(c) $\Pi = -q^2 + 500q - 22500$ (d) $q = 250, \Pi = 40\,000$
(e) $MC = MR = 2100$.
B 9.2 $\eta = 1$, $TR = 1000$, $\Pi = 900 - 0{\cdot}1q$ (linear), max. profit when $q = 0$, $\Pi = 0$ when $q = 9000$.

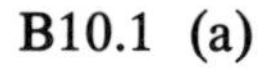
B10.1 (a)

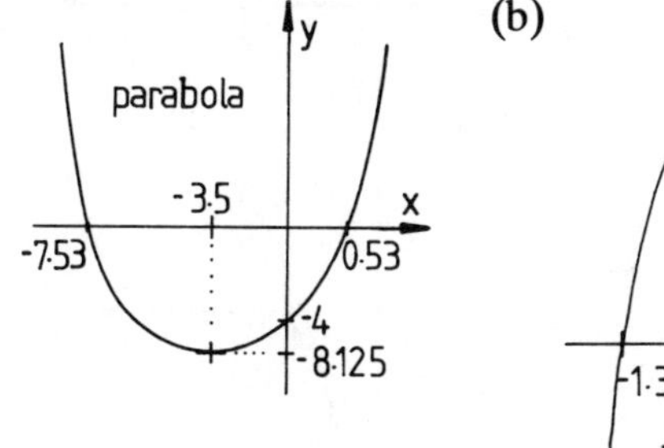

(b)

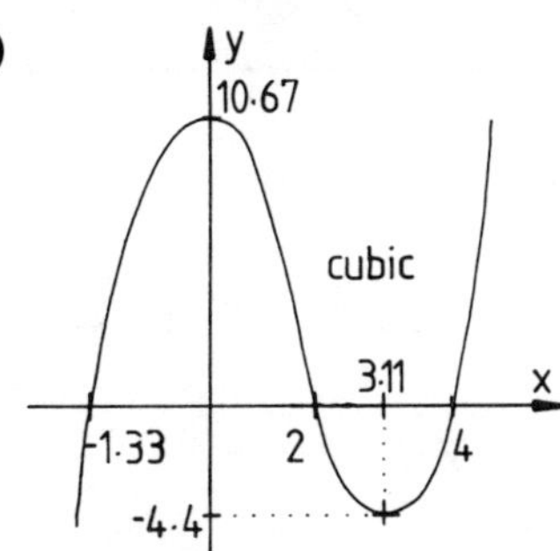

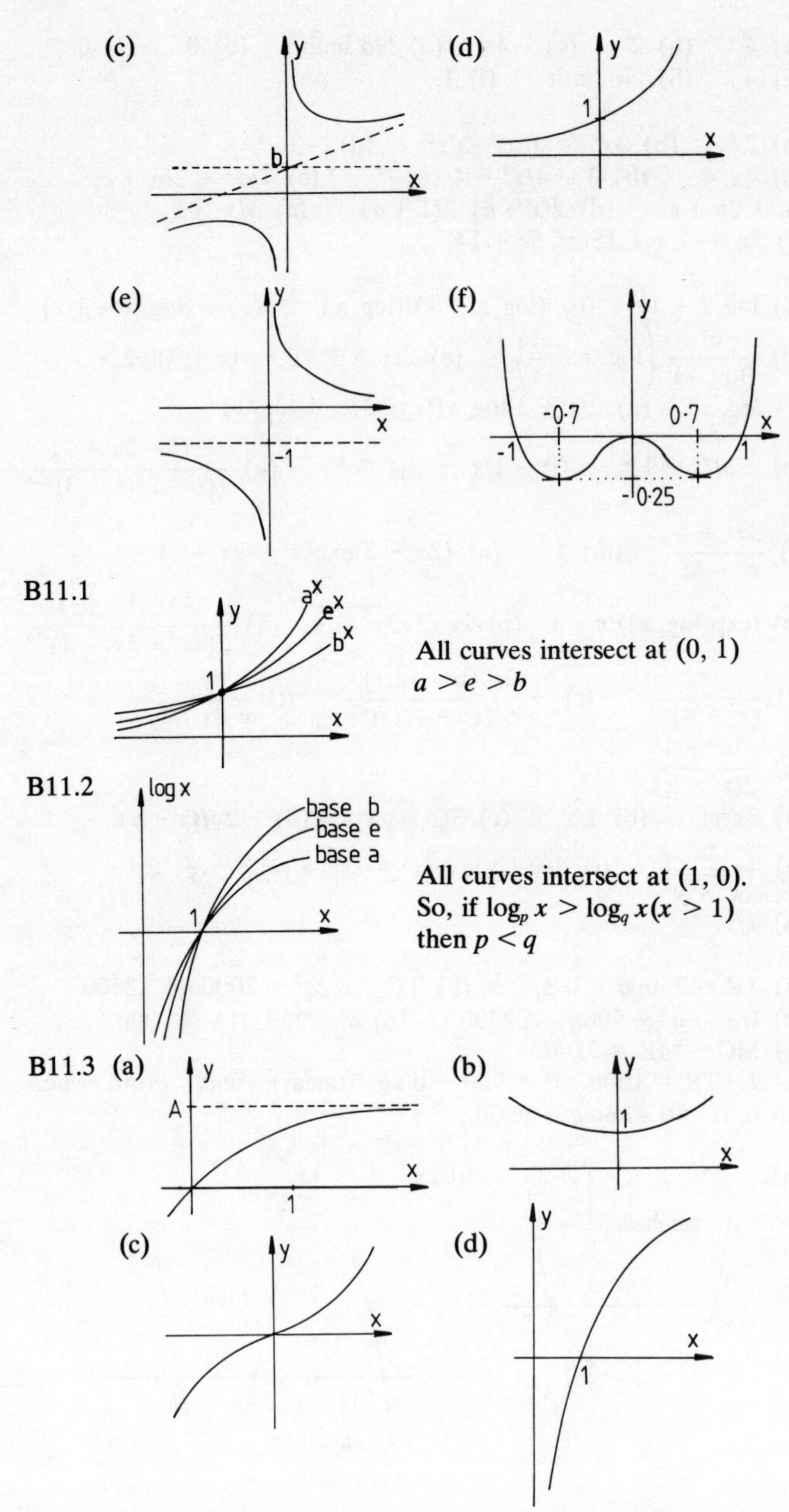

(c)
y
b
x
(d)
y
1
x
(e)
y
x
-1
(f)
y
-0·7
0·7
x
-1
1
-0·25
B11.1
y
a^x
e^x
b^x
1
x
All curves intersect at (0, 1)
$a > e > b$
B11.2
log x
base b
base e
base a
1
x
All curves intersect at (1, 0).
So, if $\log_p x > \log_q x (x > 1)$
then $p < q$
B11.3 (a)
y
A
x
1
(b)
y
1
x
(c)
y
x
(d)
y
x
1

B11.4 (a) $12x^3e^{3x^4}$ (b) $\dfrac{x}{\sqrt{x^2-1}}\exp\sqrt{x^2-1}$ (c) $2/(1-x^2)$

(d) $1/\sqrt{1+x^2}$.

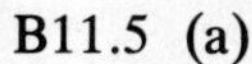
B11.5 (a)

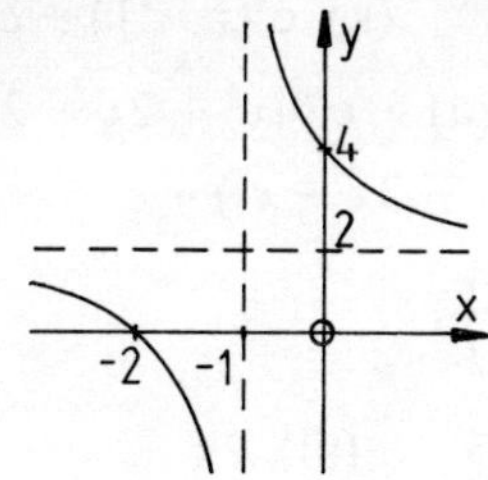

(b)

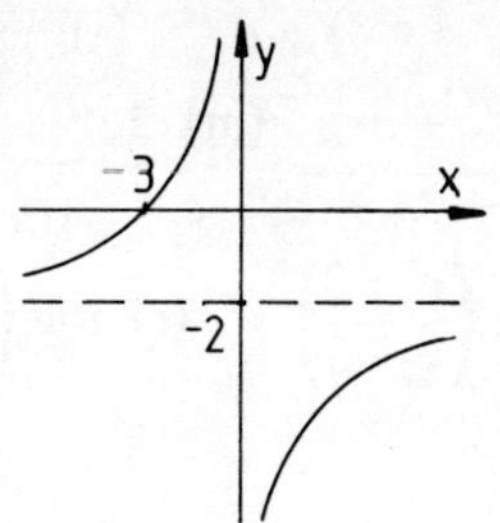

(c)

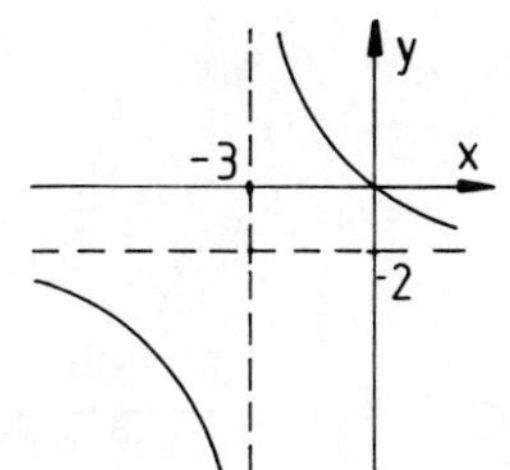

(d)

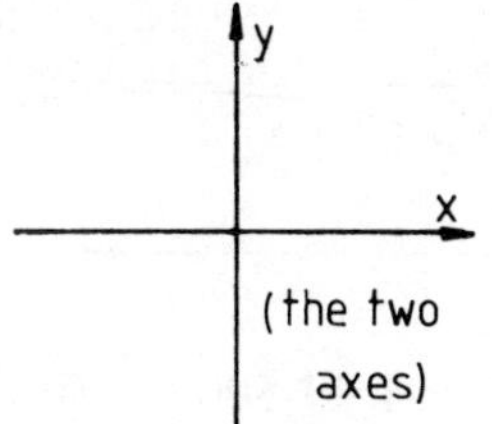

B12.1 (a) $\dfrac{x^4}{2}-\dfrac{3}{x}+e^x+c$ (b) $\log_e x+\dfrac{2}{3}x^{3/2}-e^{-x}+c$

(c) $\dfrac{x^2}{2}-x+c$ (d) $\dfrac{-1}{2x^2}-\dfrac{1}{x}+\log_e x+2\sqrt{x}$.

B12.2 (a) 36,

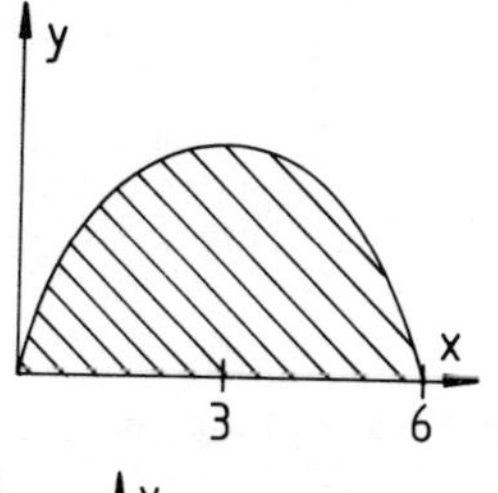

(b) 6,

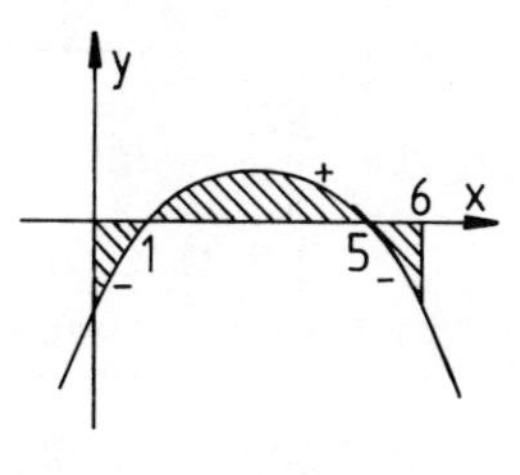

(c) −18,

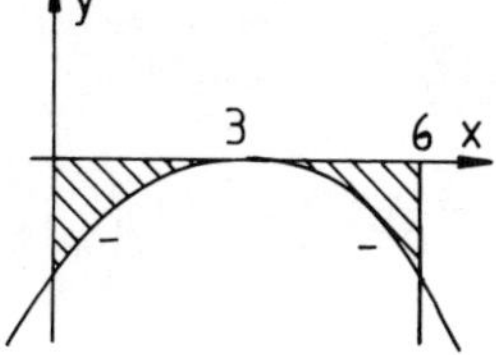

(d) 1 (e) −1.

B12.3 TC = $0{\cdot}01x^2+10x+75$

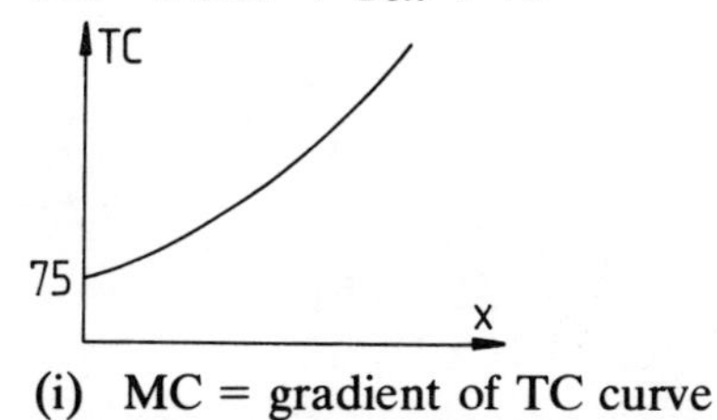

(i) MC = gradient of TC curve
(ii) TC = area under MC curve

B13.1 (a) $\frac{1}{3}(e^{3x} - e^{-3x}) + c$ (b) $\log_e(x + 3) + c$ (c) $-\log_e(3 - x) + c$
(d) $2\sqrt{e^x} + c$ (e) $(x^2 - 3)^4/8 + c$ (f) $\sqrt{1 + x^2} + c$
(g) $\log_e(x^2 - 3x + 4) + c$ (h) $\log_e(\log_e x) + c$
(i) $\log_e(e^x + e^{-x}) + c$ (j) $\frac{1}{3}e^{x^3} + c$ (k) $e^x(x - 1) + c$
(l) $-\frac{1}{2}e^{-x^2} + c$ (m) $2e^{\sqrt{x}} + c$ (n) $-e^{-x}(x^2 + 2x + 2) + c$
(o) $2\sqrt{4 - 3x + x^2} + c$ (p) $\exp(4 - 3x + x^2) + c$
(q) $\frac{1}{6}\log_e\left(\frac{x - 3}{x + 3}\right)$ (r) $\log_e\left(\frac{x - 3}{x - 2}\right)$.

B14.1 (i) (ii) 63 (iii) 24.

B14.2 (i) 20 (ii) 17 500 (iii) $\int_x^\infty y \, dx = N, y = \frac{30\,000}{x^{5/2}}$.

B14.3 (i)

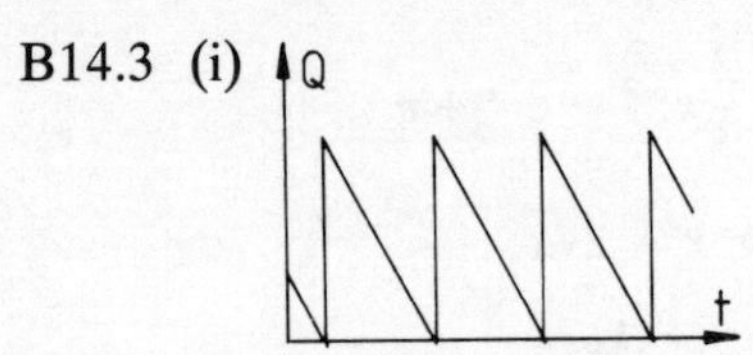

(ii) $A = \frac{SD}{Q} + \frac{1}{2}IQ$

$S = 12$, $D = 100\,000$, $I = 0{\cdot}05$ (iii) EBQ = 6928 brushes, each 3·6 weeks, annual cost £346.41.

B15.1 $3\sqrt{2} = 4{\cdot}243$. Cylinder intersected by a plane, where we wish to find the highest point of the cut.

B15.2 $2\sqrt{2} - 1 = 1{\cdot}828$.

B16.1 (a) $(5 \quad 3)\begin{pmatrix} x \\ y \end{pmatrix}$ or $(x \quad y)\begin{pmatrix} 5 \\ 3 \end{pmatrix}$ (b) $(-1 \quad 1)\begin{pmatrix} x \\ y \end{pmatrix}$ or $(x \quad y)\begin{pmatrix} -1 \\ 1 \end{pmatrix}$

(c) $\begin{pmatrix} 2 & -1 \\ 3 & 4 \end{pmatrix}\begin{pmatrix} x \\ y \end{pmatrix}$ or $(x \quad y)\begin{pmatrix} 2 & 3 \\ -1 & 4 \end{pmatrix}$.

B16.2 (a) $\begin{pmatrix} 2 & 3 \\ 4 & -1 \end{pmatrix}\begin{pmatrix} x \\ y \end{pmatrix} = \begin{pmatrix} 6 \\ 2 \end{pmatrix}$ (b) $\begin{pmatrix} 1 & -1 & 4 \\ 3 & 0 & 2 \\ 5 & 4 & -1 \end{pmatrix}\begin{pmatrix} x \\ y \\ z \end{pmatrix} = \begin{pmatrix} 16 \\ 3 \\ 7 \end{pmatrix}$.

B16.3 (a) $a_{21} = 1$, b_{32} does not exist (b) $\begin{pmatrix} 6 & 0 \\ 3 & -9 \end{pmatrix}$

(c) $\boldsymbol{A} + \boldsymbol{C} = \begin{pmatrix} -1 & 0 \\ 0 & -1 \end{pmatrix}$, $\boldsymbol{A} + \boldsymbol{B}$ does not exist

(d) $\boldsymbol{AB} = \begin{pmatrix} 2 & 6 & 0 \\ -11 & 6 & -15 \end{pmatrix}$, $\boldsymbol{AC} = \begin{pmatrix} -6 & 0 \\ 0 & -6 \end{pmatrix}$, $\boldsymbol{CA} = \begin{pmatrix} -6 & 0 \\ 0 & -6 \end{pmatrix}$,

$\boldsymbol{BA}$ does not exist (e) $-\frac{1}{6}\boldsymbol{C} = \frac{1}{6}\begin{pmatrix} 3 & 0 \\ 1 & -2 \end{pmatrix} = \begin{pmatrix} 1/2 & 0 \\ 1/6 & -1/3 \end{pmatrix}$

(f) 4, 2.

B17.1 $x = -1, y = 3, z = 0.$

B17.2 $\boldsymbol{A}^{-1} = \frac{1}{45}\begin{pmatrix} 7 & 10 & 1 \\ -10 & 5 & 5 \\ 11 & -10 & 8 \end{pmatrix}$, $x = 3, y = 1, z = 2.$

B18.1 Using the convention that position 1 ≡ he does place a bet and position 2 ≡ he does not place a bet then

(a) (1 0) (b) (0 1) (c) (0.40 0.60)

(d) 73% likely to place a bet on Wednesday

(e) 47% likely to place a bet in the long run (8/17).

With greyhounds and no double-betting, we would need matrices with dimension 3 instead of 2 (e.g. (3 × 3) transition probability matrix).

B18.2

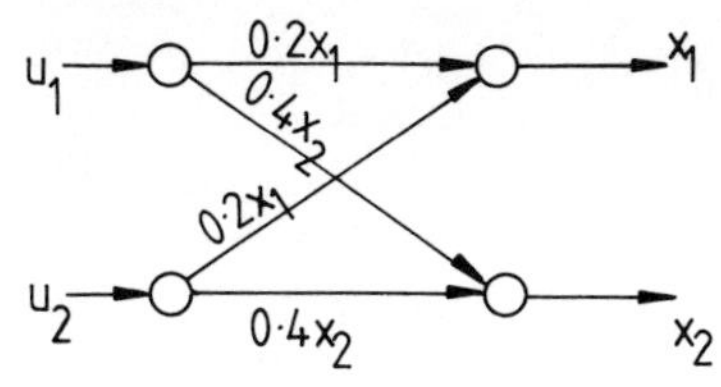

$\boldsymbol{D} = \begin{pmatrix} 10 \\ 20 \end{pmatrix}$; 35 units of G_1
45 units of G_2.

B19.1 $x = 6, y = 7, f = 32.$

B19.2 $x = 8, y = 4, f = 44.$

B20.1 (a) 0·526 (b) n^2 (c) $10(1 - (1{\cdot}1)^{-n})$.

B20.2 £4425.

B20.3 (a) £3249.50 (b) £811.60.

B20.4 £5995.

B20.5 (a) £778.40 (b) £848.80 (c) £920.

The repayments are in the ratio 9 : 9·8 : 10·6 which shows that they change nearly in proportion to the interest rates. In fact, 8000 × 9% = 720, 8000 × 10% = 800, 8000 × 11% = 880 so that the early repayments are largely interest, rather than capital repayment, and a rough guide to their magnitude can be found from simple interest payable in one year on the principal sum (£8000).

Answers to Advanced Examples

PART ONE

A1.1 $$\frac{(2x^2+xy-3y^2)\sqrt{x+y}}{\sqrt{x^2-y^2}\,\sqrt{x-y}} = \frac{(2x+3y)(x-y)(x+y)^{1/2}}{[(x+y)(x-y)]^{1/2}(x-y)^{1/2}}$$

$$= \frac{(2x+3y)(x-y)(x+y)^{1/2}}{(x+y)^{1/2}(x-y)^{1/2}(x-y)^{1/2}} = 2x+3y.$$

A1.2 $$\frac{(\sqrt[4]{xy})^{14}}{\sqrt{x^5}\,\sqrt{xy}\,(x^{1/4}y^{1/2})^6} = \frac{(xy)^{14/4}}{x^{5/2}(xy)^{1/2}x^{6/4}y^{6/2}} = \frac{x^{7/2}y^{7/2}}{x^{5/2}x^{1/2}y^{1/2}x^{3/2}y^3} = \frac{x^{7/2}y^{7/2}}{x^{9/2}y^{7/2}} = x.$$

A1.3 $$s^2 = \frac{1}{N}\sum_{i=1}^{N}(x_i-\bar{x})^2 = \frac{1}{N}\sum(x_i^2 - 2x_i\bar{x} + \bar{x}^2)$$

$$= \frac{1}{N}\sum x_i^2 - 2\bar{x}\frac{1}{N}\sum x_i + \bar{x}^2\frac{1}{N}\sum 1 = \overline{x^2} - 2\bar{x}\bar{x} + \bar{x}^2\frac{1}{N}N$$

$$= \overline{x^2} - 2(\bar{x})^2 + (\bar{x})^2 = \overline{x^2} - (\bar{x})^2.$$

(N.B. Since only one summation is involved, with $i = 1$ to N, then this has been omitted explicitly from the sigma sign, as no confusion is likely to result. All summations are taken to be for the values of i from 1 to N. This abbreviation is commonly used.)

A1.4 (a) See Section 1.14 to show $\log(x^m) = m\log x$.

(b) Using this result, we can now write the 3 curves as

(i) $y = \log x$ (ii) $y = \log x^{-1} = -\log x$ (iii) $y = \log(x^{2)} = 2\log x$.

Starting from the sketch of (i), which is given in Sections 1.14 and 11.6 of the main text, we have Figure C.

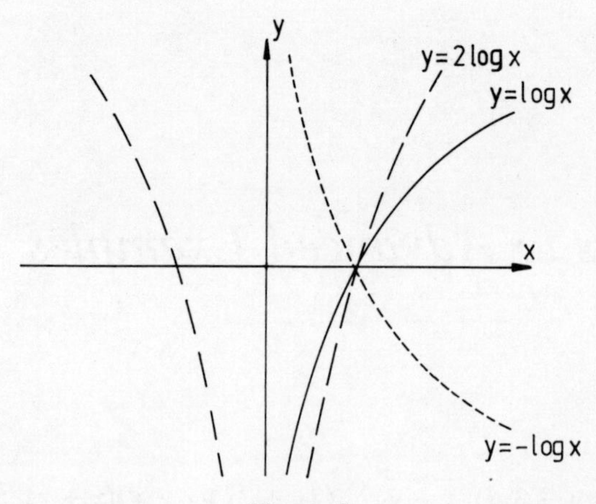

Figure C

N.B. The graph of $y = \log(x^2)$ has two branches since x^2 is always positive for real value of x, and therefore real values of $\log(x^2)$ also occur for negative values of x.

PART TWO

A2.1 The five expressions can be illustrated using Venn diagrams

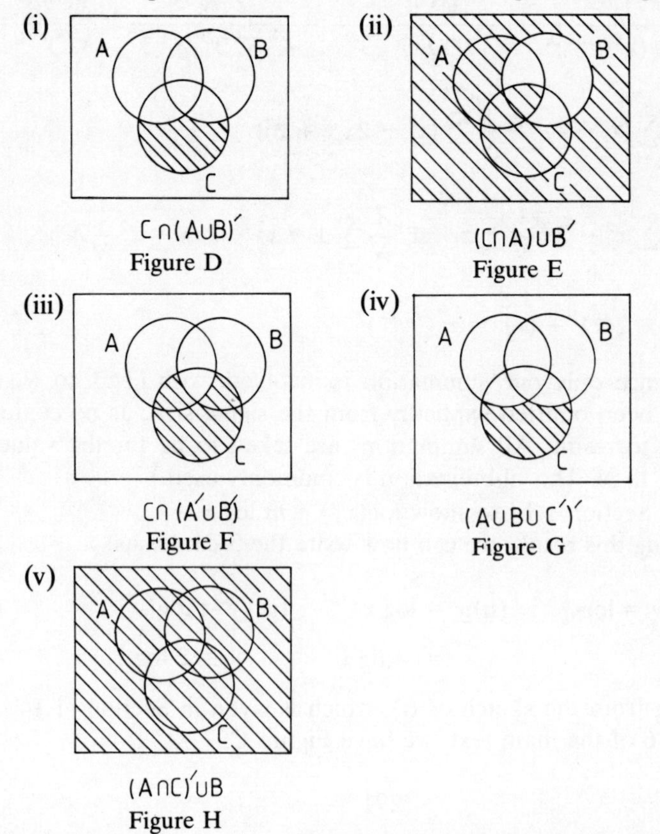

Figure D

Figure E

Figure F

Figure G

Figure H

Thus (i) and (iv) are identical. This can also be shown using the rules of set algebra:

(i) $\mathbf{C} \cap (\mathbf{A} \cup \mathbf{B})' = \mathbf{C} \cap (\mathbf{A}' \cap \mathbf{B}') = \mathbf{A}' \cap \mathbf{B}' \cap \mathbf{C}$

(iv) $(\mathbf{A} \cup \mathbf{B} \cup \mathbf{C}')' = \mathbf{A}' \cap \mathbf{B}' \cap \mathbf{C}$.

A2.2 Using the notation for the respective sets of people:

A	aristocrats	**P**	poets
M	mathematicians	**W**	wealthy people
C	academics	**I**	intelligent people

Notice that we shall treat 'poor' as the complement of 'wealthy'. This could be a case where the set definition is not watertight, and might need clarification. The five statements give us

$\mathbf{A} \subset \mathbf{W}$ or $\mathbf{W}' \subset \mathbf{A}'$

$\mathbf{M} \subset \mathbf{P}'$

$\mathbf{C}' \subset \mathbf{I}'$ or $\mathbf{I} \subset \mathbf{C}$

$\mathbf{C} \subset \mathbf{W}'$

$\mathbf{P} \subset \mathbf{I}$ or $\mathbf{M} \subset \mathbf{I}$.

This leads to the logical chains:

$\mathbf{P} \subset \mathbf{I} \subset \mathbf{C} \subset \mathbf{W}' \subset \mathbf{A}'$

$\mathbf{M} \subset \mathbf{I} \subset \mathbf{C} \subset \mathbf{W}' \subset \mathbf{A}'$

$\mathbf{M} \subset \mathbf{P}'$ or $\mathbf{M} \cap \mathbf{P} = \emptyset$.

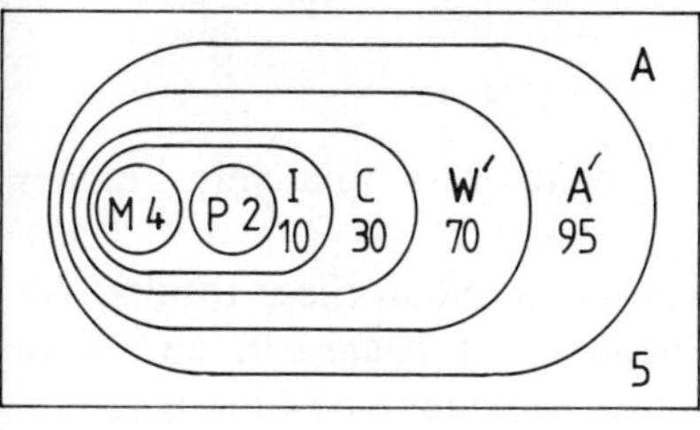

Figure I

Figure I indicates the % within each (complete) set, so that 70% fall in the set $\mathbf{W}'$, for example.

Thus, poets, mathematicians, and aristocrats are mutually exclusive groups

$$\mathbf{M} \cap \mathbf{P} = \mathbf{M} \cap \mathbf{A} = \mathbf{P} \cap \mathbf{A} = \emptyset.$$

The proportion of unintelligent academics is given by

$$|\mathbf{C} - \mathbf{I}| = (30 - 10)\% = 20\%.$$

A2.3 The figures give the following Venn diagrams for the three overlapping sets **M**, **E**, and **S**

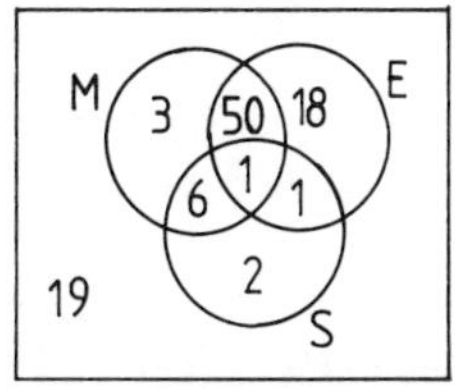

Figure J

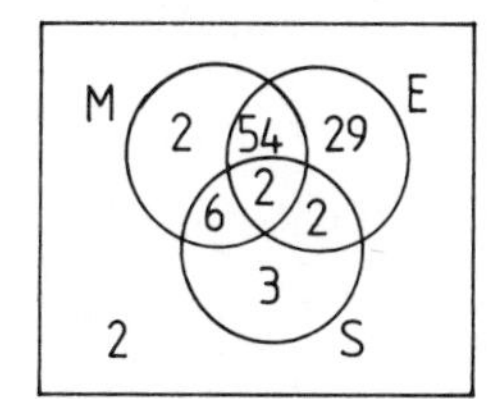

Figure K

For any group X, the simplest index i would be simply

$$i = \frac{\text{\% contribution by group X}}{\text{\% population within group X}}$$

$$= \frac{2}{1} = 2 \text{ for the group MES, for example.}$$

(i) Within the married set **M**,

If employed $i = \dfrac{56}{51} = 1{\cdot}098$

If unemployed $i = \dfrac{8}{9} = 0{\cdot}889.$

The employed group therefore make a relatively higher contribution.

(ii) Within the employed set **E**,

If married $i = \dfrac{56}{51} = 1{\cdot}098$

If single $i = \dfrac{31}{19} = 1{\cdot}632.$

The single group therefore makes a relatively higher contribution.

We have taken a very simple viewpoint to obtain these results. We have paid little attention to the meaning of *better-off*, and have therefore ignored other aspects of unemployment besides taxation, and the way income is distributed among married couples. These qualifications may, in practice, be important.

A2.4 (a) (i) We represent a successful outcome of an event A by inclusion within a set **A**. Likewise for **B**. We then define the power of each set to be the probability of a successful outcome, so that

$P(\text{A}) = |\mathbf{A}|$
$P(\text{B}) = |\mathbf{B}|$
$P(\text{A or B}) = |\mathbf{A} \cup \mathbf{B}|$
$P(\text{A and B}) = |\mathbf{A} \cap \mathbf{B}|$

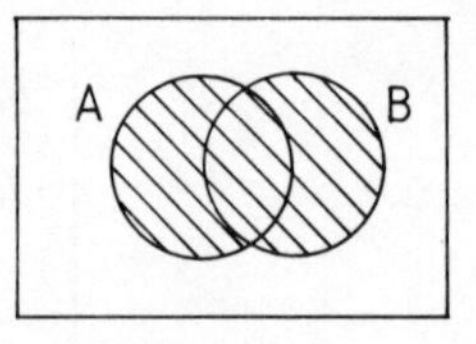

Figure L

From the usual set counting procedures (and taking care to avoid double-counting)

$$|\mathbf{A} \cup \mathbf{B}| = |\mathbf{A}| + |\mathbf{B}| - |\mathbf{A} \cap \mathbf{B}|$$

$$\therefore P(\text{A or B}) = P(\text{A}) + P(\text{B}) - P(\text{A and B}).$$

(ii) The second law can be illustrated from the following Venn diagram, where we label the outcome (A and B) by the label AB.

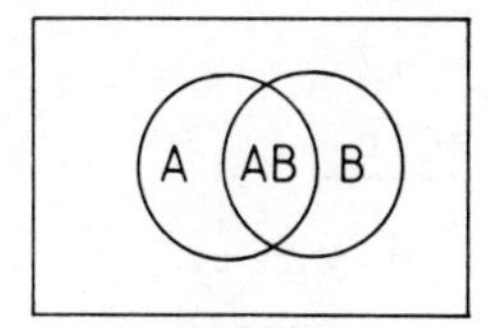

Figure M

Consider the conditional probability $P(\text{B}/\text{A})$, which means the probability of B, given that A has occurred. We are in effect considering only those outcome within the A set as our universe. Therefore

$$P(\text{B}/\text{A}) = \frac{|\mathbf{AB}|}{|\mathbf{A}|} = \frac{P(\text{A and B})}{P(\text{A})}$$

since this is the ratio of events where B is a success, to all events, *within the set* **A** *only*. Hence

$$P(\text{A and B}) = P(\text{A}) \times P(\text{B}/\text{A}).$$

(b) Consider the two sets:

R (those with a professional qualification)

D (those with a degree).

The given information leads to the following Venn diagram (Figure N).

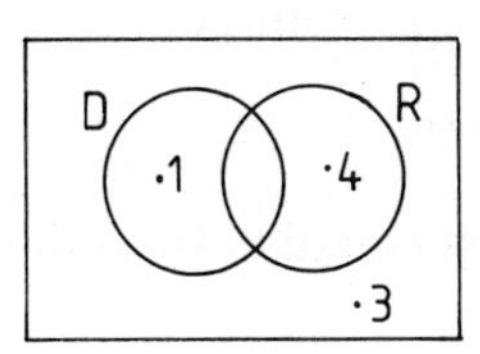

Figure N

Since the total area of the universal set must be 1, then we see immediately that

$$P(\text{D and R}) = 1 - (0{\cdot}1 + 0{\cdot}4 + 0{\cdot}3)$$
$$= 1 - 0{\cdot}8$$
$$= 0{\cdot}2.$$
$$P(\text{R}) = 0{\cdot}4 + 0{\cdot}2$$
$$= 0{\cdot}6.$$

Thus 60% have a professional qualification. (Note that this corresponds to the whole of the set **R**, including that part which overlaps **D**.)

PART THREE

A3.1 (a)

$$4x + 3y = 12$$
$$x + 2y = 4$$
$$\therefore 4x + 3y = 12$$
$$4x + 8y = 16$$
$$\therefore \quad 5y = 4 \quad y = \frac{4}{5}.$$

Figure O

$$x = 4 - 2y = 4 - \frac{8}{5} = \frac{12}{5}$$
$$\therefore x = \frac{12}{5}$$

(b)

$$x - 2y = 3$$
$$5x - 10y = -8$$
$$\therefore \quad x - 2y = 3$$
$$x - 2y = -\frac{8}{5}.$$

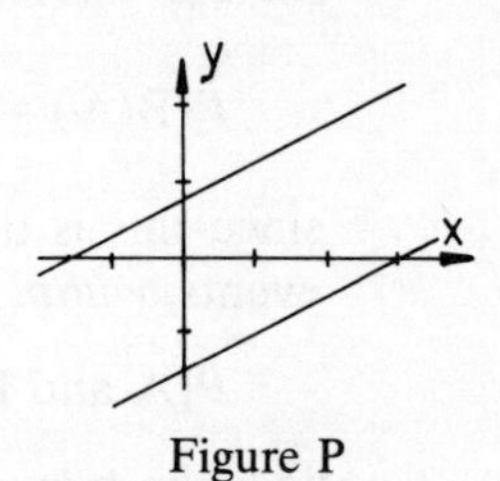

Figure P

Inconsistent, so no algebraic solution

Parallel lines, so no intersection

(c)

$$\left.\begin{aligned} x + y &= 6 \\ 3x + y &= 8 \end{aligned}\right\} \text{ subtract} \qquad \therefore 2x = 2, \quad \therefore x = 1$$
$$x + y + z = 0$$

Subtract first equation from third

$$\therefore \quad z = -6$$

$$y = 6 - x = 6 - 1 = 5$$
$$\therefore y = 5.$$

A graph would require three dimensions to demonstrate the single intersection of three planes.

A3.2 Suppose for the moment that fractional solutions are possible (i.e. partial houses and bungalows). Let x houses and y bungalows be built. The resources imply the following constraints:

Land: $\frac{1}{4}x + \frac{1}{2}y \leqslant 30$ (acres)

Tiles: $2000x + 2500y \leqslant 180\,000$ (tiles)
Cash: $8000x + 13\,000y \leqslant 1\,040\,000$ (£).

The profit yielded is given by the objective function P:

$$P = 1500x + 2000y.$$

Also $x \geqslant 0$, $y \geqslant 0$.

The equations can be tidied up, and sketched:

Land: $x + 2y \leqslant 120$ (gradient $m_1 = -1/2$)
Tiles: $4x + 5y \leqslant 360$ (gradient $m_2 = -4/5$)
Cash: $8x + 13y \leqslant 1040$ (gradient $m_3 = -8/13$)

$$\frac{P}{500} = 3x + 4y \quad \text{(gradient } m_4 = -3/4\text{)}$$

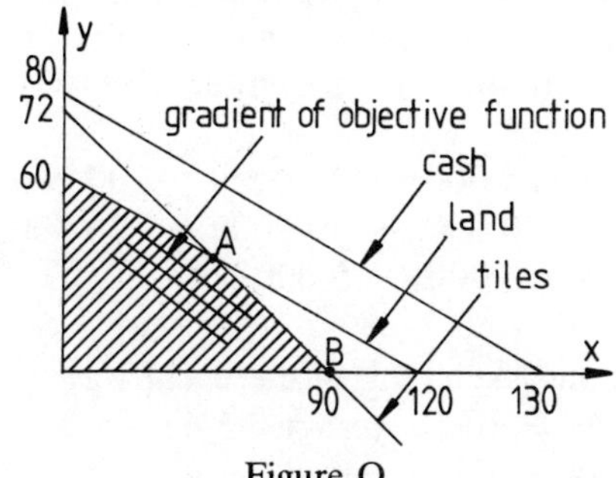

Figure Q

The feasible region is shaded. The cash constraint is redundant.

$m_1 > m_4 > m_2$.

Hence, as the objective function is moved out from the origin to increase profits, its last point of contact with the feasible region is at point A, which is the point of maximum profit. A occurs at the intersection of the tile and land constraints; that is,

$$x + 2y = 120$$
$$4x + 5y = 360.$$

Notice that *equalities* can now be used, since we know we are at the boundary of the feasible region.

$$\therefore 4x + 8y = 480$$
$$4x + 5y = 360$$
$$\therefore \quad 3y = 120 \quad \therefore \underline{y = 40}.$$

$$x = 120 - 2y = 120 - 80 = 40$$
$$\therefore x = 40, y = 40, P = £140\,000.$$

Therefore, the maximum profit will be achieved with 40 houses and 40 bungalows. (The solution has also satisfied the extra condition of whole houses and bungalows.)

If the profits on houses were to rise to H (all else remaining unchanged), then the only alteration would be that the objective function would become

$$P = Hx + 2000y.$$

Point A would remain the optimum, unless the gradient of P were to exceed that of the tile constraint. Then, as the objective function is moved away from the origin, point B would be the optimum (houses only), and the condition for this to occur is therefore

$$\frac{H}{2000} > \frac{4}{5} \quad \text{or} \quad H > 1600.$$

Therefore, he should build only houses if the profit on each rises to above £1600 (all other values remaining constant).

A3.3 Suppose x plants of type A and y plants of type S are considered. Demand constraints and the total cost (C) give us the following relations:

High grade: $25\,000x + 20\,000y \geqslant 200\,000$ (tons)
Low grade: $30\,000x + 40\,000y \geqslant 320\,000$ (tons)
Cost $C = 5\,000\,000x + 5\,000\,000y$ (£)
$x \geqslant 0, y \geqslant 0.$

Simplifying, and sketching in the usual way:

HG $5x + 4y \geqslant 40$ $\quad (m_1 = -5/4)$
LG $3x + 4y \geqslant 32$ $\quad (m_2 = -3/4)$
$C = 5x + 5y$ (£m) $\quad (m_3 = -1)$.

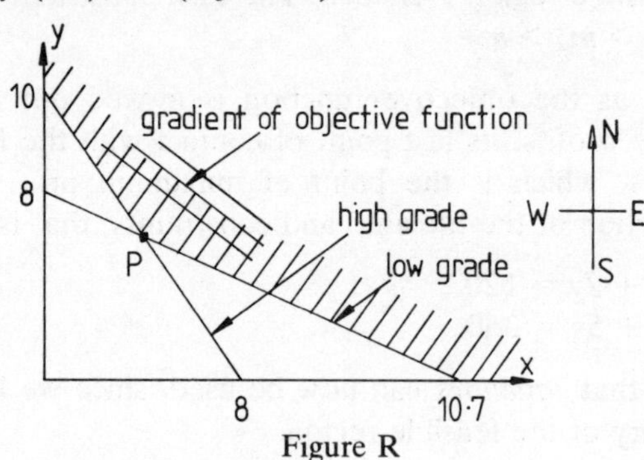

Figure R

$$m_2 > m_3 > m_1.$$

Minimum cost solution is therefore at point P when

$$5x + 4y = 40$$
$$3x + 4y = 32$$
$$2x = 8 \quad \therefore \underline{x = 4}.$$

$$4y = 40 - 5x = 40 - 20 = 20$$
$$\therefore \underline{y = 5}.$$

Therefore, if the only criterion is minimizing costs subject to providing adequate supplies, the government should purchase 4 plants of type A and 5 plants of type S.

If the demand is likely to shift towards high grade oils, and away from low grade oils, then the constraints will both shift in such a way as to push the point P on the figure towards a higher value of x, and lower value of y (in a roughly south-east direction in the navigational convention on the figure). This suggests more plants of type A and fewer of type S, but the exact choice would depend on the (anticipated) shifts in demand. This solution is intuitively obvious from an inspection of the plants' specifications. It seems likely that plant A should be favoured in the longer term since it gives an output richer in high grade oil compared with plant S, and this is consistent with the expected shift in demand.

A3.4 Suppose the student spends x minutes thinking and y minutes writing. (We assume that these are mutually exclusive activities in the following calculation, but this is not entirely obvious from the question, and would need clarification.) The student's assumptions can be expressed thus:

$x + y \leqslant 180$ (mins)	Exam duration
$x \leqslant 90$ (mins)	maximum thinking period
$y \leqslant 150$ (mins)	maximum writing period
$y \geqslant 105$ (mins)	minimum writing period
$x \geqslant 0$ (mins)	minimum thinking period.

Let M be the number of marks, in this model.

Assumption (a) gives

$$M_a = \frac{30}{60}x + \frac{20}{60}y = \frac{x}{2} + \frac{y}{3} \quad \left(m_a = -\frac{3}{2}\right).$$

Assumption (b) gives

$$M_b = \frac{20}{60}x + \frac{30}{60}y = \frac{x}{3} + \frac{y}{2} \quad \left(m_b = -\frac{2}{3}\right).$$

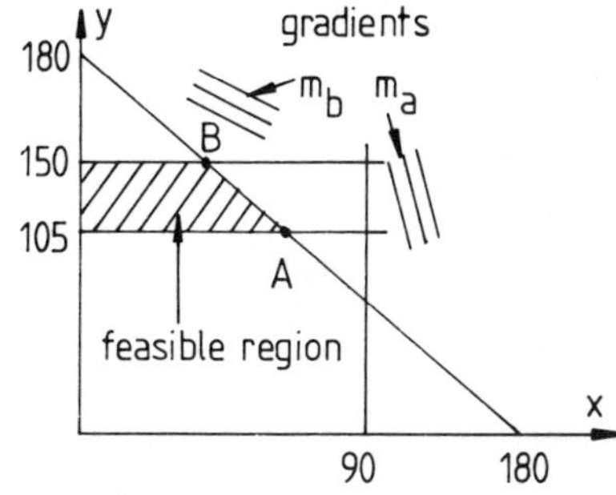

Figure S

The feasible region is shaded.

Assumption (a): As the objective function moves away from the origin to increase M, the final point of contact with the feasible region is at A, where

$$x = 75 \qquad y = 105 \qquad M = 72{\cdot}5.$$

Assumption (b): As the objective function moves away from the origin to increase M, the final point becomes B, where

$$x = 30 \qquad y = 150 \qquad M = 85.$$

The student therefore finds that the two assumptions lead to the conclusions:

(a) 75 minutes thinking
105 minutes writing
72·5 marks at most.

(b) 30 minutes thinking
150 minutes writing
85 marks at most.

Although this is a lighthearted example, it does illustrate two further points:

(a) A mixture of constraint types can sometimes occur so that the feasible region may be bounded, but away from the origin.

(b) Sensitivity to changes in the model. In this case, uncertainty as to the appropriate form of the objective function led us into trying two different cases to see how sensitive the outcome was to changes in the basic assumptions. (Different examples of this procedure were also indicated in the ends of the two previous questions.) Although we find very different results for the thinking/writing ratio, the effect on the maximum mark available is quite limited. In other words, on the basis of the two assumptions tested it seems that the maximum mark is relatively *insensitive* to the proportions of the two activity times (within the given constraints).

PART FOUR

A4.1 (a) $y = x^3(1 + x^2)$ (product)

$$\therefore y' = (1 + x^2)3x^2 + x^3(2x)$$
$$= 3x^2 + 3x^4 + 2x^4 \qquad = \underline{x^2(3 + 5x^2)}.$$

(b) $y = \dfrac{e^x + 1}{e^x}$ (quotient)

$$\therefore y' = \frac{e^x(e^x) - (e^x + 1)e^x}{(e^x)^2}$$

$$= \frac{e^{2x} - e^{2x} - e^x}{e^{2x}}$$

$$= \frac{-e^x}{e^{2x}} \qquad = \underline{\frac{-1}{e^x}}.$$

Alternatively, by simplifying first:

$$y = \frac{e^x + 1}{e^x} = 1 + \frac{1}{e^x} = 1 + e^{-x}$$

$$\therefore y' = 0 - e^{-x} \qquad = \underline{\frac{-1}{e^x}}.$$

(c) $y = \log_e(x^2 + 5x + 6)$ (function of function of x)

$$\therefore y' = \frac{1}{(x^2 + 5x + 6)}(2x + 5) = \underline{\frac{2x + 5}{x^2 + 5x + 6}}$$

Alternatively:

$$y = \log_e(x^2 + 5x + 6) = \log_e(x + 3)(x + 2)$$

$$= \log_e(x + 3) + \log_e(x + 2)$$

$$\therefore y' = \frac{1}{x + 3} + \frac{1}{x + 2} \qquad = \underline{\frac{2x + 5}{x^2 + 5x + 6}}$$

(d) $y = x^3e^{3x}$ (product)

$$\therefore y' = e^{3x}(3x^2) + x^3(3e^{3x}) = \underline{3x^2e^{3x}(1 + x).}$$

(e) $y = \sqrt{x + \log_e x} = (x + \log_e x)^{1/2}$ (function of function of x)

$$\therefore y' = \tfrac{1}{2}(x + \log_e x)^{-1/2}\left(1 + \frac{1}{x}\right) = \underline{\frac{x + 1}{2x\sqrt{x + \log_e x}}}.$$

(f) $y = \log_e 3x^4 + \log_e 5$ (function of function of x)

$$= \log_e 3 + \log_e x^4 + \log_e 5$$

$$= \log_e 3 + 4\log_e x + \log_e 5$$

$$\therefore y' = 0 + \frac{4}{x} + 0 \qquad = \underline{\frac{4}{x}}.$$

N.B. $\log_e 3$ and $\log_e 5$ are both constants.

(g) $y = \log_e\sqrt{x+1}$ (function of function of function of x)

$$= \log_e(x+1)^{1/2}$$

$$= \tfrac{1}{2}\log_e(x+1) \quad \text{(function of function of } x\text{)}$$

$$\therefore y' = \tfrac{1}{2} \cdot \frac{1}{x+1} \cdot 1 \quad = \underline{\frac{1}{2(x+1)}}.$$

(h) $$y = \left(\frac{a+x}{a-x}\right)\sqrt{\frac{b-x}{b+x}}$$

$$\therefore \quad \log_e y = \log_e\left[\frac{(a+x)(b-x)^{1/2}}{(a-x)(b+x)^{1/2}}\right]$$

$$\therefore \quad \log_e y = \log_e(a+x) + \tfrac{1}{2}\log_e(b-x) - \log_e(a-x) - \tfrac{1}{2}\log_e(b+x)$$

$$\therefore \quad \frac{1}{y}\frac{dy}{dx} = \frac{1}{a+x} - \frac{1}{2(b-x)} + \frac{1}{a-x} - \frac{1}{2(b+x)} = f(x), \text{ say}$$

$$\therefore y' \equiv \frac{dy}{dx} = \underline{f(x)\left(\frac{a+x}{a-x}\right)\sqrt{\frac{b-x}{b+x}}}.$$

A4.2 (a) $\int xe^{x^2}\,dx$ Let $u = x^2$, $\dfrac{du}{dx} = 2x$

$$= \tfrac{1}{2}\int \frac{du}{dx}e^u\,dx$$

$$= \tfrac{1}{2}\int e^u\,du = \tfrac{1}{2}e^u + c = \underline{\tfrac{1}{2}e^{x^2} + c}.$$

(b) $\displaystyle\int \frac{x}{\sqrt{1-x^2}}\,dx$ Let $u = 1 - x^2$, $\dfrac{du}{dx} = -2x$

$$= -\tfrac{1}{2}\int \frac{du}{dx}\frac{1}{u^{1/2}}\,dx$$

$$= -\tfrac{1}{2}\int u^{-1/2}\,du = -\frac{1}{2}\frac{u^{1/2}}{(\frac{1}{2})} + c = -u^{1/2} + c = \underline{-\sqrt{1-x^2} + c}.$$

(c) $$\int \frac{dx}{a^2 - x^2} = \int \frac{1}{(a-x)(a+x)}\,dx$$

$$\equiv \int\left[\frac{A}{(a-x)} + \frac{B}{(a+x)}\right]dx, \quad \text{so that } A = B = \frac{1}{2a}$$

$$= \frac{1}{2a}\int\left(\frac{1}{a-x} + \frac{1}{a+x}\right)dx$$

$$= \frac{1}{2a}\left(-\log_e(a-x) + \log_e(a+x)\right) + c = \underline{\frac{1}{2a}\log_e\left(\frac{a+x}{a-x}\right) + c}.$$

(d) $\int \frac{e^x + e^{-x}}{e^x - e^{-x}}\,dx$ Let $u = e^x - e^{-x}$, $\frac{du}{dx} = e^x + e^{-x}$

$= \int \frac{1}{u}\frac{du}{dx}\,dx$

$= \int \frac{1}{u}\,du = \log_e u + c = \underline{\log_e(e^x - e^{-x}) + c}.$

(e) $\int \frac{dx}{x \log_e x}$ Let $u = \log_e x$, $\frac{du}{dx} = \frac{1}{x}$

$= \int \frac{du}{dx}\frac{1}{u}\,dx$

$= \int \frac{1}{u}\,du = \log_e u + c = \underline{\log_e(\log_e x) + c}.$

(f) $\int \frac{x^2 - 1}{\sqrt{x^3 - 3x + 5}}\,dx$ Let $u = x^2 - 3x + 5$, $\frac{du}{dx} = 3x^2 - 3$

$= 3(x^2 - 1)$

$= \frac{1}{3}\int \frac{du}{dx}\frac{1}{\sqrt{u}}\,dx$

$= \frac{1}{3}\int u^{-1/2}\,du = \frac{1}{3}\frac{u^{1/2}}{(\frac{1}{2})} + c = \underline{\frac{2}{3}\sqrt{x^3 - 3x + 5} + c}.$

(g) $\int_0^1 (1 - e^{-x})\,dx = [x + e^{-x}]_0^1$

$= (1 + e^{-1}) - (0 + e^0) = 1 + \frac{1}{e} - 1 = \frac{1}{e} \doteqdot \underline{0{\cdot}368}.$

(h) $\int_0^\infty x e^{-x^2}\,dx$ Let $u = x^2$, $\frac{du}{dx} = 2x$

$= \frac{1}{2}\int_0^\infty \frac{du}{dx} e^{-u}\,dx = \frac{1}{2}\int_0^\infty e^{-u}\,du$

$= \frac{1}{2}[-e^{-u}]_0^\infty = \frac{1}{2}(-0 + e^0) = \underline{0{\cdot}5}.$

A4.3 (a) $y = x^3 - 7x^2 + 36$

$\therefore$ Cubic

+ve coefficient of x^3

+ve y-intercept (+36)

−ve x^2 coefficient (inverted parabola near y-axis)

No term in x (zero gradient on y-axis)

Thus far, we have Figure T

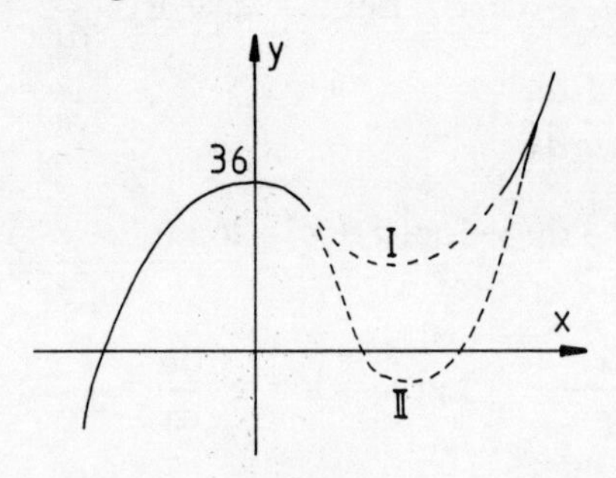

Figure T

To distinguish I and II:

$\therefore y' = 0 = 3x^2 - 14x$ at turning points
$\therefore x(3x - 14) = 0$
$\therefore x = 0$ or $14/3$ (consistent with above shape).

$$\text{At } x = \frac{14}{3}, y = \left(\frac{14}{3}\right)^3 - 7\left(\frac{14}{3}\right)^2 + 36$$
$$= -14{\cdot}8.$$

Hence y is negative at the minimum point, so that alternative II must be the correct sketch.

The x-intercepts are found most easily by trial and error here. The minimum occurs when $x = 14/3$, so one intercept lies between $x = 0$ and $x = 5$.

Try $x = 3$

$\therefore y = 27 - 63 + 36 = 0$
$\therefore x = 3$ is one intercept, enabling us to factorize:
$y = x^3 - 7x^2 + 36 = (x - 3)(x^2 - 4x - 12)$
$= (x - 3)(x - 6)(x + 2)$
$\therefore y = 0$ when $x = -2, 3, 6.$

This gives Figure U

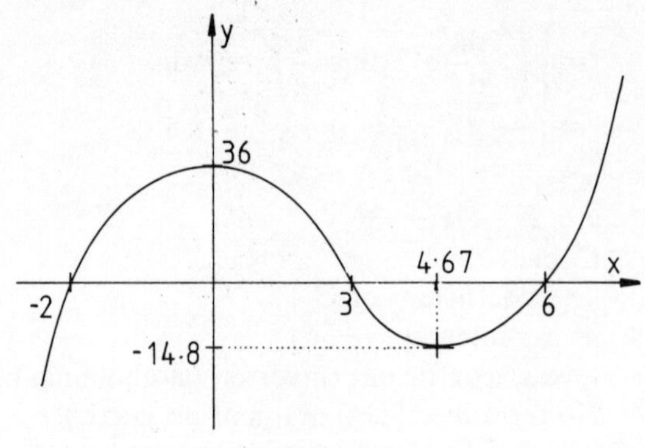

Figure U

(b) $y = x^4 - 16x^2 = x^2(x^2 - 16)$

$\therefore$ Quartic

Even powers, $\therefore$ symmetric

$y = 0$ when $x = 0, \quad \pm 4$

$y > 0$ when $|x| > 4$

$y < 0$ when $|x| < 4$

$y \to +\infty$ when $x \to \pm\infty$

$$\begin{aligned} y' &= 4x^3 - 32x \\ &= 4x(x^2 - 8) \\ &= 0 \quad \text{at turning points,} \end{aligned}$$

when $x = 0$, or $x = \pm\sqrt{8} = \pm 2{\cdot}818$
and $y = 0, \ -64, -64.$

This gives Figure V.

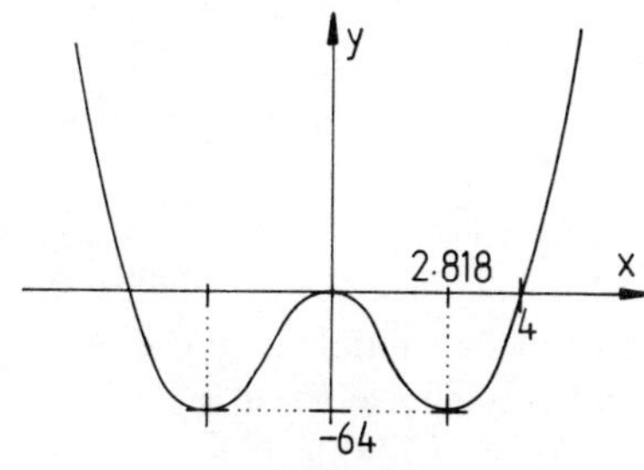

Figure V

(c) $xy = 2(x + y)$

Terms only in xy, x, y (and constants)

$\therefore$ Rectangular hyperbola.

Rearrangement gives

$$(x - 2)(y - 2) = 4$$

$\therefore$ Asymptotes are the lines $y = 2, \quad x = 2.$

Intercepts given by $x = 0, \quad y = 0.$

This gives Figure W.

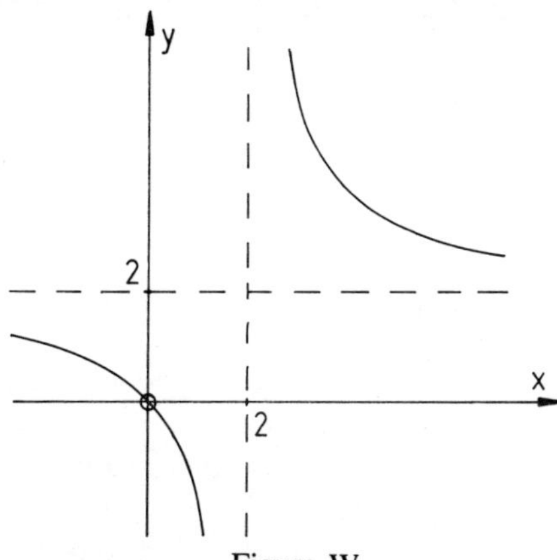

Figure W

(d) $y = \dfrac{1}{e^x + 1}$

$e^x > 0$ always $\quad \therefore y > 0$ always

$x = 0 \qquad y = \frac{1}{2}$

$x \to +\infty \qquad y \to 0$

$x \to -\infty \qquad y \to 1$

$y' = -(e^x + 1)^2\, e^x.$

But $e^x > 0$, $(e^x + 1)^2 > 0$

$\therefore y' < 0$, always a negative slope.

This gives Figure X.

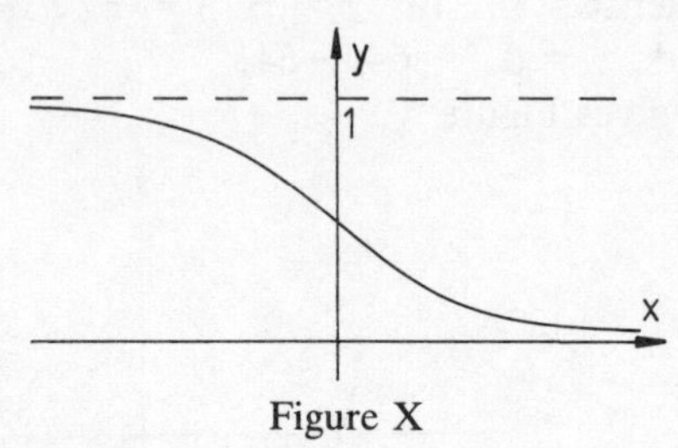

Figure X

A4.4 We develop the two models in parallel, to aid comparison. The total costs (rent and travel) are given by

Model 1: $\quad TC_1 = 15\,000e^{-x/5} + 300x$

Model 2: $\quad TC_2 = \dfrac{11\,000}{x + 1} + 300x + 3000$

Minimum total costs occur when

$$TC_1' = -3000e^{-x/5} + 300 = 0$$

$$TC_2' = \frac{-(11\,000)}{(x + 1)^2} + 300 = 0$$

$\therefore \exp(-x_1/5) = 1/10$

$\therefore \underline{x_1 = 5 \log_e 10 = 11{\cdot}51}$

$\underline{TC_1 = 4953}$

and

$(x_2 + 1)^2 = 110/3 = 36{\cdot}67$

$\therefore \underline{x_2 = 5{\cdot}055}$

$\underline{TC_2 = 6333.}$

Both model have similarity shaped relationships for TC vs. x, as in Figure Y (Rent R, Travel T).

As travel costs increase (a linear function of x), the gradient of the line (T) increases. The minimum point in the resulting total cost

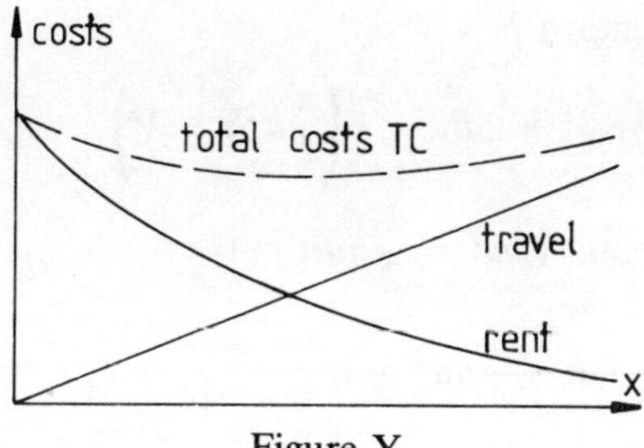

Figure Y

curve thus moves to give a *smaller* value of x. Thus, as travel costs increase relative to rents, a flat nearer the city centre is suggested, a result consistent with intuition.

If inflation affects travel and rent in the same way, we would not expect any change in the point of minimum costs.

$$\text{TC} = T + R$$
$$\therefore \text{TC}' = T' + R' = 0 \qquad \therefore T' = -R'.$$

With an inflation factor of i say

$$\text{TC} = iT + iR$$
$$\text{TC}' = iT' + iR' = 0$$
$$\therefore \qquad i(T' + R') = 0 \qquad \therefore T' = -R'$$

which is the same condition as above.

Comparing the two models directly, we find:

Model 1: City centre, $x = 0$ TC = £15 000
Minimum cost is £4953, achieved 11·5 miles away
At large distances, rent costs $\to 0$ (realistic?)
travel costs $\to \infty$.

Model 2: City centre, $x = 0$ TC = £14 000
Minimum cost is £6333, achieved 5·05 miles away
At large distances, rent costs $\to$ £3000
travel costs $\to \infty$.

The two models predict similar rent levels (and therefore total costs), nearly 6 miles from the centre.

At $x = 6$ TC_1 = £4518
TC_2 = £4571.

A4.5 Assume a can of diameter d and height h. Surface area A of a can is given by

$$A = 2.\tfrac{1}{4}\pi d^2 + \pi dh,$$

to be minimized subject to a given volume (V)

$$V = \tfrac{1}{4}\pi d^2 h.$$

Using the method of Lagrange multipliers, we find the turning point of the function F

$$F = \tfrac{1}{2}\pi d^2 + \pi dh - \lambda\left(\frac{\pi}{4}d^2h - V\right)$$

$$\frac{\partial F}{\partial d} = \pi d + \pi h - \frac{\lambda}{2}\pi dh = 0 \quad \text{(a)}$$

$$\frac{\partial F}{\partial h} = \pi d - \frac{\lambda}{4}\pi d^2 = 0 \quad \text{(b)}$$

$$\therefore \lambda = \frac{4}{d} \quad \text{from (b)}$$

$$\therefore d + h = \frac{dh}{2}\left(\frac{4}{d}\right) = 0 \quad \text{from (a)}$$

$$\therefore d + h - 2h = 0$$

$$\therefore \underline{d = h.}$$

Thus, the cylindrical can with equal height and diameter has the smallest surface area for a given volume.

A4.6 Volume of cone is given by

$$V = \tfrac{1}{3}\pi r^2 h$$

to be maximized subject to the constraint

$$r^2 + h^2 = 1 \quad \text{(from Pythagoras's theorem).}$$

Using the method of Langrange multipliers, we find the turning point of the function F

$$F = \tfrac{1}{3}\pi r^2 h - \lambda(r^2 + h^2 - 1)$$

$$\frac{\partial F}{\partial r} = \tfrac{2}{3}\pi rh - 2\lambda r = 0 \quad \text{(a)}$$

$$\frac{\partial F}{\partial h} = \tfrac{1}{3}\pi r^2 - 2\lambda h = 0 \quad \text{(b)}$$

$$\therefore \lambda = \frac{\pi}{3}h \quad \text{from (a)}$$

$$\therefore \frac{\pi}{3}r^2 = 2h\,\frac{\pi}{3}h \quad \text{from (b)}$$

$$\therefore r^2 = 2h^2.$$

But

$$r^2 + h^2 = 1$$

$$\therefore 3h^2 = 1$$

$$\therefore h^2 = \frac{1}{3} \qquad h = \frac{1}{\sqrt{3}}$$

$$r^2 = \frac{2}{3} \qquad \underline{r = \sqrt{\frac{2}{3}}.}$$

The length of the curved edge of the cone is given by

$$2\pi r = 2\pi\sqrt{\frac{2}{3}}.$$

The length of the circumference of the complete circle of paper is given by

$$2\pi \times 1 = 2\pi.$$

The angle a must therefore be given by

$$360° - a = (2\pi\sqrt{\frac{2}{3}} \div 2\pi) \times 360°$$

$$= 360\sqrt{\frac{2}{3}} \doteqdot 294°$$

$$\therefore \underline{a = 66°}.$$

A4.7 Expressing all functions in terms of q, we have

$$\text{TR} = pq = (5000 - 3q)q = 5000q - 3q^2$$

$$\text{MC} = \frac{\text{dTC}}{\text{d}q} = 4200 - 4q$$

$$\therefore \text{TC} = \int (4200 - 4q)\,\text{d}q$$

$$= 4200q - 2q^2 + c$$

$$\text{VC} = 70\,000 \quad (\text{TC} = \text{VC when } q = 0)$$

$$\therefore \text{TC} = -2q^2 + 4200q + 70\,000.$$

$$\text{Profit } \Pi = \text{TR} - \text{TC}$$

$$= 5000q - 3q^2 - (-2q^2 + 4200q + 70\,000)$$

$$= -q^2 + 800q - 70\,000.$$

(i) Break-even point occurs when

$$\Pi = 0$$

$$\therefore q^2 - 800q + 70\,000 = 0$$

$$\therefore q = \frac{800 \pm \sqrt{640\,000 - 280\,000}}{2}$$

$$= \frac{800 \pm \sqrt{360\,000}}{2} \qquad = \frac{800 \pm 600}{2}$$

$$= 700, 100.$$

The lower value of q gives the break-even point. Therefore the break-even point occurs at a production level of 100 tractors p.a.

(ii) Maximum profits occur when $\Pi' = 0$

$$\therefore \frac{d\Pi}{dq} = -2q + 800 = 0$$

$$\therefore q = 400$$

Therefore maximum profits occur at a production level of 400 tractors p.a.

(iii) At $q = 400$ we find

$$\begin{aligned} MC &= 4200 - 4q = 4200 - 1600 = 2600 \quad (£) \\ MR &= 5000 - 6q = 5000 - 2400 = 2600 \quad (£) \\ \Pi &= -q^2 + 800q - 70\,000 \\ &= -(400)^2 + 800.400 - 70\,000 \\ &= 90\,000 \quad (£) \\ \eta &= \frac{-p}{q}\frac{dq}{dp} \\ &= \frac{-(5000 - 3q)}{q}\frac{1}{(-3)} \\ &= \frac{+(5000 - 1200)}{400.3} = \frac{3800}{1200} = \underline{3{\cdot}1\dot{6}}. \\ MC(q) &= 4200 - 4q. \end{aligned}$$

This implies that the marginal cost of production decreases with increasing level of production. It is therefore implying *economy of scale*. The validity of the entire model may be limited by the range over which this economy of scale is a valid assumption. It is likely that diseconomies of scale will occur for high enough production levels (high q), unless the production process itself is altered. (This could be a new production line, for example, to avoid high overtime costs on existing lines to achieve high q.) Any such fundamental change in the production process would then, in any case, demand complete revision of the relationships in the model.

A4.8 The basic assumptions we shall use in this stock control problem are:

(a) A uniform demand rate, $D = 400$ (tons/year)
(b) Instantaneous production.
(c) Fixed cost of set-up, $S = 1000$ (£/batch).
(d) Inventory cost of $I = 80$ (£/ton/year).

Annual set-up costs ASC = SD/Q for batch size Q.
Annual carrying costs ACC = $\frac{1}{2}QI$.
Total annual acquisition costs

$$\begin{aligned} A &= \text{ASC} + \text{ACC} \\ &= \frac{SD}{Q} + \tfrac{1}{2}QI. \end{aligned}$$

Economic batch size, Q^*, is given by

$$A' = -\frac{SD}{Q^2} + \frac{I}{2} = 0$$

$$\therefore Q^* = \sqrt{\frac{2SD}{I}} = \sqrt{\frac{2.1000.400}{80}} = 100 \text{ tons.}$$

Therefore the most economic solution is to produce 4 batches per year, of 100 tons each, at a total cost of

$$\text{A} = \frac{2000{\cdot}400}{100} = 8000 \text{ £/year}$$

The relationship A vs. Q is illustrated in Figure Z, which also shows the effect of a price-break at a higher level of Q (corresponding to the lower price offered for a minimum of 400 tons).

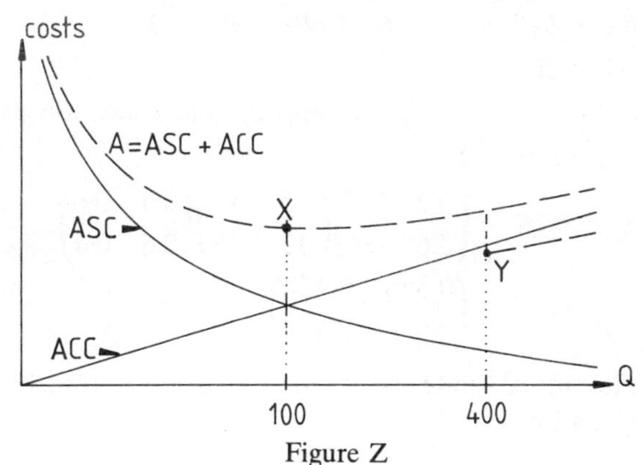

Figure Z

The choice regarding the lower-price offer depends on the value of A at points X and Y.

If we consider a batch size of 100, A = 8000. If we consider a batch size of 400, however,

$$A = \frac{SD}{Q} + \tfrac{1}{2}QI = \frac{1000.400}{400} + \tfrac{1}{2}\,.\,400.80$$
$$= 17\,000.$$

We thus incur an increase of £(17000 − 8000), or £9000, in acquisition costs with the larger batch size, per year.

This is offset by a reduction in raw material costs. The saving would be £(500 − 490), or £10 per ton, which implies a saving of £10 × 400 per year, i.e. £4000.

Thus, with the larger batch size we incur:

an increase of £9000 p.a. in acquisition costs, and

a saving of £4000 p.a. in raw material costs.

The special price offer is therefore not sufficiently attractive. We would prefer this option only if a price saving of £9000 p.a. could be achieved, that is,

$$£\frac{9000}{400} = £22{\cdot}50 \text{ per ton.}$$

Therefore we would consider a price offer of £477.50 (or less) per ton for a 400 ton batch.

PART FIVE

A5.1 **P** and **Q** commute if $PQ = QP$.

If $A = B + cI$ then

$$AB = (B + cI)B = B^2 + cIB = B^2 + cB$$

and

$$BA = B(B + cI) = B^2 + cBI = B^2 + cB$$

$$\therefore AB = BA.$$

A5.2 After three transitions, the appropriate transition probability matrix **M**, say, is given by

$$M = PQR = \begin{pmatrix} 0{\cdot}2 & 0{\cdot}8 \\ 0{\cdot}6 & 0{\cdot}4 \end{pmatrix}\begin{pmatrix} 0{\cdot}5 & 0{\cdot}5 \\ 0{\cdot}2 & 0{\cdot}8 \end{pmatrix}\begin{pmatrix} 0{\cdot}4 & 0{\cdot}6 \\ 0{\cdot}6 & 0{\cdot}4 \end{pmatrix}$$

$$= \begin{pmatrix} 0{\cdot}548 & 0{\cdot}452 \\ 0{\cdot}524 & 0{\cdot}476 \end{pmatrix}.$$

(i) The initial state is $(1 \quad 0)$, so after 3 transitions the state is given by

$$(1 \quad 0)M = (1 \quad 0)\begin{pmatrix} 0{\cdot}548 & 0{\cdot}452 \\ 0{\cdot}524 & 0{\cdot}476 \end{pmatrix} = (0{\cdot}548 \quad 0{\cdot}452)$$

(54·8% chance of being in first state, 45·2% in second state).

(ii) The long-run state $(a \quad b)$ can be found from

$$(a \quad b) = (a \quad b)\begin{pmatrix} 0{\cdot}548 & 0{\cdot}452 \\ 0{\cdot}524 & 0{\cdot}476 \end{pmatrix}$$

$$\therefore a = 0{\cdot}548a + 0{\cdot}524b.$$

Also

$$a + b = 1.$$

Hence

$$b = \frac{0{\cdot}452}{0{\cdot}976} = 0{\cdot}463 \quad a = 0{\cdot}537.$$

That is, 53·7% in first state

46·3% in second state.

A5.3 (a) We find the inverse of A using the Gauss–Jordan technique:

$$
\begin{array}{|ccc|ccc|}
3 & -2 & 1 & 1 & 0 & 0 \\
-1 & 3 & 4 & 0 & 1 & 0 \\
3 & 0 & 2 & 0 & 0 & 1 \\
\hline
1 & -\frac{2}{3} & \frac{1}{3} & \frac{1}{3} & 0 & 0 \\
-1 & 3 & 4 & 0 & 1 & 0 \\
3 & 0 & 2 & 0 & 0 & 1 \\
\hline
1 & -\frac{2}{3} & \frac{1}{3} & \frac{1}{3} & 0 & 0 \\
0 & \frac{7}{3} & \frac{13}{3} & \frac{1}{3} & 1 & 0 \\
0 & 2 & 1 & -1 & 0 & 1 \\
\hline
1 & -\frac{2}{3} & \frac{1}{3} & \frac{1}{3} & 0 & 0 \\
0 & 1 & \frac{13}{7} & \frac{1}{7} & \frac{3}{7} & 0 \\
0 & 2 & 1 & -1 & 0 & 1 \\
\hline
1 & 0 & \frac{11}{7} & \frac{3}{7} & \frac{2}{7} & 0 \\
0 & 1 & \frac{13}{7} & \frac{1}{7} & \frac{3}{7} & 0 \\
\hline
0 & 0 & -\frac{19}{7} & -\frac{9}{7} & -\frac{6}{7} & 1 \\
\hline
1 & 0 & \frac{11}{7} & \frac{3}{7} & \frac{2}{7} & 0 \\
0 & 1 & \frac{13}{7} & \frac{1}{7} & \frac{3}{7} & 0 \\
0 & 0 & 1 & \frac{9}{19} & \frac{6}{19} & -\frac{7}{19} \\
\hline
1 & 0 & 0 & -\frac{6}{19} & -\frac{4}{19} & \frac{11}{19} \\
0 & 1 & 0 & -\frac{14}{19} & -\frac{3}{19} & \frac{13}{19} \\
0 & 0 & 1 & \frac{9}{19} & \frac{6}{19} & -\frac{7}{19}
\end{array}
$$

Thus

$$A^{-1} = \frac{1}{19}\begin{pmatrix} -6 & -4 & 11 \\ -14 & -3 & 13 \\ 9 & 6 & -7 \end{pmatrix}$$

$$AX = C$$

$$\therefore X = A^{-1}C = \frac{1}{19}\begin{pmatrix} -6 & -4 & 11 \\ -14 & -3 & 13 \\ 9 & 6 & -7 \end{pmatrix}\begin{pmatrix} 5 \\ -17 \\ 0 \end{pmatrix}$$

$$= \frac{1}{19}\begin{pmatrix} 38 \\ -19 \\ -57 \end{pmatrix} = \begin{pmatrix} 2 \\ -1 \\ -3 \end{pmatrix}.$$

Hence

$$\underline{x = 2 \quad y = -1 \quad z = -3.}$$

(b) Since we already know the inverse of A, we can use it directly for the second matrix of constants, D.

$$AX = D$$

$$\therefore X = A^{-1}D = \frac{1}{19}\begin{pmatrix} -6 & -4 & 11 \\ -14 & -3 & 13 \\ 9 & 6 & -7 \end{pmatrix}\begin{pmatrix} 2 \\ 6 \\ 5 \end{pmatrix}$$

$$= \frac{1}{19}\begin{pmatrix} 19 \\ 19 \\ 19 \end{pmatrix} = \begin{pmatrix} 1 \\ 1 \\ 1 \end{pmatrix}.$$

Hence

$$\underline{x = 1 \quad y = 1 \quad z = 1.}$$

The inverse of A has been useful in solving for X very concisely, for different constants. If D were replaced by E (then F, G, etc.), the same technique becomes increasingly useful compared with any other approach.

PART SIX

A6.1 (a)
$$(1 - x)^{-1} = 1 + (-1)(-x) + (-1)(-2)\frac{(-x)^2}{2!} + (-1)(-2)(-3)\frac{(-x)^3}{3!} + \ldots$$
$$= 1 + x + x^2 + x^3 + \ldots \quad \text{for} \quad -1 < x < 1.$$

(b) $(1-3x)^{1/2} = 1 + \left(\frac{1}{2}\right)(-3x) + \left(\frac{1}{2}\right)\left(-\frac{1}{2}\right)\frac{(-3x)^2}{2!}$

$$+ \left(\frac{1}{2}\right)\left(-\frac{1}{2}\right)\left(-\frac{3}{2}\right)\frac{(-3x)^3}{3!} + \ldots$$

$$= 1 - \frac{3}{2}x - \frac{9}{8}x^2 - \frac{27}{16}x^3 - \ldots \quad \text{for} \quad -\frac{1}{3} < x < \frac{1}{3}.$$

(c) $(1+x)^{1/2} = 1 + \left(\frac{1}{2}\right)(x) + \left(\frac{1}{2}\right)\left(-\frac{1}{2}\right)\frac{(x)^2}{2!}$

$$+ \left(\frac{1}{2}\right)\left(-\frac{1}{2}\right)\left(-\frac{3}{2}\right)\frac{(x)^3}{3!} + \ldots$$

$$= 1 + \frac{1}{2}x - \frac{1}{8}x^2 + \frac{1}{16}x^3 - \ldots \quad \text{for} \quad -1 < x < 1.$$

This last series can be used to evaluate $\sqrt{10}$ thus:

$$\sqrt{10} = \sqrt{1+9} = 3\sqrt{1+\frac{1}{9}} = 3\left(1+\frac{1}{9}\right)^{1/2}$$

$$= 3\left(1 + \frac{1}{2}\cdot\frac{1}{9} - \frac{1}{8}\left(\frac{1}{9}\right)^2 + \frac{1}{16}\left(\frac{1}{9}\right)^3 - \ldots\right)$$

$$= 3(1 + 0{\cdot}05556 - 0{\cdot}00154 + 0{\cdot}00009 - \ldots)$$

$$= 3(1{\cdot}05410)$$

$$= 3{\cdot}162(29).$$

The series converges sufficiently rapidly to ensure that the largest term omitted would affect the result only in the fifth decimal place. The correct result is

$$\sqrt{10} = 3{\cdot}16228 \text{ to 5 decimal places.}$$

A6.2 (a) $1 - 0{\cdot}9 + 0{\cdot}81 - 0{\cdot}729 \ldots$ (geometric series)

$$= \frac{1}{1-(-0{\cdot}9)} = \frac{1}{1{\cdot}9} = 0{\cdot}526.$$

(b) $1 - \frac{x}{1!} + \frac{x^2}{2!} - \frac{x^3}{3!} + \frac{x^4}{4!} \ldots = e^{-x}$ (exponential series).

A6.3 If we ignore discounting altogether for the moment, the present value of an annuity would be simply the sum of the individual annual amounts, S say.

$$S = A + A + A + \ldots + A.$$

$$\uparrow \text{ 1st year} \qquad \uparrow \; N\text{th year}$$

However, each annual amount must be discounted back, each by a different factor. Each amount must be reduced by a factor $(1+X)$ for each year, to find the present value. Using the abbreviation

$1 + X = R$, we then have

$$S \text{ (present value)} = \frac{A}{R} + \frac{A}{R^2} + \frac{A}{R^3} + \ldots + \frac{A}{R^N}$$

$$= \frac{A}{R^N}(R^{N-1} + \ldots + R + 1)$$

$$= \frac{A}{R^N}\frac{(1 - R^N)}{(1 - R)}.$$

Annuity tables give the results:

(a) £1 p.a. for 25 years at 10% has a present value of £9·077. Hence £5000 p.a. for 25 years at 10% has a present value of £45 385.

(b) £1 p.a. for 20 years at 3% has a present value of £14·877. Hence £3500 p.a. for 20 years at 3% has a present value of £52 069.

(Notice the relative magnitudes of these two results. The difference in interest rates has been more significant than the total periods or annual amounts.)

A6.4 The mortgage loan may be considered to be an annuity in reverse. The original loan L corresponds to the present value, and the annual repayment is equivalent to the annual amount. Thus:

$$L = \frac{A}{R^N}\frac{(1 - R^N)}{(1 - R)}.$$

Capital Recovery tables give the results:

(a) £1 borrowed over 30 years at 12% requires an annual repayment of £0·1241. Hence, £10 000 borrowed over 30 years at 12% requires an annual repayment of £1241.

(b) £1 borrowed over 3 years at 10% requires an annual repayment of £0·4021. Hence, £15 000 borrowed over 3 years at 10% requires annual repayment of £6031.50.

For a longer term loan, with a high interest rate, each repayment in the first few years is largely interest; the principal repayment is small (but grows over the years). Thus the first repayment is approximately interest on the initial sum only; for example,

12% of £10 000 is £1200 (example (a)).

Thus the (interest + principal) repayment will be slightly greater than this. In fact, the calculation above shows it to be £1241.

This approximation breaks down for short periods since a significant proportion of principal is repaid in the first year (except at *extremely* high interest rates, when principal repayment becomes almost irrelevant anyway: most repayment is interest).

Index